THE RISE OF THE NEIGHBOURHOOD IN CANADA, 1880s–2020s

Neighbourhoods matter now more than ever before. They sustain fewer social connections, but in an era of great social inequality and high levels of immigration, they have become vital as places for homeowner investment and educational opportunity for children. *The Rise of the Neighbourhood in Canada, 1880s–2020s* traces the changing character and significance of Canadian urban neighbourhoods, city and suburban, since the 1880s.

The book highlights patterns in neighbourhood life, particularly noticeable in larger urban areas, which are especially important for the least mobile people: workers, lower income households, immigrants, women, children, and the elderly. It explores how the physical and social characteristics of neighbourhoods affect public health, crime rates, social capital, and job opportunities while shaping the lifelong prospects of children. Analysing long-term trends, the book examines the importance of communications technology in the context of rising inequality and immigration. It shows how, as homeownership rose, neighbourhoods became vital settings for investment and increasingly financialized, which has reduced affordability. Using examples from all types of neighbourhoods in cities small and large, from St. John's through Montreal and Winnipeg to Victoria, *The Rise of the Neighbourhood in Canada* argues that the current prominence of neighbourhoods will persist.

RICHARD HARRIS is a professor emeritus of urban geography at McMaster University.

THE RISE OF THE NEIGHBOURHOOD IN CANADA, 1880s–2020s

Richard Harris

UNIVERSITY OF TORONTO PRESS
Toronto Buffalo London

ISBN 978-1-4875-0063-4 (cloth)   ISBN 978-1-4875-1150-0 (EPUB)
ISBN 978-1-4875-2044-1 (paper)   ISBN 978-1-4875-1149-4 (PDF)

Library and Archives Canada Cataloguing in Publication

Title: The rise of the neighbourhood in Canada, 1880s–2020s /
   Richard Harris.
Names: Harris, Richard, 1952– author
Description: Includes bibliographical references and index.
Identifiers: Canadiana (print) 20240536177 | Canadiana (ebook)
   2024053624X | ISBN 9781487520441 (paper) | ISBN 9781487500634
   (cloth) | ISBN 9781487511500 (EPUB) | ISBN 9781487511494 (PDF)
Subjects: LCSH: Neighborhoods – Social aspects – Canada – History. |
   LCSH: Neighborhoods – Economic aspects – Canada – History.
Classification: LCC HT127 .H37 2025 | DDC 307.3/3620971 – dc23

Cover design: Val Cooke
Cover image: iStock.com/stevanovicigor

We wish to acknowledge the land on which the University of Toronto
Press operates. This land is the traditional territory of the Wendat, the
Anishnaabeg, the Haudenosaunee, the Métis, and the Mississaugas of the
Credit First Nation.

University of Toronto Press acknowledges the financial support of the
Government of Canada, the Canada Council for the Arts, and the Ontario Arts
Council, an agency of the Government of Ontario, for its publishing activities.

Canada Council   Conseil des Arts
for the Arts   du Canada

ONTARIO ARTS COUNCIL
CONSEIL DES ARTS DE L'ONTARIO
an Ontario government agency
un organisme du gouvernement de l'Ontario

Funded by the   Financé par le
Government   gouvernement
of Canada   du Canada

Canada

For Carol

Every day she teaches me about our neighbourhood.

Love thy neighbour as yourself,
But choose your neighbourhood.

– Louise Beal

Contents

List of Figures   ix

Preface   xi

1 A Historical View   3

Part 1: Eternal Verities

2 Neighbouring among Classes   17

3 Others with Less Choice   50

4 Neighbourhoods: How Varied?   85

5 The Effects of Neighbourhoods   119

Part 2: Changed Meanings and Contexts

6 Language, Meaning, and Governance   155

7 The Changing Context   195

Part 3: Reflections

8 Looking Forward, and Beyond Canada   227

Appendix   237
Tables   243
Figures   251
References   269
Index of People and Subjects   333
Index of Places   347

Figures

1 When gentrification came to Cabbagetown: Average price of residential property in Don Vale, 1945–1982   251

2 Although Montreal renters moved frequently until well into the twentieth century, their neighbours often included family   252

3 Located on Toronto's fringe, Jean Bradley's childhood stomping grounds in the 1920s and 1930s necessarily included a wide territory   252

4 In an early postwar subdivision of Grimsby, Ontario, neighbourhood children moved in packs from one unfenced backyard to another   253

5 In Vancouver's elite West End in 1908, women organized "at homes" for neighbours on adjacent blocks   253

6 Winnipeg's skid row, 1976. Although the term is not used much now, many cities still have a cluster of blocks with homeless shelters, rooming houses, and social service agencies   254

7 In 1910, facing discrimination, Chinese Canadians settled in a tight cluster in downtown Vancouver   255

8 Shunned by white colonial settler society, and with limited resources, Métis formed a small community on Winnipeg's fringe in the early twentieth century   256

9 Montreal, 1942. In the dense urban environments that once characterized many eastern and central Canadian cities, children played on sidewalks and streets   257

10 Gangs were usually male-dominated and defended a well-defined urban turf. In the 1950s, Burnsiders were an exception   258

11 Sydney, Nova Scotia, c. 1913. Even at the block scale, in immigrant districts there was a mixture of ethnicities   259

12 Calvert, Newfoundland, 1980s. Even within the smallest communities, geographical distinctions are made   259

13 In Montreal, where ethnic segregation was the norm, growth in the late nineteenth century enabled class segregation to emerge within each group   260

14 Africville, Halifax, in 1958, shortly before its demolition. Several other Black neighbourhoods existed before migration from the Caribbean picked up in the 1960s   261

15 Carol Talbot's mental map of the tight-knit Black community in Windsor in the 1940s   262

16 By the early twenty-first century, Calgary's common interest developments were distributed widely across its suburbs   263

17 A map that shows the high degree of segregation of Montreal's Jewish community. Rabbi Glazer was rarely required to travel far from home   264

18 Montreal, 1880. Reflecting the city's social geography, infant mortality rates ranged widely in the late 1800s and early 1900s   265

19 Ad hoc zoning. Over half a century, the petitions of local residents created an intricate map of restrictions that excluded non-residential land use from large parts of Toronto   266

20 Sainte-Foy, Quebec City, 1970s. Rearrangements of school catchment areas aroused widespread concern and opposition   267

Preface

I have been circling the subject of neighbourhoods ever since I came to Canada, forty-eight years ago. As a student from Britain, I found an apartment in the home of Mrs. Charitoniuk, a Ukrainian immigrant living in Kingston's working-class and (somewhat) immigrant North End. Since then I've experienced and intermittently thought about the neighbourhoods in my own life, in Sutton Coldfield, England, as a child, Columbus, Ohio, Kingston, Vancouver, Toronto, and now Hamilton. In my research I've tackled subjects from various angles, writing about segregation, residential patterns, and suburban development, as well as homeownership, schools, and home improvement. But it took David Hulchanski's request – to write a book chapter on the history of Canadian neighbourhoods and neighbourhood planning – that brought it all into focus. Does that make it sound preordained? It feels that way.

Of course, as always, I underestimated the task. Over a decade, the project grew from a chapter into an entire book. In the process, I found myself looking with new eyes at material I thought I knew, and discovering much that I should have known. I have enjoyed learning about my adopted country, finding connections, contrasts, and trends that I had not suspected and which, to my knowledge, no one else has seen – or at least recorded.

Many things have made this possible. The first is a way of thinking: historical synthesis. Particularly in chapter 7, I present some new research, but as a whole this book is, above all, a synthesis of what others have shown, or suggested. I hope that this will be useful to anyone who wants to understand more about an important part of the Canadian urban experience. But a synthesis is more than a compendium. A number of the more important topics have rarely been discussed in relation to one another. Surprisingly, for example, those who study schools, homeownership, and neighbouring have not talked very much to each other. When you bring such topics together they tend to acquire, and sometimes change, meaning: connections, contrasts, and changes reveal themselves. Some of these, I believe, emerge from the outset, but then, from chapter 6 onwards, start to come together in a narrative argument.

Historians love to tell such stories, and they have to believe that the past is interesting. But what about those of you who care much more about how we live now, and maybe how we can improve things? To you, let me make a pitch: we need to consider the past because this shows us the value of taking a historical perspective. No, that's not a tautology. The past is one thing; viewing things from a historical perspective is something else entirely. Ask Putin: he invaded Ukraine in part because – supposedly – it had "always" been part of Russia.

To understand neighbourhoods, a historical way of thinking is useful for three reasons. First, some of the most important ways in which neighbourhoods matter only become apparent after many years have passed or, in some cases, a generation or more. A snapshot cannot show us that. Second, like everything else in this world, neighbourhoods are always changing, as people, buildings, and vegetation age – or mature, if you prefer. We can try to influence change, but we cannot stop it entirely. And finally, how neighbourhoods evolve depends on a complex combination of forces, underlining the fact that single-minded interventions will have unintended, and probably unfortunate, effects. Looking at things historically can give us better understanding, coupled with much-needed humility.

Speaking of humility, let me acknowledge how much I owe to others. The project got underway with assistance from Neighbourhood Change and Building Inclusive Communities from Within (David Hulchanski, principal investigator), a partnership project funded by the Social Sciences and Humanities Research Council of Canada. This supported the work of four research assistants who surveyed local planning documents: Will Gregory (Halifax), Alexandre Maltais (Montreal), Amy Shanks (Toronto), and Emily Hawes (Vancouver). Amy Shanks also helped with the tabulation of initial results from the *Globe and Mail* and *New York Times* databases, and gathered articles by W.A. Craick from the *Globe*. Jeremy Parsons translated some French-language materials, while Geoff Rose helped with the tables.

Academic life is not without its share of competitiveness, but for all intents and purposes scholarship is a collective endeavour based on trust. Over the years I have been continually impressed by the spirit of generous cooperation among my associates. In the following acknowledgments, no doubt there are inadvertent omissions, for which I apologize. But I would like to thank the following people for their help in offering comments, suggestions, corrections, references, and as tangible expressions of intellectual community. Let me start with some non-Canadians who have offered suggestions at conferences, or through email, on various elements of the arguments that I make here. Apart from reviewers on a couple of journal articles, they include William Baer, Joe Bigott, Kevin Cox, Richard Dennis, Jack Dougherty, Claude Fischer, George Galster, Amanda Seligman, Emily Talen, and Alexander von Hoffman. Closer to home, David Hulchanski, Pierre Filion, and Richard White, not to mention three anonymous readers, offered helpful comments on all or part of an early draft. And then the following answered questions or offered invaluable suggestions on specific topics: Alison Bain, Jean Barman, Luca Berardi, Robert Boschman, Matthieu Caron, Ken Cruikshank, Brian Doucet, Jason Ellis, Peter Goheen, David Gordon, Jill Grant, John Hagopian, Dan Hiebert, Leslie Kern, Fran Klodawsky, Nathaniel Lauster, David Ley, Heather MacDougall, Phil Mackintosh, Greg Mercer, Janet Muise, Stephen Muzzatti, John Phyne, Mark Rosenberg, Howard Ramos, Chris Sharpe, Jo Shawyer, Mariana Valverde, Alan Walks, John Weaver, Carolyn Whitzman, and Trevor Wideman.

Special thanks go to Renisa Mawani, Evelyn Peters, Jordan Stanger-Ross, and Kaylah Vrabic for helping me come to terms with the historical urban experience of Indigenous Canadians, a subject on which I was – and some may say remain – regrettably ignorant. And I am pleased to take this opportunity to thank those who helped me overcome the deficiencies in my French-language skills by making sense of the francophone, and sometimes also the anglophone, literature on Quebec: Claude Bellavance, Michèle Dagenais, Lucia Ferretti, Raphaël Fischler, Michael Fox, Annick Germain, Sandrine Jean, Pierson Nettling, Sherry Olson, Thierry Ramadier, Damaris Rose, and Robert Sweeney. On that matter I owe a special shout-out to Harold Bérubé, who answered numerous queries with unfailing courtesy and knowledge.

And then heartfelt thanks to those who were there at the beginning and the end: to Carol, whose social instincts are unflagging, not only at the block scale, and on the last lap to Jodi Lewchuk and Anne Laughlin at the press, who provided helpful guidance, suggestions, and corrections, large and small It takes a village – or is it now a neighbourhood? – to make a book.

THE RISE OF THE NEIGHBOURHOOD IN CANADA, 1880s–2020s

chapter one

A Historical View

Like everything in this world, neighbourhoods have a history. That doesn't just mean they have a past: each one of them is constantly changing. Buildings age and are sometimes replaced; likewise, residents age and move on, to be replaced by others who are never exactly the same as those who left. The wider city grows, or occasionally contracts, and adjacent areas evolve. And so the character and meaning of any neighbourhood, which depend on context as well as on constituent elements, undergo steady transformation (Massey, 1995a).

These changes are commonly a source of regret. Many people, and not just older folks, are attached to their place of residence, and sometimes also to what it was before they moved in. Or what they think it was. The image and meaning of their neighbourhood are of course based on their own experiences – of making friends, of raising children, and perhaps of growing up – but often also on an understanding, impression, or idealized sense of what that area once was. In all of these ways, we associate "neighbourhood" with "community," often treating these terms as synonymous. That is why change is a source, variously, of sadness, resignation, regret, and even anger, as that community is threatened, changed, or simply gone. Thomas Nashe captured this. Acknowledging popular concern about the demise of community values, he claimed that neighbourhoods were effectively "dead and buried" (quoted in Wood, 2020: 1).

We might want to take his claim with a pinch of salt. After all, he was writing about London, England, in 1590. It would probably be possible to find comparable statements for London in every subsequent century, and in Canada for every decade since at least the 1880s, which is when we started to become an urban nation in a serious way. Of course, this could imply that the decline of neighbourhoods has been a very drawn-out process. More probably, it expresses the yearnings of every generation for a real or remembered past. I believe that the Canadian experience bears this out, and much else besides. It's true that neighbourhoods aren't what they used to be, but if they are less significant in some ways they are more important in others. That is the complicated story that this book aims to tell.

In the midst of change there are eternal verities. Even homeless people sleep somewhere, and in choosing where to lay down our heads we all pay attention to our surroundings. For over a century, a growing proportion of Canadians have been doing that in cities; "neighbourhood" is the word that anglophones, at any rate, now use when referring to their surroundings. (Francophones might use *voisinage* or *quartier*.) Ideally, it's a place to be comfortable, to make friends, to fit in, and, incidentally, to participate in the making of a community, or, at the very least, a life.

Of course, neighbourhoods, like Canadians, are not all the same. Each of us fits more easily into some places than others, depending on our income, education, tastes, ethnicity, age, and maybe sexual orientation. We pick and choose, and in the end – in varying degrees – self-segregate. Anyone who has bought a house knows this, as they ponder the attributes of the local schools, the neighbours, the public spaces. Tenants usually fret less about these things, maybe because they don't have children, don't plan to stay put for as long, or simply have less at stake financially, but they are rarely indifferent to where they live. After all, everyone cares about public safety and, following the pandemic, public health.

The problem is that many of us have little opportunity to pick and choose. Income is the obvious constraint. And, even in Canada, there's discrimination too. Visible minorities – a polite, very Canadian term – know this. So do gays and lesbians, more in the past than the present, but still. And, even when admitted to a neighbourhood, not everyone feels at home. Residents' associations often treat tenants as second-class citizens. It is easy for educated, middle-class, white Canadians like me to forget, or indeed remain oblivious to, the constraints and insults experienced by others. But they are real enough, and have always been part of the story of Canadian neighbourhoods.

But what is this thing called neighbourhood? It's elusive, existing on several scales. The most obvious is the block or, depending on where you live, the high-rise. For good or ill, we cannot help but encounter our immediate neighbours; we enjoy space, privacy, and greenery, or else we rue their absence. That is why the first section of this book is concerned with this, most intimate of scales, the one for which francophones have a name – *voisinage* – and for which everyone else should. As the next chapter argues, unavoidable and very local experiences are shaped by class: our own, and that of the larger society. After all, with few exceptions, our homes and environs have been created by developers for market segments. We insert ourselves into that environment according to what we can afford. The resulting block-scale patterns apply differently to workers than to the middle class or the socio-economic elite, especially in times of hardship, but few people have been indifferent to their immediate surroundings.

Social and economic class matters, but especially for the poor, who must scramble to find places to live, and also for those who need the support of neighbours. Sometimes, this support has come courtesy of poor people's other identities. Immigrants have routinely drawn together for community and mutual aid; that has also been true for First Nations and Métis, notably since the 1960s as many have moved off reserves. Both types of groups have also clustered because, in varying degrees, they have been treated as outsiders by dominant white society. Others, of whatever class or ethnicity, face limited social mobility, including children and the elderly. For them, immediate neighbours often count for a lot. And then there are adult caregivers. In principle, that applies to any gender; in practice, even after they entered the job market in huge numbers, it has usually been women who do most of the unpaid caregiving. In the process it is they who usually build the local social networks that go into the making of family and neighbourhood life. The character of these varied and constrained local experiences is explored in chapter 3.

Local social networks usually extend beyond the block – a lot happens at the larger scale which we usually call "the neighbourhood." Depending on one's point of view this scale is frustratingly ambiguous or richly complex. Think about the area where you live. Does it have a name? If so, what makes its identity distinct from the next one over? Where is its boundary? Would your neighbours answer those questions the same way? And would the residents of that nearby neighbourhood, let alone the folks at city hall, agree with you? If you have definite, positive answers to all of those questions your neighbourhood deserves a medal. Most of us live in places with boundaries that are multiple and blurred. As I discuss in chapter 4, that has always been true, although some persistent patterns are clear: the rich and the poor have always lived in distinct districts, with immigrants and racialized minorities often being further segregated. The scale of segregation – of working-class districts, for example, or of elite and immigrant enclaves – has always been greatest in the larger cities, and has grown since the 1880s as towns became cities and then sprawled into metropolitan areas. That segregation has also applied to gay villages, although only in the past half century, as openness about sexual orientation has become more publicly accepted. The possible combinations of class, ethnicity, and immigrant status, not to mention age and sexual orientation, are endless. Some are apparent in every Canadian city, but in differing proportions, helping to define the character of each place.

All this complexity and ambiguity is of much more than incidental interest because neighbourhoods matter (chapter 5). Exactly how is hard to pin down, but the general effects are clear, and residents know it. In the short term our mental and physical health depends on the environment, both social and physical. That was true during the Spanish Flu of 1918–19 as it was during the COVID-19 pandemic a century later. Local settings also shape lives and prospects over the longer term, especially for children. Residents, especially middle-class home-owning parents, know this and work to enhance the positive and mitigate the negative.

They prod municipalities to provide more and better services, or to prevent changes that threaten the local quality of life, not to mention property values. How governments respond depends on the power and public image of the neighbourhood, which are often shaped by newspapers, television, and social media. Neighbourhoods acrue positive or negative connotations; location and reputation matter.

So, intriguingly, does the word "neighbourhood" itself. At this point, it becomes useful to think about how things have changed, the subject of Part II. Today, everyone talks about neighbourhoods. 'Twas not always thus (chapter 6). My exploration of the *Globe and Mail* shows that a century ago this word was much less frequently used, and was often associated with low-income, inner-city, immigrant slums. The paper used other terms, notably "residential district" and then "residential area," usually when referring to more prosperous areas where residents had persuaded the city to regulate land use. These were not synonyms for "neighbourhood": they referred specifically to areas with particular types of land use regulations and were probably not in everyday use. The point is that the widespread use of "neighbourhood," with its modern connotations, has been a postwar phenomenon, albeit one associated with municipal planning.

Some of what I call the "rise of neighbourhoods" – and a convergence in meaning with the French *quartier* – can be explained by the growing importance of municipalities. The urban reform movement of the early twentieth century gave new powers to local residents, but not under the neighbourhood label. Things began to shift in the 1920s. Complementary changes in the nature of suburban development and, during the Depression, of inner city "slums" nurtured a new understanding of the housing market and, with it, of neighbourhoods and neighbourhood change. These changes bore fruit after 1945, as municipal planning emerged to shape suburban subdivisions and, less frequently but more controversially, inner-city redevelopment. From the 1960s – later in Montreal – local battles led to demands for greater participation in planning. Lately, this focus on inner-city redevelopment has triggered tensions between gentrifiers and planners, with their mantra of intensification. At the same time, governments have paid more active attention to low-income, immigrant neighbourhoods than they did a century ago. In sum, neighbourhoods and public governance – including that of schools – have become intimately intertwined. This helps to explain the growing profile of neighbourhoods, especially since the 1960s, a trend some cities have embraced to the extreme, proclaiming themselves to be "cities of neighbourhoods."

But the growing role of government cannot account for the changed significance of neighbourhoods. As I argue in chapter 7, the changes in neighbourhoods have reflected broader changes in Canadian urban society. Municipal government activity does not entirely explain the cyclical character of interest in neighbourhoods, with a moderate peak in the early twentieth century and a much higher one today. In both cases, although in varying degrees, this interest was associated with the growth of low-income, immigrant districts. These two periods saw high levels of immigration combined with great social inequality. In between, at least by Canadian standards, immigration flagged while, notably from the 1940s to the 1970s, inequality fell. Layered onto these cycles are major trends. As far as we can tell, neighbourhoods are indeed less important sources of social life – community, if you like – than they used to be. We have become more mobile, travelling farther to work and shop. Now that most women have joined the paid labour force, adults spend less time in and around the home. COVID-19 briefly reversed this trend, and for people in some occupations this shift may have become permanent, but a long-term demographic shift is clearly underway. Furthermore, it has become easier to maintain social connections remotely – to create communities of a different sort. Radio, television, and now social media enable us to keep in touch with people and events everywhere, enlarging our worlds. That is even true for the very young, once glued to *Sesame Street* and now to *Paw Patrol*. All this has surely reduced the significance of neighbourhoods for the building of community.

But I believe this trend has been outweighed by two counterforces. The first is the growth of homeownership. In 1900, two-thirds of households were occupied by tenants, and they defined the average neighbourhood. Today, two-thirds are homeowners. This matters because owners are arguably more invested in their neighbourhood: they move less often and have a greater financial stake in their homes. Although it may surprise some, the growth of homeownership has been especially notable among the middle class, who in 1900 cared more about the character and size of their home than whether they owned it. Over time these residents would be more likely to demand services and resist change. But then, notably since the 1980s, when "NIMBYism" became a byword, and as the increase in house prices have generally exceeded the rate of inflation, the financial motives behind homeownership have become dominant. Today, homeownership is (supposedly) not just a safe investment but a shrewd and necessary one, at least for those who can get on the train. And, following a similar trajectory, families with children have come to attach greater significance to education. Given that more than nine out of ten children go to public schools, mostly in their own neighbourhoods, parents struggle to get into good school districts, fight school closures, and resist attempts to fiddle with catchment areas. Because of the growing significance attached to both home investments and formal educational qualifications, neighbourhoods have never mattered more.

Some readers may wonder how useful it is to take a long historical view. After all, having struggled with the pandemic, we are now worrying about rents and house prices, and in general trying to figure out how to manage the present and near future. Will we work from home and value neighbourhoods more, less, or just in different ways? Everything seems to be up for grabs. Didn't the pandemic make the past irrelevant? It is true that we would be foolish to simply extrapolate from the recent past to the near future. It is also true that the past does not offer tidy lessons, precisely because circumstances change, and so our responses should, too. But looking to the past encourages us to take a historical *perspective*, which can in turn help us think about the present. We can go about this in three ways, which I will call the "everyday," the "analytical," and the "grand."

First, let us examine the everyday. If you are at home and in an urban place, look outside. Think about how your block has changed since you moved to it. If you have been there for years, the social environment has changed. Neighbours have come and gone; children have grown up and maybe left home. The physical environment has changed, too. Buildings have aged and either deteriorated or matured, depending on your point of view and their owners' views about maintenance; trees and bushes, if your street has any, have grown, died, been cut down, or replaced. Neighbours have carried out renovations and built additions; maybe a speculator has bought up a property only to tear it down. Some of these changes are barely perceptible, but in innumerable ways the neighbourhood is in flux.

A less visible but important way that neighbourhoods change has to do with how how they fit into the rest of the city. When we talk about neighbourhoods, we think first of older districts close to the centre. In the suburbs we are more likely to talk of "subdivisions" or "developments," perhaps using the marketing names invented by their developers. Sometimes the urban and suburban seem like different worlds. But what is now a suburban subdivision will eventually become an urban neighbourhood. Even without acquiring more density, its relative position changes; in time it becomes encircled by newer developments, absorbed into the urban fabric, and served by public transit. Suburbs have often been damned for being monotonous and bland. But this assessment overlooks the fact that every neighbourhood, even those prized by urbane gentrifiers, was once suburban. In the 1880s, Toronto's Cabbagetown was fresh-faced, with Victorian rowhouses as uniform as any of the tract homes now sprouting at the urban fringe. Nothing about a neighbourhood, or neighbourhoods in general, stays the same.

Second, we can take an analytical approach. Social scientists do this, in attempting to understand things that matter to lots of people. One of my main reasons for writing this book, and I assume that one of your reasons for reading it, is to figure out how neighbourhoods matter (the preoccupation of chapter 5). Lately, researchers have tried to quantify this, for example by tracking the educational achievement and job prospects of children who grow up in different districts but whose circumstances are otherwise similar. It turns out to be difficult to put a number on that, above all because of what social scientists call "selection bias." If they can, parents who value education will move into an area with reputedly good schools. If things turn out as they hoped, is it the parents, the school, the neighbourhood, or all three that contribute to their children's success? Or what about a recent immigrant who picks an area where others with similar backgrounds and values live? If many make the same choice, the area will attract and support stores and social institutions that provide practical and moral support to the community. But is that an effect of the neighbourhood, or a result of the needs, aspirations, and choices of those who made it what it is? The answer is surely both. We cannot disentangle this interplay of choice and environmental influence by taking a snapshot. The processes happen over time, some taking a generation or more. To understand the significance of neighbourhoods we need a historical perspective.

Third, we need to maintain a grand vision. The past does not provide answers, but it does offer perspective. It shows us that change is inevitable, that events can upend our assumptions, and that we can adapt. Emphatically, this does not mean that everything will (or should) revert to normal, whatever that is. That has never happened. New patterns – new types of neighbourhoods – will emerge. We might find that people in the past faced similar problems, and developed solutions that may be useful today. But we should be wary of looking for simplistic solutions, based on single-minded obsessions. Looking at the past, we see that life is not a controlled experiment, where one thing changes while everything else is held constant, allowing us to trace the effect of that one variable. Things are continually, simultaneously changing, so that the effects of one event become contingent on others. That is even more true in times of crisis. The past shows us concretely, not hypothetically, how complex the interplay of events can be. That is a daunting lesson, above all for those whose job it is to respond to the changes we are living through now. But it is a necessary one, because responses that take account of only one or two variables will get things wrong, as has happened too often, for example with attempts at urban renewal in the 1960s. And if changes are carried out with forceful conviction the results can be regrettable, or worse. The past is messy, and so is the future: planning for it, we must expect the unexpected, and exercise humility.

Any book about urban neighbourhoods that covers more than a century must generalize, or else drown in detail. The problem is that even today, and obviously in the past, there is so much we do not know, and what we know varies over time, with the changing interests of academics and the preoccupations of the wider society. With the partial exception of chapters 6 and 7, my account relies on the work of others: historians, academics in other fields, writers of fiction and non-fiction, and reporters. In that sense, it is a work of synthesis that bridges several divides: between social scientists and historians, and between those with different interests, notably housing, education, local politics, public health, and everyday social life. It's important to take a comprehensive view of these subjects, especially in an urban context, because each affects the other, sometimes profoundly. It is possible to talk about rural neighbourhoods, and some people still do. But the urban context condenses social relations to create a uniquely complex dynamic. When the elements are assembled and put into motion they tell their own story.

Unavoidably, this account also relies on a written record which has biases – social, historical, and geographical. It privileges a literate middle class, the people whose ideas and views most easily find their way into print. Even then, it also leans towards those experts and bureaucrats who worked for public agencies, for it is their job to record, think, and write about cities. And so the experience, views, and actions of most urban residents appear, if at all, at second hand. Semi-autobiographical fiction, memoirs, and oral histories can give us insights into everyday experience. More systematic evidence is available in newspapers, including the *Globe and Mail*, archived digitally from 1844. A commercial operation, the *Globe* reflected the changing concerns and experiences of its readers, at first locally in Toronto but then nationally.

This makes it difficult to recapture the experiences of many Canadians, including children and first-generation immigrants, many of whom lack fluency in English or French. Settler Canadians are only just beginning to understand the urban experience of Indigenous peoples. Given that it is only in recent decades that they have moved in large numbers to urban areas, their experience is in some ways comparable to that of immigrants. But in others it is profoundly, and indeed morally, different.

Another type of bias, equally vexing, is geographical. We know far more about large cities than about the smaller centres in which most urban Canadians have lived. Big cities have always figured disproportionately in newspaper reports, in fiction, and in the writings of social scientists and historians. The problem is that these big cities are atypical. They have attracted a higher proportion of immigrants, while enabling more types of segregation and on a larger scale. Toronto and Vancouver, in particular, have experienced the greatest pressures for growth and redevelopment, of the sort that triggers local opposition. And then within major metropolitan centres, a disproportionate amount of writing has been devoted to a handful of neighbourhoods: Cabbagetown, the Annex, and Don Mills in Toronto; the Downtown Eastside and the West End in Vancouver; the St. Urbain corridor in Montreal. Perhaps none has received as much attention as The Ward, recently billed (not quite accurately) as "Toronto's First Immigrant Neighbourhood" (Lorinc et al., 2015). Studying such high-profile neighbourhoods can offer insights into the wider urban experience but often at the expense of neglecting the "ordinary." And then, to complete this litany of challenges, even for well-documented places, there are substantial holes in our understanding of the processes of change. Some of these can be filled, or at least bridged, with evidence from the United States. In important ways, the experiences of our two countries have been similar: we are nations of immigrants that prospered and urbanized on a similar trajectory; we have seen parallel waves of social inequality, of trends in homeownership, and in the salience of formal, public education. On some of these topics, including the everyday practice of neighbouring, the American literature is abundant, and it would be obtuse to ignore it. That said, the particular areas where I have relied most on that literature do suggest an agenda for Canadian research, an issue to which I return in some concluding reflections.

I have also drawn on American evidence for two other reasons: to compare and to generalize. Canadians have always compared themselves with their American cousins. We delight in touting our health care system – although now with more qualms than previously – and our more extensive transit systems, while deploring their gun culture, race relations, and polarized politics. Sometimes we have envied their prosperity, influence, and self-confidence. All of these elements, and more, have shaped the character and dynamics of neighbourhoods, and American comparisons can help us to understand the distinctiveness of our national experience. Conversely, comparison allows us to speculate whether US metropolitan areas have also seen a "rise of neighbourhoods" (Harris, 2024). This book is primarily addressed to Canadians – hence my use of "we" and "our" – and is about the Canadian experience. But, as I also suggest in the conclusion, there is a strong case to be made that that experience has parallels, and indeed louder echoes, south of the border.

One of the comparisons is intranational, concerning Quebec and English Canada. These are two cultures and sometimes, it seems increasingly, two solitudes. Each has had its own ways of acting in, and speaking about, urban space. One of the greatest challenges in telling the national story of Canadian neighbourhoods is that of figuring out how Quebec's experience has compared with that of English Canada. Some differences have been obvious, but I have also been struck by the parallels and indeed the convergences. Postwar, these include the rapid growth of homeownership in Montreal and the evolution of *quartiers*. Whatever the name, I argue that the rise of neighbourhoods has been a national experience that has transcended borders, both provincial and national. But that intellectual journey of a thousand sentences starts with a step towards our next-door neighbours.

PART 1

Eternal Verities

chapter two

Neighbouring among Classes

Not only do we need to learn who our neighbor is, but also how we can help him ... we must learn to be neighborly.

J.S. Woodsworth

The Bible tells us to love our neighbours and also our enemies. Probably because generally they are the same people.

G.K. Chesterton

It is natural to care about who our neighbours are. After all, like our coworkers, they are hard to ignore. We run into them on the street or, in a high-rise, in the hallway or elevator. If you live in a house, you might find yourself mowing the lawn or shovelling the sidewalk at the same time. If they live next door you might have to deal with the fallout from their renovation project. You will know when they (or their adolescent children) are having a party, what sorts of music they like, and maybe whether they argue a lot. My wife and I will never forget our ex-neighbour's chained and relentlessly barking dog, kept outdoors, and a fellow apartment-dweller who worked nights and sometimes left his stereo on full blast when he went out. Jack Batten's history of Toronto's Annex neighbourhood used various sources and covered different subjects and time periods, but the personal connections he drew on were with the people who lived on his block on Albany Avenue (Batten, 2004: 13–16).

Those of you who live in townhouses or high-rises will be especially attuned to those on the other side of your walls, ceiling, or
floor. Indeed, South Korea, where most city-dwellers live in high-
rises, has a public agency that handles noise complaints. The Korea
Environment Corporation reported that in 2020, as people spent
more time working from home, complaints went up by 60 per cent.
Heavy footsteps were the leading irritant, followed by hammer-
ing, furniture moving, and door slamming. Residents of some
of Canada's less-well-built condominiums can probably relate –
apparently, the existence of condo rules encourages complaints
(Lippert, 2019: 138; McGinn, 2013a). Wherever we live, whether
we actually get to know our neighbours, beyond polite cursory
exchanges, is up to us (and them). But one way or another we can't
avoid them.

James Lorimer and Myfanway (Mif) Phillips have advised
neighbourly caution in what is still the most sustained, if unstruc-
tured, investigation of neighbouring in a Canadian city. In 1966, the
couple moved onto a street he calls "Minster Lane," on the fringe
of what had been Toronto's Cabbagetown neighbourhood, "East
of Parliament" to locals. Lorimer had an eighteen-month grant to
undertake an ill-defined study of the area, and the couple's joint
goal became to "live the life of ordinary residents, but as people a
bit more curious about their neighbours and neighbourhood than
people usually are, and certainly with more time to spend talking
to people"; the methodology was that "we were mutually anxious
to meet people" (Lorimer and Phillips, 1971: 4, 16). In the book
about their experiences, with photographs provided by Phillips,
Lorimer made many telling observations, some of which I will
mention later.[1] The most general of these concern the nature of
the neighbouring relationship. Close friendship, he suggested, is
unusual, or at least risky, because "once someone is friendly with a
neighbour it is far more difficult to withdraw from friendship than
from family or friends who live in different locations" (45). That

[1] It is clear from his use of the first person singular that Lorimer wrote the book.
Accordingly, when quoting, I attribute all text to him alone even though the work
cited is co-authored.

is one reason why "relations among neighbours are of a special, rather limited type" (46–7).

That said, the stereotype is that, at least for minor everyday matters, neighbours are there when we need them; that is why "neighbourhood" and "community" have been so often linked. It is easy for me to think of examples. After snowstorms, Ron Adams, a neighbour of mine who lives at the end of our block, typically gets out his snowblower and does the entire sidewalk on our side of the street. One of the houses on Ron's route was once occupied by a family by the name of Plant. When, aged 91 and suffering from Alzheimer's, Polly Plant had two strokes and was confined to bed, her daughter Monica, who had been looking after her, reached out to a couple of neighbours. They in turn contacted others and, in the end, for a month almost every household on the block was making meals, buying groceries, doing yard work, responding to night calls, or undertaking respite care, sitting with the invalid in her second-floor bay window (Plant, 2016, 2017). An urban barn-raising, albeit one undertaken mostly by women.

Drawing on such anecdotes, most of us tend to overstate the case. We value community, hoping to find it near at hand. In general, Canadians do seem to be aware of the value of neighbours. A large Canadian survey showed that three-quarters of urban residents saw their neighbourhood as a place where people helped each other; one-fifth had received a neighbour's help in the preceding month and one-sixth had offered help (Ray and Preston, 2009). Connections were considered beneficial. Two-thirds felt a strong sense of belonging to their area, and almost nine out of ten trusted their neighbours a lot … or at least somewhat. Trust and familiarity went together, as a smaller Toronto survey showed: "by far the most important influence on social trust is knowing one's neighbours" (Toronto Foundation, 2018: 26). This sounds good. But how durable and significant are these connections? I have lived on my block for thirty-six years and, in my experience, when people move away social connections are (more or less quickly) dropped; that was even true for Monica after she sold her mother's house and moved on. Most of the time, people go about their lives in mutual ignorance, placidly indifferent to the lives of those around them. In the large survey, only two-fifths of respondents knew many neighbours, and more than 10 per cent did not trust their neighbours. Only 2 per cent had been helped by a neighbour in finding a job. Helping out usually goes only so far. Neighbours matter, but not in the same way, or as much, as close friends or kin.

And there can be conflicts. In the nineteenth century, when municipalities were busy making muddy streets more passable, they sometimes gave residents a choice of surfacing materials. On each block the majority ruled, often in favour of the cheapest option, thereby angering the minority. Exaggerating only a little, Phil Mackintosh (2020: 113) reckons this produced a "near-social dissolution of neighbourhoods." During the Depression, one resident of the East York suburb of Toronto recalled that "a neighbour across the road was a bit on the nosy side and she reported my husband and me to the welfare department" (Schultz, 1975: 10). Surveillance with an edge! Today, on my block, probably quite typical, conflicts have arisen about parties, the use of power tools and, most recently, a proposed back-lane redevelopment. (This really stirred some people up!) Dissension among residents in condos can also be strong, although often more muted because many are tenants and attendance at monthly meetings is restricted to board members. There is a social media site for immediate neighbours, appropriately called Nextdoor. Started in the United States, it launched in Canada in fall 2019. The company's Canadian head observed that those who use it "tell … me that I'm surrounded by people who could use a hand – and who would probably lend one back if I asked" (Grainger, 2020). But he also conceded that there are feuds and pettiness online "as in any real-life neighbourhood."

Occasionally, some neighbours do become friends, and not just in the social media sense. Other neighbours may be kin, bringing its own benefits and tensions. But most are neither and, as Suzanne Keller (1968) has explained, our relationship with them is unique. The ones who matter most live nearby: next door, across the road or, like Ron, just down the street. That was Lorimer's (1971: 5) experience: the four families that he and Mif got to know best all lived within a block. As the author (Roberts, 1993: 38) of an outstanding British oral history project showed, this is the scale of

"real significance" to most people. French Canadians have a specific term for it, *le voisinage* – ironic, since in 2003 a survey revealed
that the average Quebecker knows fewer neighbours than did residents of Ontario and British Columbia, and was less inclined to
trust them (Ray and Preston, 2009: 233).

Local trust is important in relation to crime. The current thinking
about crime is that it is highly concentrated in just a few, narrowly
circumscribed locations. A study in Minneapolis found that half of
all crimes happened in 4 per cent of all "places," the latter being so
tiny that researchers defined 115,600 of them (Weisburd et al, 2016:
19). In Montreal and Toronto, too, the block is the intimate setting
at which crimes are concentrated (Boivin and de Melo, 2019). In
practice, that is the scale at which Neighbourhood Watch groups
are most effective. Assuming that the block was the most important scale of social life, Jane Jacobs (1961) famously argued that
safety is helped by having "eyes on the street" (c.f. Grant, 2011).
It turns out that women are better at recognizing strangers and
potential disorder, in part because they know their neighbours better (Gaub, Wallace, and Hoyle, 2022). My wife reckons she knows
almost everyone, and every car, on our block, and will notice if a
strange vehicle is parked on the street. The block, then, is a good
scale at which to begin to think about the ways in which, and for
whom, neighbours matter.

Physical and Temporal Settings

It may seem obvious that the physical setting – detached home or
high-rise, city or suburb – affects neighbouring activity. Indeed it
does, but it can be difficult to tell whether we are observing the
impact of the environment or whether people have chosen places
that suit their ways of life (a matter of "selection bias"). Both can
be true simultaneously, and usually are.

Think, for example, of a high-rise apartment, an unpromising
setting for neighbourly relations. Tenants move frequently and so
have less opportunity, and less reason, to build local connections.
The setting is not even conducive to maintaining connections that

already exist. In Singapore, the government has resettled villagers from the same kampongs into the same high-rises, but "the kampong spirit ... the trust and neighbourliness of village life ... was hard to transplant and often fizzled" ("The last holdout," 2021: 23). As one ex-resident recalled, "we only met neighbours along lifts or corridors and then sometimes it's hardly a hi or bye" (ibid.)

In many apartment buildings, there are few pretexts for connection. Hired help clean windows, mow lawns, clip bushes, and clear sidewalks, reducing the opportunities and pretexts for casual conversations. Not surprisingly, then, in a comparative study of Toronto's middle-class families Bill Michelson (1977: 146, 156) found that apartment-dwellers perceived their neighbours to be relatively "withdrawn," while they themselves were unlikely to make meaningful social connections nearby. Michelson suggests that self-selection, rather than the built environment, could account for much of this difference: some people simply prefer the anonymity of apartment living. Shawn Micallef is a fine example. In the course of a survey of Toronto's inner suburbs, Micallef (2017: 133) comments: "I don't know my neighbour. We say hi. She smiles at my dog. I hear occasional noises from her apartment that remind me she's alive. That's enough." But Micallef's neighbour might not feel the same. In Canada today, the single-person household is the most common type (comprising almost one-third of all households), a recipe for loneliness. As I write, a front-page article in the *Globe and Mail* reports the emergence of "companionship services" (Xiao, 2024). Once upon a time, escorts had a dubious reputation. Now, for an hourly fee, you can hire someone who will accompany you for the purposes of shopping, eating out, walking, or working out. There are plenty of people who have not chosen to ignore their neighbours.

High-rise living does not have to be anonymous. Design can make a difference. In Vienna, for example, multi-storey buildings encircle large green courtyards where children can play and adults socialize. Currently, a smaller project in Port Moody, BC, aspires to combat loneliness by doing something similar (Bula, 2024). Reputation can also help. Where buildings have acquired a reputation for community, self-selection can sustain a high incidence of neighbouring. Gerda Wekerle (1976) documented this in an upper-income cluster on Chicago's Near North Side. Such situations are unusual because most buildings contain at least a modest social mix. A recent study of British high-rises has shown that micro-segregation is quite common, with division based on floor or position in the building (Maloutas and Karadimitrou, 2022). More common in Canada are immigrant communities in rental apartment buildings, as M.G. Vassanji (1991) has described in his fiction, and Amy Harris (2010: 158, 198) has discussed in her study of a Toronto inner suburb. What has become more common are high-rise condominiums. Because buyers move less frequently than tenants and have a stake in the property, close neighbouring in condos probably falls between that in large rental apartment buildings and in blocks of single-family homes.

Many condo units are owned and occupied by single women; in her study of these residents Leslie Kern found a spectrum of practices and respondents. At one extreme: "for Ava, the hallways are like the neighbourhood streets, where she runs into friends and makes plans for further socializing," while in her pet-friendly building Samantha found that that "really brings people together" while Naomi reckoned that there is "a great deal of neighbourliness" (Kern, 2010: 112, 113, 114). But, at the other extreme, "April" declared, "I like being anonymous … I don't like people knowing my business because, you know, you don't know what creep lives next door"; meanwhile "Chloe" was very blunt: "I don't want to be bothered by people that much … I take the stairs to avoid small talk" (117). This is consistent with Randy Lippert's (2019) finding that few condo residents are interested in getting involved in their building's governance. Of those that Kern quotes, Sandra is average: she liked the seasonal barbeque but in general "I kind of like the fact that it's not like everybody on your street knows exactly what you're doing all the time" (117). Evidently, Shawn Micallef had a kindred spirit. And, significantly, the responsibilities of collective governance did not bring people together because "participation was seen as onerous" (123). Before the pandemic, many young single women had active social lives in downtown Toronto, but this did not routinely involve their immediate neighbours.

Along blocks of detached homes, a contrast is sometimes drawn, or implied, between older neighbourhoods and newer subdivisions. A British study undertaken in the early 1950s shows how misleading this can be. Studying two suburban council (public housing) estates near Liverpool, sociologist Peter Mann made a useful distinction. In Estate B he found that *most* residents wanted to be friendly and helpful; in Estate A, *the majority* preferred neighbours to be "friendly but [to] mind their own business" (Mann, 1954: 166). There were aspects to life in both areas unique to postwar Britain (and indeed Liverpool), but Mann's distinction between "manifest" and "latent" neighbourliness, respectively, crosses borders. One emphasizes openness and mutual aid, the other a "respect for the privacy of other peoples' lives" (Mann, 1954: 164).

But the operative words emerging from Mann's study are "most" and "the majority." There was no consensus on either estate, and that has always been typical, everywhere. People do not have identical preferences. Undertaken at much the same time, a study of a single block of homeowners in Bloomington, Indiana, found that residents differed enormously in terms of how many, and how well, they knew people on their block (Sweetser, 1941: 525–33). Canada's own Alice Munro provides a striking example. In the mid-1950s, she and her first husband, James Munro, moved with their two young children to North Vancouver. Her eldest daughter, Sheila, later recalled that her mother "hated" the neighbourhood: "the most serious threat to her writing didn't come from being a housewife and mother, it came from the neighbourhood women who were always dropping in for coffee unannounced or congregating in her backyard without invitation" (S. Munro, 2001: 29, 33). Alice later dramatized this in a short story, "The Shining Houses," featuring a block on which "any gathering together of people who lived there was considered a healthy thing in itself" (A. Munro, 1968). The Munros soon moved to a house in West Vancouver where, Sheila recalls, "the yards were bigger, the fences were higher, and there was more respect for privacy" (S. Munro, 2001: 31). Alice Munro's needs and preferences were unusual. But everyone differs in how they wish to connect with neighbours.

Mann's distinction is helpful, because meaningful, but we should beware of stereotyping any person or place.

The usefulness of his study extends beyond the concepts of "manifest" and "latent." He showed that the mutuality associated with inner-city neighbourhoods could exist in the suburbs. In fact, suburban sociality has often been reported. Alexander Von Hoffman (1994) has shown the rich associational life that developed in Jamaica Plain, a new neighbourhood of Boston in the late 1800s. Half a century later, such busy community-building had become a media cliché. What many missed was that there was a temporal aspect to this. Drawing on both American and British research, Suzanne Keller (1968: 68) suggested that the character of neighbouring varied with the phase of settlement, as the "eager interaction and mutual helpfulness" of first settlers gave way to "selectivity and withdrawal." Canadian sociologist S.D. Clark had made the same point two years earlier, in a wide-ranging study of suburban Toronto. For him, the intense socialization that some American writers had observed and labelled as the "suburban way of life" was limited to the settling-in period. Certainly, this was the experience of geographer Chris Sharpe. He recalls that, growing up in suburban Ottawa in the 1950s, "my world was largely constrained by my block," where he saw and experienced "a strong community spirit" (Sharpe, 2024). But "the camaraderie didn't survive." With such experiences in mind, Clark reckoned that in short order "the suburbs ceased to be suburbs" (Clark, 1966: 222). He surely overstated the point. Bill Michelson (1977: 194–5) later concluded that, as at the scale of the building, there was an element of self-selection: more of those moving to the suburbs were looking for family- and child-centred sociality, creating what they sought. It is likely that both influences – selectivity and recent arrival – were at work.

Timing of settlement also has an effect in inner-city districts that have been redeveloped. Lately, most have been high-rise condominium projects, but the best-documented example is Regent Park, Canada's first public housing project. This involved the redevelopment of much of Cabbagetown, with many residents being rehoused. By the end of Phase 1 in 1957, Albert Rose (1958: 188, 189) noted the "considerable" use of community facilities, including a playground and baseball diamond; he reckoned that it had "become a community … because these people have interests in common." But he hinted that this feeling of community would not last. Rose (1958: 150) noted that many residents were "indifferent" while others were "directly opposed to community participation." Even those who "recognize[d] the need to come together" did so because that was the only way to solve "common problems" (ibid.) Within a decade, a different, discouraging narrative had developed (Johnson and Johnson, 2017).

Keen socializing affects established areas too, and is notable when there is rapid, wholesale turnover. The area that Lorimer and Phillips lived in for three years turned out to be one of these and, by coincidence, was the subject of a study for the Canada Mortgage and Housing Corporation (1974). The study described the experience of middle-class incomers, who were still known locally as "whitepainters," a term that Harry Bruce had coined in *Maclean's* in 1964, but who soon adopted the term that was going global: "gentrifiers" (Batten, 2004: 13). The CMHC study found that these incomers knew more neighbours than did the area's long-term working-poor residents (CMHC, 1974: 43–4). Lorimer argues that this difference was not attributable to class. He observed that "during the first two years these people lived on Minster Lane, they formed an informal but quite intensive social circle," of which he and Mif were a part (Lorimer, 1971: 27). But "this intensive neighbourhood life of people who saw themselves literally as a colony in hostile territory did not, however, last," as some began "to withdraw from the steady round of visits" while others, having renovated their home, sold out and moved on, having made a tidy profit (ibid). Indeed, after 1970, speculation had become rife. Between 1970 and 1972 the number of building permits issued to residents who lived in and around Minster Lane jumped from 32 to 154, mostly for building alterations or new heating and plumbing; meanwhile, more than two-thirds of owners sold within two years of taking out a permit (Sabourin, 1994: 278–9, 281). They made a tidy profit; it would have been even tidier if they had waited a little longer (figure 1). It is easy to assume that suburbs are different sorts of places. In some ways they are: since at least 1945, suburbanites have relied more heavily on automobiles to get around. But at the block scale there are fundamental similarities. Significantly, a large survey in Toronto found very little difference in how well city, as opposed to suburban, residents knew the neighbours on their street or in their building (Lu, 2010).

And, of course, suburbs do not remain suburban forever. Harlem, once a Jewish suburb, became famous as a Black, inner-city ghetto (Osofsky, 1963) before its more recent gentrification. As Carolyn Whitzman (2009) has shown, Toronto's Parkdale underwent a similar transition, from 1880s suburb through so-called slum to self-styled urban village. The process of becoming urban takes decades. It took half a century for blocks of suburban homes to become inner-city Edmonton, prompting redevelopment (Smith and McCann, 1981). Duberger, a postwar suburb of Quebec City, attracted its first residents with its spaciousness and tranquillity (Ramadier and Deprés, 2004). Today, incomers enjoy a relatively central location and access to services. It takes effort to see yesterday's suburb in today's neighbourhood (McManus and Ethington, 2007). But, doing so, we can see that here, too, lines are blurred, and shifting. In their different ways Lorimer, Clark, Keller, and Mann – the Canadians, the American, and the Brit – remind us that the dynamics of neighbouring and neighbourhoods exist across the urban landscape.

The stereotype of frenetic suburban socializing that Keller and Clark viewed as temporary sits beside a more enduring one: that "manifest" neighbouring has been, above all, a feature of older, working-class districts. Here, mutual aid arises out of the constraints of "need and limited opportunity, isolation and relative poverty" (Keller, 1968: 51). Keller (ibid.) contrasts this with the "more selective" socializing favoured by a middle class that can afford to be choosy, deciding who to associate with on the basis of "personal compatibility." This is a good place to start making sense of patterns of neighbouring.

Neighbouring arrangements are always personal, and sometimes quirky, but systematic patterns often emerge, depending on the experience and outlook of the people involved. A relatively unusual example were the few thousand Canadians who, between the 1930s and early 1960s, joined building cooperatives. These were pioneered in Nova Scotia, but soon spread, notably to Newfoundland, Quebec, and Ontario (Harris, 2001: 123–4; Sharpe and Shawyer, 2016). Sometimes, the homes ended up being owned by individual families; sometimes the cooperative had title. Either way, the bonds formed during the building process endured. Ronald Chafe (n.d.: 21), one of the 10 men who formed the Grandview Co-op Housing Group in Gander, Newfoundland, later recalled how "friendship was solidified through the building stage and was continued for many years after the completion of the houses." Operating on a much larger scale, for many years the Coopérative d'habitation de Montréal, which built 655 single-family homes in Saint-Léonard, supported a wide range of collective endeavours. These included establishing a co-op store for building materials, a restaurant, a consumer co-op, and a Caisse Populaire, along with sports, recreational and cultural activities. It held well-attended monthly meetings, demonstrating a level of interest that was "unusually high" (Collin, 1998: 478), and mobilized politically. It was not the typical suburban subdivision.

The bonds of friendship often lasted a lifetime, but increasingly they separated from their roots. In 2007, members of the Bonaventure Co-op in St. John's got together to fund a fiftieth anniversary plaque. But by 1986, when geographer Chris Sharpe moved into one of the homes, all 10 of the original families had moved on. On a larger scale, in 2016 the 81 families of a building co-op in Grimsby, Ontario, celebrated their sixtieth anniversary. Their activities were documented in an exhibit that I attended at the Grimsby Museum, with informative text and images now available on the web (Grimsby Museum, 2021). Here too, however, and except in the memories of those originally involved, the bonds are no longer anchored in place.

Cooperators have always been a small minority. Much larger groups, defined by class, gender, and age have found, and made, neighbourhoods important. Those with low incomes have hardly ever been served by builders and so have made homes wherever they could. The same has usually been true of immigrants, themselves often poor but with compelling reasons to congregate. Arguably, however, close neighbours matter most to children and, in different ways, to the elderly. In most residential settings, women have tended to be more neighbourly than men. The experiences of these groups that have shaped Canadian neighbourhoods will be considered in the next chapter. But first we need to consider class. More than anything, its divisions have shaped the residential landscape. Most housing has been provided by the private sector, and builders target buyers or renters above all on the basis of income. Since World War II, the main exception, accounting for about 5 per cent of housing units, is public housing, but that, even more explicitly, is designed for a particular income group. And so, above all, the residential environment of Canadian cities expresses the differences of class. The rest of this chapter looks at the ways close neighbours matter to those in different class groups.

The Differences of Class

Social class is one of the most fraught and ambiguous terms in common use. Occasionally, social distinctions are clear, as is the case in the traditional "company town," where many employed residents are either workers, supervisors, or managers/owners. But in larger centres the diversity of occupations, coupled with social mobility, blurs lines to the point at which trying to draw them – is it occupation, education, or status that matters most, or does it just come down to income? – seems impossible. This was apparent in the mid-1960s, as most Canadians rode a wave of postwar prosperity. At that time, John Porter (1965: 3) opened his influential study of class in Canada, *The Vertical Mosaic*, with the suggestion that "one of the most persistent images that Canadians have of their society is that it has no classes." He himself believed no such thing – otherwise he would not have written the book – and I doubt that many of his contemporaries did either. As a starting point, many would probably have accepted B.K. Sandwell's epigram, "Toronto has no social classes – just the Masseys and the masses," a distinction now referred to as "the one per cent." But even then, as we do today, they would have drawn another line, between a large, amorphous middle class and the marginalized poor.

Even this won't do, however. I would argue that we need to speak about at least four classes: the rich and the near-rich; the ill-defined middle class; a large working class, once defined by manual labour but which now includes precariously employed service workers; and the very poor and the homeless. To be sure, group boundaries are shifting and hard to define. Especially during economic slumps when many are laid off, absolute poverty has afflicted segments of the working class. But most people are intuitively aware that such class distinctions exist, whether at work or at home. Half a century ago, after hanging out with his new neighbours for several years, Lorimer reported that "the class difference which is always implicitly and often explicitly recognized … is that between working people and middle-class people" (Lorimer and Phillips, 1971: 46). Similarly, drawing on a long-standing distinction, many, perhaps especially those on low incomes who fear being lumped together with undesirables, have often drawn a line between themselves and the "undeserving" poor. Alice Munro (1978: 69) had one of her characters express this succinctly. Rose, just returned from university, is speaking with her mother, Flo:

"This would have to be the last part of town where they put the sewers," Flo said.

"Of course," Rose said coolly. "This is the working-class part of town."

"Working class?" said Flo. "Not if the ones around here can help it."

This distinction is alive and well, as seen in a September 2024 *Toronto Star* headline: "'Get off your a-s-s,' Doug Ford scolded, telling homeless people to get jobs."

Since Porter's time, fewer Canadians have been inclined to identify as working-class (Livingstone, 2023: 52). In part, this may be because of the stigma that has come to be associated with manual work. More substantially, it reflects deindustrialization and the associated loss of well-paid, unionized jobs in manufacturing. Meanwhile, the middle class has acquired new cultural variants, ranging from the precariat (e.g., baristas) to the elite (e.g., tech CEOs). It has taken on a meritocratic character, which was predicted, described, and satirized more than 60 years ago by Michael Young (1958). More recently, Richard Florida (2005) has characterized a chunk of it as the "creative class." Cutting across income groups, this segment of the middle class places a premium on education and aligns itself with socially liberal values. This "new middle class" seeks the ambience and walkability that are to be found in gentrified neighbourhoods (Ley, 1996). And because it includes many academics, we have plenty of studies of such areas, in Canada and beyond.

No one appreciates the character of this culturally defined group more than those who have crossed its boundary. An articulate, witty spokesperson of this group is Shawn Micallef, who grew up working class in Windsor and acquired his middle-class credentials in downtown Toronto in the 2000s. In *The Trouble with Brunch: Work, Class, and the Pursuit of Leisure*, a brilliant commentary on the larger social meanings of "brunch," he insists that "being or feeling middle class" is "a world view, a sensibility, a sense of self ... that cuts across incomes," with the "creative class" being a "subset" that has lately received much attention (Micallef, 2014: 29–30, 43). Income, of course, is relevant, but formal education can be just as important, and so are – here he draws on the century-old argument of Torsten Veblen – particular forms of "conspicuous consumption." (Significantly, Veblen was writing in the era when the cartoon "Keeping up with the Joneses" was launched.) Brunch is Micallef's chosen example of conspicuous consumption but, as I argue in chapter 7, the branded neighbourhood matters more. Lorimer and Micallef leave no doubt that a line can be drawn between the working and middle classes, and hardly anyone can question the existence of classes above and below these two. For all their ambiguities, four class categories can usefully serve as our frame of reference: the working and middle classes, the elite, and those living in poverty. The first three have some, albeit varying, ability to choose where they live.

The Working Class

Many writers have offered accounts of local, tight-knit "communities" in urban working-class districts, ones that offer "mutual aid" (Comacchio, 1999: 41). To move meant losing a network of social connections. The son of Hungarian immigrants in Winnipeg's North End in the 1910s and 1920s, John Marlyn (1957: 26) recalls that, after one man got a good job and the family prepared to move into a better home, his wife began to distance herself: she "slowly stopped borrowing lard and sugar and things from her neighbours." Descriptions of such local social networks, particularly for Montreal and Toronto, abound. In francophone Montreal in the 1800s family ties strengthened such networks. Sherry Olson and Pat Thornton found that in some areas almost half of all households lived within 500 metres of a relative (figure 2). Even in this city of tenants, "despite the intensity of the first-of-May shuffle, kinfolk remained neighbours ... little knots of neighbouring kin were crocheted along the street fronts and threaded through the alleys, testimony to the organizing power of kinship" (Olson and Thornton, 2011: 78–9). Kin networks supported local businesses. To get, or renew, a tavern licence the publican submitted a petition signed by at least six people; most petitioners were close neighbours or family, or both (Poutanen and Olson, 2020). Landlords, too, were in the mix, typically living close to their rental properties (Sweeney, 2014). This arrangement largely survived into the twentieth century. In *The Fat Woman Next Door Is Pregnant*, his semi-autobiographical novel set in Plateau-Mont-Royal in the 1940s and 1950s, Michel Tremblay (1981) describes a dense web of family, friends, and close neighbours. Looking at a similar area, but in a darker mood and as an outsider, *The Tin Flute*, by Gabrielle Roy ([1947] 1989: 95), describes how, during the day, "women, thin and sad, stood in evil-smelling doorways" while "others, indoors, set their babies on the windowsill and stared out aimlessly." But there is a silver lining. In the evening, after the men return, "there is this village life, when chairs are pulled out to the sidewalk, or people sit on door-sills and talk passes from one threshold to the next" (290). Also during the 1940s, Richard Valeriote (2010: 16) recalls that on the street where he grew up in The Ward – that is, St. Patrick's Ward in Guelph – most houses had a porch or verandah, "and these were almost always occupied in warm weather by neighbours who were characters in a play called *Alice Street*." The people interviewed by Steven High (2022) also recall that this type of community persisted in Pointe-Saint-Charles until the area was hit hard by deindustrialization, beginning in the 1970s.

Allowing for cultural differences and more limited family networks, accounts from Toronto, notably Cabbagetown, are similar. Hugh Garner grew up there after being brought from England in 1919 at the age of six. He, too, reports dire conditions during the Depression, these being leavened by mutual supports. The latter could make a big difference when a tenant was threatened with eviction. In a magazine article, Garner ([1936] 1972: 146) reported that "sometimes gangs of neighbours come around and carry the furniture back into the house as fast as the bailiffs carry it out." In a later semi-autobiographical novel, he creates the character of Myrla Patson, a young woman who is kicked out of her parents' home. A neighbour, Mrs. Plummer, steps in: "You come to my house. Elizabeth is working in Detroit now you know. I've got lots of room" (Garner, 1971: 180). This, at a time when, according to Garner (1971: vii), Cabbagetown had become "the largest Anglo-Saxon slum in North America." Although working class from the outset, as the name suggests, the area had not always been so rundown. The historian Maurice Careless portrays it as flourishing in the late nineteenth century. He sees a "well-knit" community, with "mutual aid and accepted, bonding obligations" that created "cohesiveness, vitality and" – setting it apart from the 1930s – "confidence" (Careless, 1985: 26, 40). According to Careless and Garner, then, in working-class areas local community manifested itself through thick and thin.

This sort of bonding extended beyond the inner city. In the early twentieth century, thousands of British working-class immigrants settled in Toronto's suburbs. Many men built their family's home, commonly with help from neighbours, whom they in turn helped (Harris, 1996: 200–32). Their wives also looked out for each other. One woman I interviewed for my study of the northwestern suburb of Earlscourt commented simply, "everyone helped one another" (Harris, 1996: 212). Those to the northeast of the city, in East York, were hard hit by the Depression. Work returned as war approached. When East York resident Nell Binns finally got a job, a neighbour brought round "a precious tin of salmon she had been saving for years and a tin of peaches"; meanwhile, because Mr. Binns had sold his tools, the neighbour loaned him a hammer so that he could start work (Schultz, 1975: 55).

Jean Bradley (1996) has provided the fullest account of a suburban working-class family's experience at this time. She recalls the (sometimes extended) geography of the area as well as the everyday details (figure 3). Born to British immigrant parents in 1928 in Newtonbrook, between Finch and Steeles in what was then outer suburban Toronto, Bradley reports that several neighbours helped her father to rebuild the family home. (The first had burned down.) One neighbour, Charley Gordon, had moved from Cabbagetown some years previously. Soon, however, arthritis compounded a war injury, which was aggravated by a failed experimental treatment. He became bedridden during the Depression. According to Bradley, this "was a time of neighbour helping neighbour" (34), and in all sorts of ways. A neighbour provided extended daycare for the young Jean, organized birthday parties for her, and taught her her "numbers" and how to read (81). Another shaved her father; still another brought Sunday dinners to give his wife a break; in winter, a local man would "crawl under the house and thaw pipes with a blowtorch"; others occasionally "assumed the role of medical practitioner" (31, 34, 50, 68). Given that few families could afford a doctor, that was crucial. Similarly, the mother of another respondent, Wilf Royle, who grew up nearby in suburban York Township, volunteered her services as an untrained midwife: "she must have delivered most of the babies in Silverthorn" (Bailey, 1980: 32). Especially during periods of hardship, working-class community was apparent in city and suburb alike.

This solidarity persisted in working-class areas in more prosperous times. In the Grimsby building co-op, the men, mostly employed in Hamilton's factories, built homes in the evening after work. Meanwhile, women such as Helen Komadoski raised the children, fed everyone, and cleaned up. Former resident Avryle Wilson recently commented, "we raised each other's kids" (Grimsby Museum, 2021). Without fences between yards, a hockey rink straddled several. Steve Cheverie recalls that as a child growing up

there, it felt like "we could walk into any home in the neighbourhood and without hesitation it would be opened to you." As the
online museum exhibit states, "these 81 families built more than
just homes: they built a strong community and everlasting friendships" (ibid.). Co-ops were rare, but other, more conventional, settlements also fostered community. As immigration revived after
1945, new ethnic enclaves formed, some of which nurtured strong
localized loyalties. One was Goose Village – "Village-aux-Oies" – in
Montreal's Pointe-Saint-Charles neighbourhood. Marise Portolese
(2023: 40), daughter of a couple of who lived there from 1953 to 1964,
insists that "residents had unwavering support for one another."

Later, more problematic, examples of communities are the public housing projects built between the late 1940s and early 1970s.
Their public image has suffered over the years, but they still foster mutual aid. One of the largest projects in Toronto, now slowly
being redeveloped, is Lawrence Heights. In an ethnographic study,
Luca Berardi (2021a) notes pervasive concerns about public safety,
along with "local wisdom" about how to handle it. But in conversation he also emphasizes that "people take care of each other"; for
instance, they offered loans, childcare, and even furniture, when
an apartment fire had damaged the interior of one of the buildings (ibid.). Similar mutual assistance has existed in Regent Park,
Canada's first housing project, the one that replaced much of Cabbagetown. From interviews with residents of the area, Martine
August, too, noted local concerns about violent crime but emphasized the "strong sense of community" based on "dense networks
of friendship and support" (August, 2014: 1317). She reports that
"when asked for examples [of mutual aid] most residents readily
described supportive actions, including: carrying shopping bags
and opening doors; picking up mail or watching the house when
a neighbour is away, giving rides, lending a credit card, lending tools, sharing food, watching kids and picking kids up from
school," and in two cases taking care of "home and family (cook-
ing meals and cleaning) when they were unwell or hospitalized"
(1325). Shades of Jean Bradley.

But Mann's "manifest" neighbouring has not always characterized working-class areas, and even in the Plateau and Cabbagetown

it may have been rarer than published accounts imply. Memory plays tricks, especially when coloured by nostalgia; even eye-witness accounts of an event that happened yesterday can be inaccurate. This is compounded as the years pass, so that nostalgia colours memories, making them selective (Blokland, 2001; Massey, 1995a). That is especially true when descendants are involved. Montreal provides many examples. Matthew Barlow (2016) has shown how, in the early 2000s, generations after their ancestors first settled in Griffintown, Irish-Canadian Catholics were claiming the area as their historic home. Marisa Portolese (2023) heard stories about Goose Village, near Griffintown, later an Italian-Canadian enclave, from her parents and spoke with some of their neighbours. She did not grow up in the area, but she visited what is left of it many times, took photographs and, in a labour of love, wrote a book. But how reliable is her account, or indeed that of anyone who has cared enough to write a neighbourhood history?

Even among those of us who like to think of ourselves as objective, there is the problem of confirmation bias, making us selective in particular ways (Kahneman, 2011). To tell a good story, we imagine communities, invent traditions, (Harris, 2024; c.f. B. Anderson, 1991; Hobsbawm, 1983). Romanticized stereotypes of working-class community can shape where researchers look, what they find, and how they interpret it. A classic English study of life in a working-class neighbourhood is Michael Young and Peter Wilmott's (1957) portrait of Bethnal Green, in London's East End in the 1950s. Both worked, and one lived with his family, in the area for a while, and both undertook interviews with its residents. Their book documents, and waxes eloquent, about the supportive intimacy of neighbours. But, reviewing their interview notes, Jon Lawrence (2016) has argued that they overstated their case. He might be right. The timing of the original study is significant. The French sociologist Christian Topalov (2003: 476) has pointed out that "sociologists and other researchers ... 'discovered' working-class neighbourhoods" in the 1950s and 1960s ,"at the very time they were about to vanish." Young and Wilmott, left-leaning middle-class outsiders, arguably played up the contrasts with their own experience. Perhaps their work, and that of other sociologists like Herbert Gans (1962) in the United States and of historians such as Maurice Careless, overstated the strength of local working-class community, even in the areas they studied. And how typical were those areas anyway? London's East End was understood to be unusually tight-knit, as was Cabbagetown and Gans's West End of Boston in their heyday. Confronted, over a period of three years, with a more typical area, Lorimer rationalized his disappointment with Minster Lane: "during its recent past it seems to have lacked much of the community that characterizes most neighbourhoods in large cities" (Lorimer and Phillips, 1971: 12, and see also 8, 46–7). "Most neighbourhoods"? That was his assumption, not a fact.

We should also listen to the sceptics, and to evidence that probes that assumption. Kenneth Scherzer (1992) has challenged the belief that, even in the nineteenth century, many working-class families in New York City made significant connections with neighbours. Most were tenants, and moved frequently. In a similar vein, David Ward (1976) suggests that the stereotypical Victorian "slum," implying a cosy if impoverished village, was more myth than reality. James Gray (1970: 11) recalls that in 1920s Winnipeg his own family "never stayed long in any one neighbourhood," adding "we moved on average once a year." That was common. In the city's North End, "the week seldom passed when a new immigrant pupil or two failed to show up for classes" because the family had moved (Gray, 1970: 11). The point was echoed by W.E. Mann for the Lower Ward, a poor inner-city district in Toronto in the 1950s. A local school experienced a 100 per cent turnover rate in one year, with some families moving multiple times (Mann, 1961). Under these circumstances it must have been difficult to get to know, less trust, immediate neighbours. After looking at another low-income district in Toronto after 1945, S.D. Clark (1978: 155) commented that "these were people who appeared to have developed virtually no roots in the community in which they lived."

In reality life in these communities was surely more nuanced than the foregoing accounts would suggest. One has only to look to the accounts of working-class life in Hamilton, Canada's most industrial city in the twentieth century. Craig Heron (2015: 174) reports that, before World War II, "people who grew up in working-class

Hamilton later recalled ... neighbourliness all around them," but he adds that "relatives and neighbours could usually help out only for short spells or in limited ways." Jane Synge's (1976: 46) interviews with Depression-era Hamiltonians amplify this. One woman recalled that "we were very close with our neighbours. If they were in trouble they'd come to mother" (Synge, 1978: 103). But this was unusual. On an everyday basis "there was a well-established code of behaviour. There was little visiting in others' homes, but considerable conversation over back fences, and from porch to porch" (Synge, 1978: 103). "Respectability" was key. In her brilliant and evocative semi-autobiographical novel about a girl's wartime life in east end Hamilton, Sylvia Fraser (1972: 33, 35) insists that although Oriental Avenue was "just a working-class street ... certain standards are scrupulously maintained,"; she then elaborates on the wider world view: "the residents ... are good neighbours united by the tough virtues of honesty, industry, sobriety and the certain knowledge that life is short, difficult, unpredictable and likely to get more so." Respect required a certain distance. What mattered was family; it was they who you turned to in times of trouble. Stuart Crysdale (1970) found the same thing in 1960s Riverdale, then a mostly working-class area across the Don Valley from Cabbagetown. And then, a decade later and a mile northeast in East York, Barry Wellman (1979: 1210, 1219) found that immediate neighbours had only limited significance for most residents, whose "close intimates" lived elsewhere.

Far from reaching out, working-class men and women often erected social fences. Denyse Baillargeon interviewed women with families in Montreal's working-class districts during the Depression, another place where households moved a lot. She reported much mutual aid, not least because neighbours included relatives. But she also found a strong desire for privacy. This was a tactic to "camouflage their problems and their poverty," but there was more to it (Baillargeon, 1999: 160). She suggests that, "while it was generally desirable to 'get along' with the neighbours, relations with them were limited to greetings on the street or a 'bit of chat' on the balcony in the summertime" (ibid). Women were reluctant to open their doors to neighbours, or to seek help. Her conclusion is that relationships were cautiously balanced: the prevailing spirit was "not interference, but not indifference" (Baillargeon, 1999: 161). This is something like what Suzanne Morton found in Richmond Heights, a working-class area in Halifax developed in the 1920s. Women, especially, worried about "respectability" and how to build the "social 'fences' of distance that created the proverbial 'good neighbours'" (Morton, 1995: 39). Expectations were changing. This was, after all, the decade when "children's behaviour and development were apt to be judged by the standards of middle-class professionals" (Strong-Boag, 1982: 178). Pride could trump need.

The balance shifted towards pride, and privacy, when conditions improved. When incomes rose after 1945, needs and mutual assistance became discretionary. Donald Foley's (1952) study of Rochester, New York, suggests that neighbouring soon begun to wane, a trend echoed in Rotterdam (Blokland, 2003). Shawn Micallef (2014: 16) recalls that in 1980s Windsor his family was "quite friendly" with their "Croatian next-door neighbours – he a General Motors line worker, she a hairdresser with a home shop" and that "when they had their driveway done, we did ours too, sharing their Croatian crew." Relatives also helped out each other to save money. That was nice, but not essential in the way it had been in Cabbagetown or Newtonbrook during the Depression. Several of the Toronto suburbs that S.D. Clark studied in the early 1960s were solidly working class; at first, they saw some of the same sorts of neighbouring that Jean Bradley described, but as incomes rose so did people's possibilities and expectations. The impact of rising prosperity is apparent in Dale Gilbert's (2015) study of the parish of Saint-Sauveur, Quebec. There, the postwar exodus eroded the old neighbourhood, and people made newly private, consumerist lives. This shift fits the changing emphases of Quebec writers. *Alley Cat* and *Juliette*, the semi-autobiographical novels of Yves Beauchemin (1986, 1993), take place in working-class Longeuil, on Montreal's South Shore in the 1980s. They portray lives that lack the local intimacy apparent in Michel Tremblay's *Fat Woman Next Door Is Pregnant,* set a generation earlier. In that respect, and as mobility has increased, neighbouring has become more discretionary. In Vancouver of the 1970s, Alan Hob-kirk (1973) found that the "immediate neighbourhood" did matter more to residents of working-class Renfrew Heights than to those in middle-class University Hill. Class still mattered. But even those in the Heights were mobile to a degree unknown half a century earlier. And then, with more women holding down full-time jobs, they simply have had less time to connect with their neighbours on a regular basis.

The Middle Class

It is even more difficult to speak about middle-class views and experiences. We each have our own experience and that of friends – here I am making assumptions – but little in the way of objective information to expand or verify our understanding. When reformers like Sir Herbert Ames, or organizations such as the Bureau of Municipal Research, began to undertake surveys in the late nineteenth and early twentieth centuries, they saw no rea-son to tell people about what was happening in their own back-yards. Social reformers, many based in the Protestants churches, were concerned with "'educating the popular mind'" (quoted in Christie and Gauvreau, 1996: 183). Since World War II, social scien-tists have generally taken the same point of view. As the American sociologist Junia Howell (2019a) has pointed out, ethnographers have favoured working-class communities and the marginalized. Studies have been carried out by people like James Lorimer, who are, and think of themselves as, middle class and who write for people like themselves. The same is true of historians, who have used neighbourhood studies as a way into working-class culture (Bérubé, 2016). Implicitly, both treat middle-class life as the norm and perhaps assume that it is not worth writing about because its features are obvious and unremarkable. Significantly, the two neighbourhoods that have received the greatest per capita cover-age were working-class communities, settled by Métis and Black people respectively: Rooster Town (pop. 190 in 1916) on Winnipeg's fringe and Africville (pop. 394 in 1959) adjacent to Halifax (Burley, 2013; Peters, Stock and Werner, 2018; Clairmont and Magill, 1999;

Africville Genealogy Society, 2010; Loo, 2010). A rare middle-class portrait is John Seeley et al.'s *Crestwood Heights* (1956), of what is obviously a pseudonym for Forest Hill in the 1950s, after its incorporation in 1924 and before annexation by Toronto in 1966. But its being a Jewish enclave makes it atypical. As Erna Paris (1976: 102), who grew up there, recalls, "our lives in the Forties and Fifties were insular and unreal – unconnected to the WASP reality of Toronto."

There are many semi-autobiographical novels and memoirs set in poor and working-class districts. In David Arnason and Mhari Mackintosh's (2005) literary history of Winnipeg, one neighbourhood gets its own section: the poor, immigrant, working-class North End, also the subject of a fine book of photographs (Paskievich, 2017). As Amy Harris's (2010) exhaustive survey shows, Toronto fiction has a similar bias towards places like Cabbagetown and the Ward. Similar biases are apparent for every major Canadian urban centre. For Quebec City there is Roger Lemelin's *The Town Below*; for Montreal, Gabrielle Roy's *The Tin Flute*, Mordecai Richler's *Apprenticeship of Duddy Kravitz*, Michel Tremblay's *The Heart Laid Bare*, Jovette Marchessault's *Like a Child of the Earth*, and Kathy Dobson's *With a Closed Fist*; for Toronto, Hugh Garner's *Cabbagetown*, Jean Bradley's *A Home across the Water*, Catherine Hernandez's *Scarborough*; M.G. Vassanji's *No New Land*, and Louisa Onomé's young-adult novel *Like Home*; for Hamilton, David Baillie's *What We Salvage*; and for Winnipeg, John Marlyn's *Under the Ribs of Death*, James Gray's *The Boy from Winnipeg*; and Katherena Vermette's, *The Break*. Even Prince Albert has Robert Boschman's *White Coal City*, set in the working-class Flat rather than the middle-class Hill. Halifax's Africville has also been fictionalized (Colvin, 2019). There are almost no works that portray neighbourhood life in middle-class districts. Phyllis Young's *The Torontonians*, apparently set in Leaside, is the exception. The implication is that, because most readers are familiar with such areas, they are not much interested in seeing them described in print.

But the greatest difficulty in speaking about middle-class neighbouring is that it is diverse, both geographically and in substance. For workers, and especially the working poor, reaching out to neighbours has been a survival strategy. Those with higher incomes can be, as Keller notes, "more selective": they may build connections for "personal compatibility," social status, or networking (Keller, 1968). Greater selectivity has been reflected in, and enabled by, changes in domestic environments. Beginning in the 1920s in some suburbs, lot sizes grew larger, so that, as Peter Ward (1999: 132) has observed in his history of the Canadian home, "the dwelling migrated back from the road and in from the sides of its lot." The verandah or porch had served as a quasi-public space from which neighbours could be observed and greeted, but Cynthia Comacchio (1999: 80) suggests that for the middle class in the 1920s "porch-sitting, street play, and chatting with neighbours" became "not quite respectable." Since 1945, as the automobile enabled a sprawl of larger lots, "the suburban home has turned its back on the street" (Ward, 1999: 142). As Christopher Grampp (2008) has shown in his history of America's "home grounds," backyards ceased to be functional and became places for recreation. Front porches were abandoned. In summer, people barbequed on backyard patios. Until the early 1960s, back as well as front yards were unfenced and so, as Anna Andrzcjewski (2009: 50) has described in suburban Madison, Wisconsin, "kids moved in packs (nearly every household has a child) and everyone watched out for one another's children" (figure 4). One woman who Barbara Lane (2015: 199) interviewed recalled that "we all played together, in the back yards, on the sidewalks, in the streets and on the building sites where new houses were in construction. We rode our bikes on the sidewalks and around the block." This could be a scene from any one of a hundred Hollywood movies or sitcoms set in the 1950s. But such loose sociality did not last. As more women went out to work, fewer were around to watch out for each other's children. From the 1960s, homes grew while lots have shrunk. Yards, once open, have been enclosed. On a recent walk around a typical 1950s suburb in Hamilton, I counted only three unenclosed backyards among over a hundred homes. Social life has become more private. Larger homes, air conditioning, television, and then the internet have encouraged everyone to spend more time indoors. Households have turned away from the street.

If socializing has become private, the assertion of a public image and social status requires something larger – the neighbourhood. For Toronto's postwar suburbs, S.D. Clark (1966) made what was becoming a conventional distinction between planned and unplanned subdivisions. Of course, they were all "planned" in a minimal sense: a developer had laid out streets and decided whether to set aside land for non-residential use. But subdivisions now thought of as planned had framing arterials that emphasized the distinctiveness of the area, while a primary school, community centre, and/or church were built as hubs (Harris, 2004). (Until the 1960s, Christian places of worship were overwhelmingly the norm.) Ever since, the status of such subdivisions has been communicated by the size and design of the homes and also by the way they are named and marketed. Clark (1966: 105, 176) found that families who moved into planned subdivisions were self-consciously "mindful of the advantages of a good address'" while "for the women and children it was in the associations of neighbourhood that the supports and values of social class had to be found." As a result, "a certain, carefully calculated, amount of time, money, and effort was directed to the support of [local] associations" (177). The significance and selection of larger neighbourhoods are explored further in chapter 5. I mention it here because it matters for the way the middle classes view their immediate neighbours: as with whom to invite for dinner, they can be choosy.

The chosen neighbourhood is not necessarily a planned suburb. Clark noted that some people – like Karen, the protagonist in Young's *The Torontonians* – chose to move back to the city, while others never left it. Indeed, he suggested that they, not new suburbanites, were the people "with a strong commitment to a way of life" (Clark, 1966: 223). The city-suburban contrast in preferences was nicely – if predictably – captured by Bill Michelson. Those who chose the suburbs cared most about having more private domestic space; city-dwellers valued transit and accessibility. In sum, people "chose different forms of environment to represent different needs and wants" (Michelson, 1977: 182).

In recent decades, the cultural variant of the middle class has chosen the urban option, developing a wave of gentrifiers that began with those filtering onto the proverbial "Minster Lane" in the late 1960s. They are leaving their mark on neighbourhoods, and neighbouring, across the country. In Montreal, Sandrine Jean shows this by comparing Vimont-Auteuil, in suburban Laval, with a city neighbourhood, Ahuntsic. She reports that in V-A large yards encouraged private lifestyles, while a mother expressed the Ahunstic spirit: "the street is an extension of my house, the park is my back yard, my neighbourhood is the whole city" (Jean, 2016: 2573). Alan Walks (2006: 49; 2008) found much the same in Toronto, where residents in the urban Beaches neighbourhood claimed to value community while suburbanites led more privatized lives. These differences were in turn reflected in political party support: urbanites vote NDP, suburbanites Conservative, with Liberals sprinkled in both places.

There is no simple conclusion about middle-class neighbouring. In an environment where family, friends, and children can be well-entertained within the confines of the private home, the household's core requirement may be that they be "left alone" by their immediate neighbours. As one woman told Clark, "I think the neighbourhood life around here is about ideal, friendly but don't bother me" (Clark, 1966: 159). Shawn Micallef's declaration suggests that that sentiment is even more common in rental and condominium high-rises. But in some middle-class neighbourhoods, people have sought community. Clark (1966: 104) himself declares that in planned suburbs "good fellowship and neighbourhood friendliness were [seen as] virtues." To be sure, that urge faded after a year or two. But for many it is more than a mere, or temporary, gesture. In her study of the middle-class New Urbanist and gated communities that have gained momentum in recent decades, Jill Grant (2007: 494) found that buyers were looking to "make a statement about valuing heritage, community, and sociability," this being part of their "self-image." In New Urbanist developments such values are expressed by historic designs, front porches, and streetscapes where cars are restricted; in gated or common interest developments they are baked into associations, although it is unclear whether the "imagined community" of such developments is often realized. Do people actually sit on those porches or even attend association meetings (Grant, 2007: 483)? But the goal, and the social message, is clear.

In city and suburb, the middle class has almost always had more power than the working class to choose where to live, creating for them more residential options and engendering a certain self-consciousness. It may be possible to generalize about their neighbourhood selection process, but not about the outcomes.

The Elite

But what about those who, in theory, have the most choice, and those who have little or none? At one extreme there are the truly wealthy, a group whose neighbouring preferences are not so much a closed book as one that has yet to be written. At best we get glimpses, as if through a tall, dense, and well-clipped hedge.

Curiously, we know more about the local experience of elites a century ago than we do about those today. Before World War I, Vancouver's "one per cent" lived in the West End. In 1908, 86 per cent of those listed in the city's Elite Directory lived there, and the children had little reason to venture beyond (Robertson, 1977: 31–2). There were two private schools, both established in the 1890s: the Granville and Miss Gordon's. The Granville's curriculum circular noted that "special attention is paid to the manners and general bearing of the pupils," while instruction included French, music, and Latin, as well as math, natural science, and a "sound English education" (Robertson, 1977: 47). Robert McDonald (1996) has described the lives of the adults in the West End, but its neighbourhood life remains elusive. One element was that, as in other Edwardian enclaves, women organized a ritual of "at-homes" and, in summer, garden parties. These catered to a self-enclosed world. At Mrs. William Sully's garden party, 38 of 48 guests were named in the Elite Directory (Robertson, 1977). The at-homes were more intimate, bringing together immediate neighbours (figure 5). These comprised "subworlds" where, for example, those on Melville Street met on Fridays, but "other streets had their own social orbits of like members within the elite hierarchy" (Holdsworth, 1977: 200). These local networks produced a tight geography of intermarriage.

Among the elite, marriage was, and indeed remains, a significant event. The day itself assembles people who matter, a fact that Jean Rosenfeld (2000) exploited. Using the guest lists for two weddings, in 1892 and 1901, combined with the overlapping membership list of the Hamilton Club, she constructed an inventory of Hamilton's turn-of-the-century elite. She concluded that "through the location and form of its houses" this class shaped "a small section of the city," thereby controlling "entry into, and membership in, a simple social pyramid" (1). Simple because it was small. The houses were mansions; almost all contained live-in servants, usually two or more (74). The "small section" was the south-central area on the lower slope of the Niagara escarpment, roughly what is now Durand; 85 per cent of the elite lived there. That arrangement was convenient, and it also made a social statement. Clustering meant that "ladies [could] reinforce the social structure and conveniently flit from drawing room to drawing room in the seemingly never-ending, but socially essential, ritual of teas and social calls" (4–5). They could have walked, though doubtless many preferred to create an impression by taking the carriage that was housed round the back, sometimes with servants' quarters attached. Hamilton's elite area, like Ottawa's Sandy Hill a few years later, was small, giving it a cosy feel (Taylor, 1986: 94).

Even in larger cities there was a tight – if sometimes fraught and competitive – intimacy. A century ago, Canada's most prestigious elite area, in its largest city, was Montreal's Square Mile, and adjacent Westmount, the first jurisdiction in North America to have a comprehensive zoning regulation (Fischler, 2016; Germain and Rose, 2000). Mansions sat on estates "so that," as historian Margaret Westley (1990: 26, 70, 90) observes, "the whole area was a huge park," a place where the children who were not being supervised by nannies were taught in private schools. But it was compact enough that "everyone who lived there knew everyone else ... like a small town" (90). There, as in Vancouver and Hamilton, at-homes kept social networks alive, and the day after a dinner or dance one was expected to drop by to "leave a card." As one woman recalled, however, "you hoped the lady was not at home when you came to call" (34). It was all very WASP. The British connections – and architectural styles – were prominent; the community lived in "self-segregation" while, even though most Montrealers were francophone, "everyone … encountered was English or spoke the language well" (38, 93).

That local coherence surely no longer exists, but exactly what has replaced it, and when, is a mystery. In the postwar period we get little more than glimpses. Toronto is a case in point. In the 1950s, life in Toronto's Forest Hill was described in pedantic detail by John Seeley and colleagues (1956). Some elements resonate with descriptions of those earlier elite areas. There is a strong sense of community, of relationships created in the community centre and schools, the churches and synagogues, and in the clubs and summer camps, but whether and how residents related to their immediate neighbours is unclear. In any case, then, as now, Forest Hill was merely upper middle class, with staged dinner parties to confirm and extend social and business networks. These did not reach stratospheric social heights, and there has always been some social mix. In the 1950s, teachers were "well aware" of the children who came from a small, poor "southernmost enclave" (Seeley, Sim, and Loosley, 1956: 442n9). In the same enclave, in the 1980s the area still accommodated the sort of modest apartment buildings like the one where Aubrey D. Graham, a.k.a. Drake, was raised by a single mother. The area has never been the preserve of the one per cent, any more than was Phyllis Young's fictional version of Leaside or, for that matter, the exclusive Kingsway Park area, which was developed between the wars with strict restrictions on the value, quality, and design of homes (Paterson, 1985).

The prime elite area in Toronto for over a century has been Rosedale, but I know of no work, factual or fictional, that illuminates its neighbouring life. We see glimpses through servants. Hugh Hood (1975: 23) speaks of the area's working-class northern fringe during the Depression, which was "in some degree the dormitory of the servants of the south." By the 1960s, some of those servants were immigrants. In *The Meeting Point* (1967) Austin Clarke imagines two friends from Barbados, Bernice who works in Forest Hill as a nanny and Dots, a Rosedale maid. Later, an outsider's impression is sketched by John Miller in *A Sharp Intake of Breath*, whose narrator is taken by sister Bessie to see the McNabb mansion in Rosedale, where she works as a maid. Toshy declares "I'd never seen houses as big as the one we walked by that day ... there was a quiet on Glen Road that gave the neighbourhood a smugness, a self-satisfied calm" (Miller, 2006: 228). The only street life was a "couple that strolled by silently" and "a man who drove by in a fancy car," with his "stiff back and smirk" (Miller, 2006: 228–9). There were no children playing hopscotch or young men hanging around on the corners.

In Montreal in the postwar period, outsiders had added "Golden" to the local's term "Square Mile," while Westmount continued to guard its privilege (Westley, 1990: 25). In 1951, when the average assessed value of homes in the City of Montreal was $4,191, those in Westmount were an impressive $9,374 (Choko, 1998). It was one of the owners of those homes, F.R. Scott, a lawyer, dean of law, and influential poet, who gave us a clue that intimate neighbourly ties were no longer the norm. One day, a laundry truck careened down the hill and hit a maple tree in his front yard. In "Calamity" he describes what happened next (Scott, 1981). "Normally," he writes, "we do not speak to each other on this avenue / but the excitement made us suddenly neighbours / People who had never been introduced / exchanged remarks / and for a while we were quite human." After the tow truck and police had left the scene, "Order was restored / The starch came raining down." His poem speaks eloquently of the dignified – some would say snobbish – aloofness of the rich. They might throw dinner parties for friends and business associates, but they ignored the neighbours.

As for French Canadians, much the same was true in Outremont, Westmount's poor cousin, where the average home was assessed at $6,516 in 1951. In *The Heart Laid Bare*, Michel Tremblay (1989) presents it as a symbol of achievement, "not a place to live but a way of life, a social status." He shows us the life of Jean-Marc, a university professor. Strolling there one evening as a visitor, Jean-Marc comments that "walking through the streets of Outremont at night makes you ambitious ... I saw myself emerging from my castle to turn on the magic sprinklers." There are cultural differences in the ways that elite anglophone and francophone Montrealers live and entertain, but in neither case is there much evidence of active neighbouring.

Like other elite districts, Westmount was guarded by covenants, land use bylaws, and building regulations that ensured that only the right sort of homes were built, and for the right people (Bérubé, 2015). That put a floor on prices, defining residential settings. Kingston's Alwington Place, in the city's west end, is not much larger than a block; in Hamilton's Westdale, Mayfair Crescent and Oak Knoll Drive are the most prestigious blocks in a middle-class area. Racialized restrictions were sometimes used as well, for example in Victoria's Uplands (Forward, 1973; McCann, 2017) and Edmonton's Mount Royal (Foran, 1979). Until declared illegal in the late 1940s, they targeted Jews in Ontario, and Chinese in British Columbia (Walker, 1997: 190). In Westdale, developers built a legal wall to exclude more than a dozen groups (Weaver, 1988). For many decades, because of exclusion or active preference, elite areas have rarely shown much ethnic diversity.

But lately that has been changing as affluent immigrants have arrived, notably from Hong Kong and mainland China, creating upmarket districts of their own. Within the frame of class landscapes produced by developers and builders, there have always been concentrations. These include those of the poor, the fourth class, which have often included lower-income immigrants, as well as clusters of the young, and the young at heart. These vital elements contribute to the complexity and diversity of the social geography of Canadian cities.

chapter three

Others with Less Choice

> It is the sentiment that inheres in ... proximity, the common bond of shared space, that defines the neighbourhood's essential social bond.
>
> Albert Hunter, "The Urban Neighborhood"

Builders and developers have catered to the elite and the middle classes. They have also created districts for workers too, as the east ends of Montreal, Hamilton, and Vancouver, or the north ends of Halifax and Winnipeg, testify. In so doing, they produced environments that could support the "manifest" neighbouring of working-class families, or the latent versions of those who could afford to be choosy. But other types of people have clustered in particular areas, or at least made as much as they could of their immediate surroundings, often because their mobility was limited.

The people that builders have always neglected are the poor, who have ended up living where and how they could. Typically, this has meant older housing in rundown areas. The effects of the filtering process, a term used since the 1940s, have often been described. In Montreal, after his mother had fallen on hard times, Paul, the main protagonist in Hugh MacLennan's *Two Solitudes*, ended up in a place that had declined: "It had been a good street once ... but the old families had long ago sold out to rooming-house proprietors" (MacLennan, 1945: 123). As Nicholas Lombardo (2014: 6) has shown, in the 1880s Jarvis Street was the "premier address in Toronto," home to the Masseys and the Gooderhams. By the 1920s, however, large homes had been chopped up into rooming houses, eventually to be redeveloped. Given sufficient size, the right location, and building restrictions, elite areas such as Victoria's Uplands have retained their character, becoming effectively "immortal" (Forward, 1973). But that is not always the case, and whole areas can decline. In Vancouver, West End mansions were soon subdivided for lower-income families and singles before being redeveloped after 1945 (McAfee, 1972). The same happened to parts of inner Edmonton (Smith and McCann, 1981). But redevelopment only happens when the inner city thrives. This has not yet happened on any scale in some cities, notably Winnipeg. David Burley and Mike Maunder (2008), in what is perhaps the richest account of the filtering process, have shown how fine homes on Furby Street in Winnipeg wasted away, taken over by a succession of poor and migrant families, many of First Nations origin, struggling to build community while making a living and a life. Homes have been rescued and improved but, so far, only a few.

The poor, of course, have often included a disproportionate number of immigrants, Métis, and First Nations. In varying degrees, all of these groups have faced discrimination when arriving in urban areas. In most of Canada, WASP culture was long dominant and everywhere "white" was regarded as a cultural norm – indeed many would argue that it still is. Immigrants have had to fend for themselves, having brought too few savings or marketable skills to be able to afford new homes in ethnic subdivisions. In recent decades, as rural-urban migrants, First Nations and Métis have also had to insert themselves into a pre-existing dominant culture and urban landscape as best they could. But at least immigrants, and to a lesser extent lower-income households and First Nations, have been able to congregate with neighbours with whom they share a cultural identity. There are other groups, for whom social isolation is an everyday reality. Seniors, and single mothers and children, for instance, gain much from neighbourly association yet are often scattered across the urban landscape. Paradoxically, though, these groups have themselves valued and benefited from neighbourhood connections most of all.

Poverty

Although poor people are defined by their lack of money, it is never easy to say who is living in poverty. Beyond absolute destitution, the worst form of poverty (Frankfurt, 2015), there is relative poverty. In any event, decent shelter and sufficient food are absolutely basic to human health and happiness. Recent surveys suggest that a quarter of a million Canadians experience homelessness at some point during the year, a number that has grown since the 1980s (Gaetz et al., 2016). In addition, there are about three times as many who make up the "hidden homeless," couch-surfing with family or friends, or living in empty houses or abandoned buildings (Wellesley Institute, 2010). As many again experience conditions that we now regard as overcrowded, while double that number occupy substandard housing and/or have to set aside an inordinate proportion of their income for rent. To different degrees, all of these households are considered poor.

But, beyond that, and underneath those percentages, much is open to debate. In the nineteenth century, many urban households lacked hot and cold running water, while sharing rooms. Today in Canada these deficiencies are marks of poverty or substandard housing. But that is not how those people were seen – or viewed themselves – at the time. In some measure, then, poverty exists according to prevailing norms. There have long been wide differences in income and wealth. Those in, say, the bottom tenth in terms of income are likely to be seen as poor, regardless of their absolute standard of living. Just as important, they are likely to think of themselves that way, and experience psycho-social consequences of poverty (Payne, 2017; Wilkinson, 2005). If, in absolute terms, poor people have always been with us, in relative terms they always will be.

And of course many people move in and out of poverty. Households that were solidly working-class in the 1920s descended into desperate poverty during the Depression, only to emerge into the postwar middle class. The same has been true during other economic depressions, notably the 1890s, as well as short-term recessions. Were those households poor? Temporarily, yes of course.

But some had material and social resources that enabled them to weather hard times, including supportive family and neighbours. They had advantages – albeit modest – denied to those who never had steady work, or who had no prospect of employment. In their survey of the history of Toronto's poor, Bryan Palmer and Gaétan Héroux (2016) use an expansive definition of the poor that implicitly includes all of the unemployed. I would suggest that, for present purposes, a narrower focus makes sense: those with durable challenges, who Palmer and Héroux (2016: 4) call "the dispossessed."

How do close neighbours matter for people like this? In most cases the answer seems to be not much. Many poor people, and poor households, have always been geographically scattered, finding shelter wherever and however they could. This was apparent during the Depression; Palmer and Héroux (2016: 180) observe that, at that time, the housing crisis "was by no means limited to the well known and long recognized 'slum districts'." However, the poor have been quite dispersed, although less obviously so, even in better times. Jordan and Hildy Stanger-Ross (2012) conclude that the majority of the poor live in socially integrated neighbourhoods, albeit in basement apartments, older multi-unit buildings, or crowded together. The converse is also true: half the people living in low-income areas have middling or better incomes.

That said, there have always been concentrations of poverty. The most visible have been in and around skid rows. "Skid row" came into use during the Depression, but such areas had existed for decades. The upper streets of late Victorian Halifax, with a core of "vice" along Barrack Street, was a well-known haven for repeat offenders, many Black or of Irish heritage (Fingard, 1989: 17, 19). The term "skid row" fell into disuse in the late 1970s (Ford, 1994). Its implied reference to alcoholism, drug abuse, transience, and homelessness came to be seen as too critical, or dismissive. But the reality persisted, 1970s Toronto being a case in point (Whitney, 1970). Where- and whenever they have existed, skid rows typically extended several blocks and included flophouses, rooming houses, hostels, shelters, and, as Gwyn Rowley (1978: 214) described in his ethnographic account of Winnipeg's, "random unkempt business premises" (figure 6).

Winnipeg has been the subject of some of the fullest accounts. By the 1970s, the majority of its skid row residents were First Nations or Métis men, living in a "state of limbo between the reserve ... and the reality of a modern Western city" (Rowley, 1978). Three decades later, David Burley and Mike Maunder (2008: 87, 99, 101) described life on a nearby block where people still "moved back and forth between the reserve and the city," where transience was the norm – one resident, Margaret Bonnette, moved 23 times in 28 years – and where "boozy, blurry sociability ... could erupt into anger, rage and violence," with the Manitoba Warriors gang active nearby. The experience of women could be especially fraught (Klodawski, 2006). Another study found that even recent refugees found the area troubling. To be sure, as one said, "everything is close. Easy to get a bus to go everywhere. It is walking distance to all you need" (Carter, Polevychuk, and Osborne, 2009: 305). But his dominant impression was negative. Others commented, "the only reason I live here ... is because I can afford it. Nothing else is good about it," "not a good place for kids," and "there are gangsters on the next street" (ibid.). After one year, three-quarters of the residents wanted to move away; after another year or two, the proportion had risen to 90 per cent.

Was community possible on skid row? Rowley (1978: 220) suggested it was. After eight years of research, Christopher Hauch (1985: 70) concluded that "the most powerful influence was the fluidity of the people," so that "very little social organization exists," but a type of community had developed. Racialized and cultural differences were obliterated by shared experience: "gestures of seeming kindness and sharing, among men so evenly impoverished, are routine" (41). Curiously, however, "unlike [in] our own society such acts are never accompanied with sentiment": no thanks were given, nor mutual obligations implied (41). The bleakest form of neighbourliness imaginable.

But I wonder whether such indifference is the norm. The pandemic saw a surge of overdose deaths. Marcus Gee's (2021) report from Oshawa hints at how some responded. The city's downtown became "home to a floating population of struggling people who live in shelters, rundown apartments or clusters of camping tents." Although the term itself is not much used now, this amounted to skid row reborn and, according to other reports, it was – and remains – hardly unique. In Oshawa, addicts carried Naloxone kits to save others on the street. One, Ken Chopee, reckoned that he had saved 20 lives. As Gee (ibid.) comments, "many take turns saving each other, forming a kind of buddy system to stay alive." Hardship can bring out the best, as well as the worst, in people.

As various accounts suggest, skid rows have commonly bled into larger areas of poverty, such as Montreal's Plateau-Mont-Royal and Pointe-Saint-Charles, Vancouver's Downtown Eastside, Kingston's Near North, Calgary's Greater Forest Lawn, and Hamilton's North End. Some have detected real neighbourliness in these places. In Greater Forest Lawn, Erik Meiji and co-researchers (2020: 211) found that poor people "exchanged stories and experiences ... which instills a sense of solidarity and fosters reciprocal emotional support"; they shared "local knowledge" about resources, services, and which businesses allowed them to use washrooms. Those who stuck around became well known, sometimes affectionately. In Hamilton, Denise Davy (2021) found that a troubled regular who shuffled in and out of shelters, became well-known as "Princess." On Winnipeg's Furby, rooming-house apartments became "home and gathering place for many more relatives and friends" with a "continual coming and going." Especially where families were involved, networks were maintained.

But, in such areas, neighbourly connections and trust have been thin on the ground. Yovette Marchessault, herself of mixed Cree and French-Canadian background, grew up on Boyer Street in Plateau-Mont-Royal in the 1940s. She recalls the rats indoors and the violence beyond (Marchessault, 1975: 124). A generation later, Kathy Dobson (2011) described in graphic detail her girlhood in Montreal's Pointe-Saint-Charles, "Canada's toughest neighbourhood." It has had competition, from Toronto's Jane-Finch corridor, and Vancouver's Downtown Eastside, both of which have been framed negatively in the media (Richardson, 2014; Liu and Blomley, 2013). Less well known, except to locals, the fringes of downtown Hamilton have presented a troubling picture, one which David Baillie's (2015) *What We Salvage* has evoked from the teenage protagonists' point of view. The closing of institutions for the mentally ill led to the creation of a service ghetto for patients across the Niagara peninsula. As Michael Dear and Jennifer Wolch (1987: 110–38) have shown, rooming houses and short-term care facilities became concentrated in certain areas, and so, inevitably, was the service-dependent population – those with mental and physical disabilities. The authors do not suggest that bleak living conditions were mitigated by supportive neighbours. Two decades later, Erin Mifflin and Robert Wilton's study of rooming-house residents found that little had changed. To be sure, a few social-ized with other tenants, sitting around the kitchen table with a few beers, but most were troubled by noise and lack of public safety. Currently sleeping in a homeless shelter, one respondent told them that once "I had my own apartment … I had good neighbours, never bothered me, I never bothered them" (Mifflin and Wilton, 2005: 413). Peaceful anonymity was the best he could hope for, and he could not even depend on that.

What to make of these varied vignettes? The fullest and prob-ably most accurate picture of the neighbourhood experiences of poor people in a modern Canadian city is that provided by John Ecker and Tim Aubry. They interviewed over three hundred peo-ple in Ottawa living on the street, in single-room occupancy apart-ments, or in rooming houses. They found that location mattered, especially in terms of access to shelters, cheap accommodation, and social resources. A minority liked the places they lived, find-ing them walkable and quiet. Some said that they "fit in" (Ecker and Aubry, 2017: 536). But most felt uncomfortable and unsafe, especially where drugs were traded. Today's street drugs are more addictive, and socially more disastrous, than alcohol, which has long been associated with poverty. But recent evidence surely speaks to a long-standing situation. For the very poor, transience has been a fact of life. Neighbours can be helpful, but oftentimes are indifferent or worse. Even at its best, neighbourliness cannot reverse a generally bleak existence. The experience of immigrants, however, even those who started out poor, has routinely been much more encouraging.

Immigrants, First Nations, and Métis

Living side by side, workers and the working poor have been available to help each other in hard times. They share a culture that includes certain tastes in food and drink, in entertainment, and everyday speech. This is especially true of ethnic and racialized minorities, and for those whose first language is not English or, in Quebec and parts of Northern Ontario and New Brunswick, French.

Immigrants have all sorts of reasons to congregate. That has even been true of wealthy and well-educated recent arrivals from India, Hong Kong, and mainland China. But it has been doubly true for the poorer majority, who have had to resort to crowding into older dwellings and neighbourhoods, perhaps taking in lodgers and boarding families, including extended families or relatives from back home. This clustering brings the material advantages of having people close by who could help you out, with whom you could communicate and learn how to survive in a strange place. Apart from these practical aspects, there is what Robert Harney (1975: 208) has called the "*ambiente* of a social and cultural transition." Harney was speaking about Italian Canadians, but environmental adjustment is needed by most, if not all, migrants.

Even the British needed kindred spirits nearby in the instances they were in the minority. In early twentieth-century Trois-Rivières, there was no "English neighbourhood," but "the members of the anglophone community generally tended to favour certain neighbourhoods or streets" (Bellavance and Normand, 2014). The same could be said of British immigrants in other Quebec cities, including Montreal. In the early 1900s, many settled in the west end of Winnipeg, near the Canadian Pacific Railway shops, and in Elmwood. One street was populated by families from Leicester, England, while a Scottish woman recalled that her neighbours "were all so friendly; we were all like one family," offering help in times of need (McCormack, 1984: 367). But in general the British did not self-segregate. They shared a language and cultural assumptions with most native-born settler Canadians, themselves of mostly British heritage. From interviews with British immigrants who grew up in Hamilton's East End during the Depression, Jane Synge

(1976: 44) reports that there was less reciprocity than among other immigrants. Robert Harney (1985: 10–11) has observed that "some groups obviously required a 'little homeland' more than others."

If the British lie at one extreme of the aggregation continuum, Jews and the Chinese, targets of discrimination and worse, lay at the other. Victoria's Chinese Canadians formed a tight-knit community in the early 1900s, with almost 3,500 people living and working in just a handful of blocks by 1911. Congregation in Canada's first "Chinatown," as white settlers called it, was for "convenience and mutual assistance" and, given that Chinese immigrants were ostracized and sometimes attacked, self-protection (Lai, 1973: 113). The result was "a fully self-contained community" where homes, workplaces, stores, Chinese schools, churches, temples, opium dens, fraternity associations, and even a hospital sat cheek-by-jowl (115). At greater length, and with a subtle account of the dynamics of racialization, Kay Anderson has shown that much the same was true in Vancouver, whose Chinese population only exceeded Victoria's in the 1920s (figure 7). The early Chinese community was largely confined to a few blocks, choice playing little part: the solicitor for the city's Chinese Board of Trade observed that "the Chinese live in aggregation, but this is more a matter of necessity than choice" (K. Anderson, 1991: 69). Indeed, Anderson argues that, expressing "the frames of mind of the West" as much as the "ethnic attributes of the East," "Chinatown" was in part "a European creation" (9, 31). In time, most of the community dispersed; by 1931 barely a third lived in Chinatown, but block-scale concentration still held advantages (Barman, 1986: 113). It persisted in much the same area into the 1960s, proving effective in mobilizing opposition to redevelopment, something that was threatening many Chinatowns in Canada and the United States at that time (Vitiello and Blickenderfer, 2020). Ethnic festivities, along with "block-based, multi-lingual, neighbour-to-neighbour communication systems" gave life to "informal networks" (Lee, 2007: 395). The block scale continued to matter to Chinese immigrants.

As it did for Jews, a similarly targeted group. Harold Troper (1987: 56) judges that, among all immigrants until mid-century, Jews were placed by WASPs "at the bottom" in terms of desirability.

In Toronto, those who prospered moved to Forest Hill, but they had first concentrated in the Ward, one of the city's poorest districts until it was demolished. Choice barely came into it. As Dan Hiebert (1993: 209) puts it: "Jews were not particularly welcome in many other parts of the city." Here there was a tight intermingling of homes, stores, and factories, so that men and women could walk only a block or three to do their shopping or get to work. But, even within this community, who lived next door mattered. Cynthia MacDougall (2015: 163) recalls that, having grown up in Toronto's Ward district, her "uncles and aunts said Walton Street was a neighbourhood all of its own."

And then there were those, such as the Italians and Portuguese, who faced discrimination of a milder kind and who also found that tight clustering was helpful and comforting. The early wave of men lived in boarding houses or rented from compatriots. In Hamilton in 1911, on the 100-block of Sherman Avenue, there was a large concentration of Armenians and Italians, many boarding with fellow immigrants. Fifteen years later, Italian tenants were fifteen times more likely to rent from Italian landlords than from anyone else (Doucet and Weaver, 1991: 386). The proportion for East Europeans was even higher. Women soon joined them, and helped expand the enclaves. Michele Stella arrived in Toronto in 1930, settling in Little Italy on College Street, where three-fifths of her "immediate neighbourhood" was Italian (Sturino, 1978: 43). Later, when families arrived in large numbers after World War II, homes were purchased and subdivided, cantinas and second kitchens were built in basements, vines were planted in backyards, and families hung out on the street (Zucchi, 1985; Iacovetta, 1992; Stanger-Ross, 2009).

Those immigrants usually squeezed into two- and three-storey houses. Recent arrivals from more diverse places have often found themselves in different kinds of spaces. Sutama Ghosh (2014) has shown how, in Crescent Town and Scarborough Village, Bangladeshi immigrants made high-rise buildings into neighbourhood networks, converting a ground-floor apartment into a grocery store and other units into daycares and beauty parlours. In Thorncliffe Park, floors and corridors serve much the same function as city blocks. Adinah Heqosah, a recent arrival from Afghanistan, was initially envious of "the Pakistanis and Indians, who have taken over multiple floors of her building" (Saunders, 2010: 313; c.f. Vassanji, 1991). Immigrants make do however they can.

Any generalization is complicated by differences within each ethnic group. Outsiders speak about "Italians" but northerners and southerners perceive of themselves as different from each other; indeed, identities, affiliations, and distinctions can matter down to the scale of the village. These can be accentuated by patterns of settlement. Even in the early twentieth century when the Italian community was quite small, Zucchi (1988: 67) reports that Toronto's three Little Italys looked down on each other: "'The Ward' was considered a slum by the other two neighbourhoods. The Ward residents found the College-Grace district rough. Meanwhile, those two neighbourhoods viewed the north Dufferin Little Italy as downright dangerous." Even more obviously, to outsiders, the label of "Black" hides cultural differences between American, Caribbean, and African immigrant traditions. This is especially true in districts like Montreal's Saint-Antoine – later renamed Little Burgundy – which was settled by African-American porters and their families but which attracted immigrants from the West Indies after World War I (Williams, 1997: 44). Most recently, geopolitical developments have intensified divisions within the Chinese-Canadian community. Even before China's takeover of Hong Kong, the settlement patterns of those born in Taiwan differed from those born in mainland China (Hiebert, 1999). A Hindu from Mumbai might find more in common with a Protestant from Barbados than with a Muslim from Kolkata.

Immigrants care about who lives on the block, but much depends on the context. True, some researchers may overstate the significance of the immigrant enclave, but here again, as much as anywhere, residents and outsiders may be inclined to imagine the presence of a community that is tighter, and more complete, than it ever was, is, or could be (Qadeer and Kumar, 2006; c.f. Massey, 1995a). But enclaves clearly have mattered a lot. The only way that one Thorncliffe Park immigrant, Maryam Formuli, could start building a personal network was to hang around a local kebab counter, "nervously approaching fellow Afghanis" (Saunders, 2010: 312). To make small local businesses viable, and to attract specialized services such as language training, "finance, travel, and media institutions," neighbourhood-scale numbers of people are required (317). Working men and women have needed affordable stores and restaurants, along with churches or other religious institutions, that require more patrons or members than a single block can provide. As discussed in chapter 7, the middle classes have increasingly paid attention to school catchment areas. But no group values neighbourhoods as much as newcomers. They look for affordable stores, but ideally these sell the food, clothing, furniture, and sundries with which they grew up. They look for informal venues – cafés, restaurants, clubs, or bars – that have the right ambience and/or music. They may need specialized immigration services in their own language, along with community halls and religious institutions, each requiring the support of hundreds or thousands of people. That is why, when speaking about the immigrant experience of the city, it is usually the scale of the neighbourhood that matters most.

The same has not been true in the same way for the Métis and First Nations. (The Innu still have little urban presence to speak of.) Their experience has been different from that of immigrants for several reasons. First, and most obviously, they were here first. It was European settlers and their successors, like myself, who were the immigrants, and the cities they created have expressed their culture, broadly capitalist and liberal-democratic in character (c.f. Hugill, 2017). For many decades, this fact was publicly ignored, or suppressed. And so there is a deep irony in conflicts such as those described by David Ley. Vancouver's Shaughnessy neighbourhood was developed in the interwar years as the city's elites migrated away from the West End (Nader, 1976: 393). Guided by building restrictions, homes were large and estate-like, and buyers favoured architectural styles that evoked their British origins, including Scottish baronial and Tudor. Gathering momentum in the 1990s, however, affluent Chinese immigrants began buying up and tearing down homes, replacing them with boxier, modernist "McMansions." Residents resisted, in part because some purchases were speculative, with new homes being left vacant for months or years. But their objections were also cultural and aesthetic: heritage was being destroyed and, along with it, stately trees. As one resident declared, "we want to stress that this is a place to live not just to make money out of" (Ley, 1995: 197). That, of course, is what First Nations people had been telling settlers for generations.

The experience of Indigenous people has also been different from that of immigrants because, for decades, their presence in urban areas has been neither anticipated nor accepted. For a while, First Nations actually made up the majority of residents of Victoria, BC, with a great concentration on Johnson Street, and then on Herald Street (Lutz et al., 2014). But, being pushed away, their numbers declined rapidly from the 1860s. In the Lower Mainland, the Coast Salish had established a village at Xwáy̓zway (Whoi Whoi). As European settlement grew in the late 1800s, there was discussion about creating a reserve for them there but Gilbert Sproat, head of the Indian Reserve Commission, reckoned that "white settlers did not appear to desire that Indians would be concentrated" (quoted in Kheraj, 2013: 54). They wanted "Indians" to be dispersed, available as labourers in the sawmills, or simply gone. In a less planned manner, the Métis were also displaced from a number of urban centres in Western Canada (Kermoal, 2022).

Although 46 per cent of Canadians were living in urban areas by 1911, only 2.7 per cent of First Nations and 10.6 per cent of Métis did so at that time (Goldman, 2014). Such disparities often exist in other settler-colonial nations, including Australia, the United States and, in a different way, Kenya and South Africa (Hugill, 2017; Peters, 2005: 340–52). Unlike immigrants, First Nations people did not arrive in cities by boat or plane. Instead, they were dispossessed of their land and pushed onto rural reserves where, for many decades, they were expected to remain – and indeed, under the pass system, were compelled to do so. As Evelyn Peters (1996: 60) says, based on her interpretation of literature on Canadian cities, "Aboriginal people are confronted again and again with explicit or implicit messages that cities are not where they belong" (see also Peters, 2005: 340–52).

And so, when they came, at least some Aboriginal people have sought "place[s] of cultural safety within the hostile, white settler, urban environment" (Peters, Stock, and Werner, 2018: 7), in effect reclaiming space. An early example was the small community of Rooster Town, which grew up on the outer fringe of Winnipeg in the first half of the twentieth century (Burley, 2013) (figure 8). It was socially tight, and almost entirely Métis. Many of the families were related, visiting was common, and for entertainment they enjoyed dancing and fiddling (Peters, Stock, and Werner, 2018: 133–4). But the group was scattered and never large: at its peak around 1940, it consisted of only 50 households and about 250 people, and soon eroded.

There is another way in which the urban experience of First Nations, especially, has been distinctive. Once they began to migrate in numbers, to the point that today about half of Canada's First Nations people live in urban areas, it appears they have not consistently tried to congregate (Peters, 2005: 356–9). "Appears" because, although since the year 2000 more has been written about their urban experience than in the previous century, relatively few have been from a First Nations' perspective. It hasn't helped that many Indigenous peoples were dispossessed not once but twice: after being relegated to a few reserves located in or near urban centres, they were later removed. In the Lower Mainland, the Squamish lived year-round on the south shore of what settlers later called False Creek (Macdonald, 1992: 10). In 1869, the area was made into a reserve, but displacements eliminated the settlement, and the associated rights to nearby lands – at least until a recent court decision (D. Harris, 2017; Stanger-Ross, 2008; Wade, 1994: 50–1). The Songhees also suffered displacement. Allocated a site across the bridge from Fort Victoria in 1850, they were later removed wholesale to Maplebank, near Esquimalt (Mawani, 2003).

On the other side of the continent, a small group of Mi'kmaq on the Kings Road outside Sydney, Nova Scotia, were "given" a reserve in 1882. Following complaints from an "occasional" member of Parliament, and after a court hearing in 1915, the Mi'kmaq were displaced (Walls, 2016: 544). In such ways, several urban neighbourhoods were nipped in the bud. Others were non-starters.

In 1877, a site was set aside for a Papaschase Indian reserve in Edmonton but because it was not immediately occupied its status was annulled in 1888 (Shields, Moran, and Gillespie, 2020).

A few reserves survived and became urban. In 1869, the Musqueam were assigned territory on the north bank of the Fraser River, where they had a winter village (Macdonald, 1992: 10, 14; Weightman, 1978). In the early 1900s it became important for market gardening, farmed by Chinese immigrants, and today is home to about a thousand Musqueam (Stanger-Ross, 2008). That is similar to the population in Membertou, south of Sydney (Mercer, 2021). Beginning in the 1930s, the Department of Indian Affairs offered more financial assistance to people in such urban reserves, assuming that the rural ones were self-sufficient; by 1935, disbursements to the Wendake reserve in Quebec City were 74 times greater than in 1900 (Gettler, 2020: 161).

Reserves have drawn attention for various reasons. Even though the Mohawk built an important railway bridge to the city, Kahnawake was long ignored by historians of Montreal. Becoming a settlement of nine thousand, it eventually won media coverage for its militancy (Rueck, 2011). In Quebec, Wendake has drawn tourists as well as a mix of Indigenous peoples (Iankova, 2010). Indeed, many urban reserves have piqued the curiosity of settlers. Barbara Weightman (1978: 204) spoke with a number of neighbouring residents who had visited the Musqueam reserve, finding that many were curious to know "what an Indian reserve was like." I did that myself, while living in Marpole, south Vancouver, in the early 1980s. In Kelowna, Westbank has boomed, growing from a trailer park and housing development to include a variety of commercial operations which bring in business from outside the community (Flanagan, 2019). On a smaller scale, after losing land when the new town of Oromocto, New Brunswick, was planned, almost two hundred of the Welamukotuk First Nation ended up in an urban neighbourhood on its fringe (Gordon, 2021). Other, nonresidential, urban reserves have been established to promote economic development (Tomiak, 2017). In this way, at different scales, clustered neighbouring within and among First Nations people takes place in an urban setting. But such experiences are the exception, not the rule.

Other Indigenous enclaves, like Rooster Town, just outside of Winnipeg, emerged with rural-urban migration. Today Winnipeg overall has the largest Indigenous population of any Canadian city; 12 per cent of the city's population are First Nations or Métis, a figure that has doubled since 1996. The proportion in the North End is almost double that, and in Lord Selkirk Park, within the North End, self-identified Indigenous comprise a majority of the population (Distasio and Zell, 2020: 224; Canadian Centre for Policy Alternatives, 2005: 21–9). Not surprisingly, as early as the 1970s a nearby 8–10 block stretch of Main Street, extending north from the city centre, despite its reputation for single-room occupancy hotels and pawnshops, had become "an important cultural, social and political hub for the city's growing indigenous community" (Hugill, 2022: 251). Although many Indigenous people have had little choice about where to live, such clustering must have reflected the desire of at least some to live near kin and clan members. In Winnipeg more than in any other Canadian city, Indigenous residents have worked to create service centres akin to those designed for immigrants in the early twentieth century. A notable example is Neeginan (Cree for "Our Place"), which has been functioning since 1990 (ibid.).

Neeginan has depended in part on government support, but in general Indigenous settlement in urban areas is hardly a tribute to the hospitality of urban settler society. Indeed, many Indigenous migrants have felt ambivalent about urban life, especially with the risks of living in areas with concentrated poverty. There is a class dimension to this. Unlike many first- and second-generation immigrant communities, the Indigenous middle class – such as it is – has always been geographically integrated (Peters, 2014: 198). For David Newhouse, growing up on the Six Nations reserve in southern Ontario, "the city ... was an escape ... where I could be myself, where I could live in relative obscurity away from the peering eyes of neighbours" (Newhouse, 2011: 25). Settling in Peterborough, and having become a university professor, he found "our families and homes are here ... along with a sense of community, albeit not in neighbourhoods" (33).

More typical is a North End Winnipegger interviewed by Jim Silver and colleagues; he did not view his move as an escape and did not enter the middle class. "Growing up on the reserve," he observed, "there was a community feeling that I don't feel here. Sure you know a lot of people, you meet a lot of people, you see them every day going down the street, you say 'hi'," but the community of the reserve "is not, like, here" (Silver, Hay, and Gorzen, 2006: 47). Silver and colleagues report the similar experience of an older woman in Winnipeg's North End who lacked connection with neighbours. Although a long-time resident who had raised children in the city, she said "I never bother anybody, the only one I talk to here on this street is these people next door here, maybe every once in a while just say hi" (53). The man perceives a lack of community; the woman does not even seek it.

For her, as for many others, safety was the issue. Alan Anderson (2017: 143, 145) provides telling accounts for Pleasant Hill, Saskatoon, an area akin to Lord Selkirk Park in its high proportion of First Nations residents. He found some "ad hoc advocacy and support" among local residents, but an overwhelmingly concern about crime, with pervasive gang activity, drug deals, prostitution, and alcohol abuse. At that time, there were 31 gangs active in the area. One resident commented that "children, youth, and women cannot walk around the neighbourhood or even sit on their front doorsteps because 'johns' approach them," while "seniors are afraid to walk to the bus for fear of being attacked or robbed" (Anderson, 2017: 147–8). Perhaps because of the crime, and because they saw their move to the city as temporary, most of the people Anderson spoke with "did not state a preference when it came to socializing with Aboriginals" while "a large majority did not have a preference when it came to living in an Aboriginal neighbourhood" (Anderson, 2017: 128).

In districts like Pleasant Hill and Winnipeg's North End, many Indigenous residents found themselves living near kin, or others from their reserve. For some this was surely comforting, and useful. Richard, one of the men whose account is reported by Jim Silver, Parvin Ghorayshi, and colleagues (2006: 27) recalled that, growing up in the North End, "in a block radius I had a couple of aunts, an uncle, some cousins, and then very familiar people who also became in some sense extended family." As a result, Richard felt "very safe" (ibid.). But this was unusual. Many First Nations and Métis people have ended up in low-income areas because of poverty, not cultural preference. Indeed, those cultures have been eroded, notably through the residential schools, and so the *ambiente* on streets occupied by First Nations and Métis has not been that of the typical immigrant neighbourhood.

The Young, and the Young at Heart

People care about neighbours in different ways, and some people care more than others. Children and seniors – or at least retirees – are the obvious examples (Buffel and Phillipson, 2018). This has been brought home to me – literally – when the pandemic struck. I had always enjoyed walking: to get exercise, experience the out-of-doors, to clear my mind, and organize my thoughts. (I drafted this paragraph during a neighbourhood ramble.) But during lockdowns I got to know some neighbours better. Alan from next door, Paula further down, and Bill across the road, each walking their dogs, not to mention David and Laura, being exercised by their puppy. All, like me, were retirees out in the fresh air. And then, next to Bill, there were Nicola, Thomas, and Miles, shooting hoops and, yesterday, shovelling the sidewalk before making snowmen and angels. These neighbours, young and old(er), and the boys up the street who played street hockey, are the people who spend the most time outside, and for whom the neighbourhood – or at least the block— matters most. Since the pandemic one of the older couples has moved away but, in greater numbers than before, the dogs and sidewalk greetings remain.

The Young at Heart

My older neighbours are typical. A huge American survey showed that the elderly did more neighbouring than anyone (Guest and Wierzbicki, 1999: 105). The same holds true in Canada. Ray and Preston's large survey of 2003, mentioned earlier, revealed that those over 60 knew more, had helped more, and had been helped by more neighbours than had younger adults; they had a stronger sense of local belonging and were more likely to trust their neighbours (Ray and Preston, 2009: 232). Richer, if anecdotal, evidence confirms this. In 2011, Paula Gardner interviewed older people in the High Park area of Toronto. Everyone reported that local people and places played a large part in their lives. Some, like a respondent named Allan, went looking for strangers: "Down at the sports park," he said, "you can have a coffee and just sit around and always someone will come up and talk to you" (Gardner, 2011: 266). Others, like Ed, expected to connect with immediate neighbours: "Oh if I want to talk to anybody all I need to do is go out into the yard and stare at a tree … and Sam or one of the other neighbors will be over in a heartbeat" (Gardner, 2011: 266–7). Still others, more typically women like Jane, used an established social network: "I looked out for their children for years, now they look out for me, they bring in my garbage and recycling bins, shovel my driveway" (Gardner, 2011: 267). Shovelling is a recurrent theme; this is Canada, after all.

And then there are less fortunate ones, those who live alone, perhaps with physical or mental disabilities. The frail and those in wheelchairs value curb-cuts at the corner, a level sidewalk without trip hazards, and a surface cleared after winter storms (Kerr, Rosenberg, and Frank, 2012). Even shut-ins, whose closest encounter with neighbours is from their porch, or through the front window, care about the people next door: How noisy are they? Do they look out for me? They will also be fretting about the area: Are break-ins common? Is it safe? Those in granny flats might be able to call upon their landlord in times of need, or simply to chat, but those in multi-unit buildings can get lonely. In 2019, the City of Vancouver launched "Hey Neighbour," a program for such people. It sought resident "animators" to bring people together. One senior commented "I like to talk and I'd get into the elevator or meet someone in the hall and try to talk to them and there was just a wall" (Keillor, 2019). He wanted neighbourly connections, but the interest was not reciprocated.

Well-heeled seniors have avoided this problem by buying into age-defined communities, both high- and low-rise common interest developments (CIDs). The "interests" in question are legal and financial, in that all residents are charged for shared facilities, but they are also social. Part of the attraction of these areas is that maintenance is outsourced, for a price (Walks, 2014). In addition, these projects guarantee neighbours of a similar age, the absence of noisy children, and often a clubhouse and recreational facilities (Grant, Greene, and Maxwell, 2004). As Jill Grant (2005: 297) reports from a national survey, many projects also have their own Neighbourhood Watch programs, while one official commented that a clubhouse is "a real positive thing which has spin offs for the community. With people meeting each other and getting involved in projects." The concern is as much to exclude undesirables as it is to include those with similar views about property maintenance and sobriety. Grant (2005: 299) quotes one developer: "people buy on the basis of security and living with people they are comfortable with." These are relatively affluent communities, so mutual aid is not much needed. Instead, as Jill Grant (305) observes, paraphrasing and quoting one resident, "everyone knows each other, he said, 'it's a real community'." Residents have time on their hands and, measured by the hours spent in one another's company, CIDs are among the most neighbourly places in Canada.

Older adults in immigrant communities are often especially invested in their neighbours. In Vancouver recently, Catherine Tong and colleagues spoke with men and women from various visible minorities. (This Canadian term begs the question: "visible to whom"?). Raveena (age 67) said she went out "for the fresh air and I want to meet the peoples [*sic*] and I want to make my feelings better," while for Jasmeet (age 78) "these few blocks, these are my village" (Tong et al., 2020: 643). Making and affirming social connections were vital for women. Only two of the fourteen women interviewed had a driver's licence and, as Ka-Lee (age 81) observed, "he [her husband] can't drive me everywhere ... it's too much for the old man" (644). Limited mobility framed their geography.

Immediate neighbourhoods have always been important for older people, but this may have become more so over the past several decades, at least for social interaction at a time when, generationally, elders' significance for others has declined (Guest and Wierzbicki, 1999). We live longer and move less often than our grandparents did. Many residents of High Park had not moved for decades; that is how they had come know many neighbours. Theirs was what Michael Hunt and Gail Gunter-Hunt (1986) have called a "naturally occurring retirement community" (NORC) – an area which acquires older people by chance. Such areas are becoming more common. In Toronto in 2006, only 9 per cent of all census "dissemination areas" (average population 700) were NORCs, where "over 55s" made up 40 per cent of the population or more. In 10 years the proportion had doubled, to 20 percent (Donnelly et al., 2020). Aging in place is more drawn-out than 50 years ago. In the 1920s in Richmond Heights, a working-class suburb of Halifax, few people lived beyond their 60s, and most were poor. This constrained mobility and compelled many to move in with their children, away from old friends their own age (Morton, 1995: 51–66). One effect of the longer life expectancy – and state pensions – is apparent from a British study. Long-term residents, especially women, have become "neighbourhood keepers" (Phillipson et al., 1999: 741). Fifty or, even more so, a hundred years ago there were fewer keepers to care. And they care more because they have long memories, memories that are notoriously selective, encouraging them to idealize the past. They value their neighbourhood as much for what it was as for what it is (Blokland, 2001).

Not all residential areas are equally friendly to the elderly. In another study of Vancouver, Thea Franke and colleagues (2013) looked at the experience of "highly active older adults." Not surprisingly, they found that those who were more resourceful were more active. But two other factors made a difference: local social connections and the character of the built environment. Areas with green spaces and scenery, where it was easy to avoid busy roads and which had good sidewalks, encouraged people to get out and about. These features surely mattered more to older adults than to their grown-up sons and daughters – unless, of course, they had children.

The Young

The neighbourhood features identified above would have mattered to all of those adults when they were the age of Nicola, Thomas, and Miles. Curiously, not much has been written about how the everyday "culture of childhood" is rooted in place (Sutherland, 1997; c.f. Onusko, 2021). Sutherland (1997: 224) argues that, apart from home and school, the neighbourhood has been much more important than indicated in the psychological and sociological literature. The neighbourhood "provided more scope for parents and children to differ over the boundaries of place," and "in it, as well, most children found a wider opportunity to explore forbidden places, to spend time with forbidden people, and to partake of forbidden pastimes" (237). In other words, more freedom.

Of course, freedom always varied with age. Some preteens are given – or take – much more latitude than others. Until at least the 1960s, children were told to "go out and play" (Lewis, 2002), perhaps supplemented with "and don't come home until supper." In Depression-era working-class Hamilton, Glynn Leyshon (1999: 32) recalls that "there was little or no adult supervision, little or no equipment, and the games were played out on the pavement, in the alleys, and on the back lanes." She played marbles in the gutter and made primitive scooters from orange crates and roller skate wheels. But most young children stayed close to home. Much the same was true in early postwar middle-class suburbs. Sheila Munro (2001: 67) recalls that, "almost as soon as we moved to West Vancouver I discovered a neighbourhood gang of friends." She did not stray far. The important friends were "Bruce next door whom I adored, Mark at the end of the block who adored me, and Karla and Linda across the street" (ibid.). Unsupervised, they made their own fun: "We played in ditches by the side of the road … we built little bridges and sailed boats in sandy streams. We climbed trees …" (ibid.).

From his account of American urban neighbourhoods in the early 1900s, David Nasaw (1985: 32) concluded that "the block was the basic unit of social organization for city kids." Canadian cities were hardly different. For Toronto at that time, Philip Mackintosh (2017: 180) describes how, depending on their age, "children splashed, toddled, and whiled away hours … the roadway provided space for … sports and myriad group games." Memoirs, oral histories, and social-realist fiction indicate that versions of this experience persisted into the postwar era.

This was especially true in working-class districts, where domestic spaces were crowded and few homes had front or back-yards of any size. Evoking Montreal's Saint-Henri ward in 1940, the outsider Gabrielle Roy (1989: 95) conjures "a group of ragged children … playing on the sidewalk among the litter." Denyse Baillargeon (1999: 116) reckons that rear lanes were "particularly lively" being "playgrounds for the children" and, as such, the "quintessential area for the women and children." They still are (Zask, 2021). In the 1930s, there was little alternative. Parks were few, and in poorer areas public schools offered little beyond a paved playground. As an adolescent in Montreal, Paul, the protagonist in Hugh MacLennan's *Two Solitudes*, had to adjust when his family's household income fell. He was moved from a private academy to the local public school where "no games were provided" (MacLennan, 1945: 248). For exercise and fun, young children played, above all, "on streets and sidewalks, in back alleys, city parks, vacant lots" (Lewis, 2002: 6) (figure 9).

Lucky children who lived on the urban fringe had other options: "undeveloped woodlands, and along creek and river banks" (Lewis, 2002: 6). Bill Bailey's 1980 oral history included accounts of individuals who grew up in suburban York Township in the 1910s and 1920s. According to Wilf Royle, "the bush was a wonderful place for us kids" (Bailey, 1980: 19). He elaborated: "there were lots of places where we made swimming holes with a little work" and in the winter "a pretty good rink. We all skated in those days and played hockey or shinny" (20). Frank Fisher recalled that "in summer we made kites of sticks and newspapers" (64). Such experiences were not the norm for city kids. Robert Boschman (2021: 139) recalls that, in Prince Albert, Saskatchewan, in the 1950s and 1960s, playing shinny was the big deal "in every season but summer." But this was not done on the large concrete pad behind the King Coin Launderette where his family lived and worked. Young children did not have much opportunity to stray far. Attuned as Lorimer and Phillips (1971) were to the local children, Lorimer commented that "the kids who live nearest are the most likely candidates for the circle of one's kids' close friends" (47). Neighbourhood space was children's domain.

That became briefly even more true after World War II. Children had fewer jobs or responsibilities and more free time. Since the 1970s, however, especially in middle-class families, they have been subject to closer supervision and guidance, including organized sports and cultural activities. In their remaining free time they play video games, watch TV, scroll through social media, surf the internet; when outside, they are discouraged from venturing far from home. And so, as a recent study of 9- to 13-year-olds in London, Ontario, has shown, their geographical spaces are "very small," mostly "the environment within and immediately surrounding the home" (Loebach and Gilliland, 2019: 11). Some never walk beyond their block. Those that do go in pairs or groups, heading for parks, playgrounds, malls, or convenience stores.

Gender has always been an important factor: girls have had more limited freedom. Veronica Strong-Boag (1988: 12) points out that during the interwar years "little girls were expected to be mothers in the making." Chores kept them indoors; boys delivered papers, ran errands, helped in the garden or with other outdoor projects (Sutherland, 1997). In inner-city St. John's, the boys could venture as far as the wharf, but girls stayed in or close to home, where they were taught the "domestic arts" before being shepherded to church events (Phyne and Knott, 2018; Sharpe and Shawyer, 2021: 85). A man who grew up in Banff Trail, Calgary, in the 1950s recalls, "I was expected to cut the grass and shovel the walks. The girls were supposed to help Mom and help with the dishes" (Onusko, 2021: 82). Among working-class Ukrainians in Depression-era Sudbury, boys ran errands while girls washed floors (Zembrzycki, 2007). It was the boys and young men, for example, who fathers called on when the DIY craze emerged after 1945 (Harris, 2012a: 51). The gendered division of labour was not perfect. Boys might help in the home, especially when there was no sister, but this was a temporary thing, not training for life. Perhaps the most significant point is underlined by Neil Sutherland (1997: 115): "families divided [children's] work into two … categories. In one category was work done mostly by girls, and in the other was work done by both girls and boys." Girls spent more time indoors.

Once outside, however, the young ones had a range of play options. Speaking about the early twentieth century, Norah Lewis (2002: 10) suggests that the line between boys' and girls' games was "not clearly drawn." All youngsters could enjoy hide-and-seek, ball games, and tag, as well as "Red Rover, hopscotch, skipping and Ring around the Rosie," which Cynthia MacDougall's mother played while growing up on Walton Street in Toronto's Ward (Sutherland, 1997: 233; MacDougall, 2015: 163). But it was easier for girls to be "tomboys," playing catch or "scrub," "the neighbourhood version of baseball," than for boys to risk being called "sissies" by enjoying Double Dutch, jacks, or – heaven forbid – dolls (Lewis, 2002: 10–11, 124). That would have been as risky as for a man to do housework: one of the women interviewed by Baillargeon (1999: 126) recalled that "men who helped their wives might become the neighbourhood laughingstock." But, until recently, the serious organized sports such as "football and baseball were 'the' games of boys," along with conkers and marbles (Sutherland, 1997: 234). And, hockey, the quintessential Canadian game was off-limits to girls. Robert Boschman (2021, personal communication) would have been astonished to see a girl playing street or backyard hockey in Prince Albert in the 1970s. Like adults, children could mercilessly enforce gender norms. Fortunately, in the past generation, things have started to change. Notably, through its victories and demonstrable team spirit, the national women's soccer team has helped to change public attitudes as to the sorts of games that girls and women can play.

The same has become true for those who cross gender norms. Growing up in the 1990s, at first Stephanie Nolen (2017) played games outside with trans men, "bois," but she got the message that she wasn't "the right kind of queer" and so dropped out (79). She made up for this as a young adult. Living in Toronto, west of Spadina, away from the gay ghetto, she played football with other lesbians in Trinity-Bellwoods Park, "something many of us had not had the chance to do growing up" (78).

There is a game that, when they reach adolescence, most boys and girls have always been happy to play together, whether in dark alleys, dance halls, clubs, or the back seats of cars or movie theatres.

Here, too, local peers taught each other the rules, and sometimes the consequences. Cynthia Comacchio (2006: 78) recounts how one girl learned the biology of sex from her older brother, who had got it from "the boy across the street," who had found out from some girls who had learned it from their mother. Today, adolescents get much of their sex education from the internet, but it is unclear whether this modern version of the telephone game is more reliable. The consequences, planned or unplanned, might be more children and hence marriage. Marisa Portolese (2023: 40) reports that, in early postwar Goose Village, it was "customary" for neighbours to marry neighbours. An American study hints at just how common this sort of thing once was. Using 5,000 marriage licences from 1932, researchers found that in Philadelphia – an industrial, working-class city –a third of partners had lived within four blocks of each other or less before getting hitched, of which 17 per cent had lived on the same block (Bossard, 1932).

Middle-class girls were discouraged from hanging out on the street, especially towards nightfall, to avoid the perils of sex and other temptations. Karen, Phyllis Young's fraught heroine in *The Torontonians*, felt this prohibition as a girl in the Annex neighbourhood of Toronto in the 1930s. This area, once middle class, was becoming diverse as some homes were subdivided into rooming houses (Moore, 1982: 31). Karen recalls that during summer evenings "you could see [girls] playing jacks or hopscotch under the street-lamps until after ten o'clock at night" (Young, [1960] 2007: 104). But Karen herself could only observe this from her bedroom window: her parents forbade such behaviour. There was rarely cause for concern. A woman interviewed in Vancouver by Peter Sutherland (1997: 227) recalls that "we had a gang if you want to call it that … we would hang out under the lamp on the corner. We didn't do any harm. Mostly we just talked." But, when girls became teenagers, Karen saw (or sensed – it isn't clear) that they might now run giggling in that same street "into a darkness from which they might not come back" (Young, [1960] 2007: 105). In middle-class households, when they had free time, girls were more likely to play indoors, maybe at a friends', or elsewhere under adult supervision.

Girls rarely indulged in organized gang activities. The Burnside teenagers that hung out in territory between Victoria and Nanaimo in the 1960s (figure 10) were an exception. Eight of the core group of 25 were girls (Porteous, 1973). And some went "bad," even in the suburbs (Iacovetta, 1999). More typically, they hung out with the boys. Cynthia Comacchio (2006: 34) quotes a Toronto journalist writing about adolescents in 1919: "they drift to sidewalks, to gatherings under street lights or to shops and dance halls, anywhere were they may find space and light." A century later, in Mississauga, Louisa Onomé's 16-year-old protagonist describes how she "like[d] to hang out on street corners" where "everyone here knows each other, even if you're up to no good" (Onomé, 2021: 37).

Formal and informal gangs have been a male preserve, an opportunity for insecure adolescents to assert their masculinity, perhaps while making money. This could take innocuous forms. Don Kerr (1989: 9) grew up in Saskatoon after his family moved in 1932: "there were back alleys full of garages and weeds and we played can-the-can and baseball and ten other games ... We knew whose yards we could and couldn't invade for hide-and-seek and war games." Kerr doesn't indicate the gender of his playmates, but we can make a plausible assumption. In Winnipeg in the 1910s, John Gray (1970) delivered newspapers and not just in his own area. He recalled that "every neighbourhood had at least one boy who played 'chicken' with all the delivery boys who came around" (58). In Banff Trail in the 1950s, "boys perpetrated much" of the crime and delinquency (Onusko, 2021: 150).

Similarly in Montreal, the titular hero of *The Adventures of Duddy Kravitz* led a couple of informal gangs before moving on to more profitable pursuits (Richler, [1959] 1970). Duddy called one of his gangs "The Warriors." Gatherings of teenage boys and young men were long thought to spell trouble. Male "waifs and strays" caused much hand-wringing in late nineteenth-century Toronto: "you can scarcely walk a block without your attention being drawn to one or more of the class called street boys" (Houston, 1982: 139). Almost a century later, local youth regularly caused trouble at Marge's Lunch, a greasy spoon in Kingston's inner north end (Harris, 1988: 74). The most dramatic gang incident occurred in Toronto when the antisemitic Pits gang triggered the Christie Pits riot on 16 August 1933. But the Pits was exceptional. Fred Sharf, one of its victims, emphasized that most gangs were not antisemitic, instead defining themselves by their turf, age, and (implicitly) gender (c.f. Comacchio, 2006: 35). Turf could express race or ethnicity, as in Montreal's Little Burgundy and some modern public housing projects (Berardi, 2021a; High, 2022: 12; Richardson, 2014), and in western cities, notably Winnipeg, some gangs define themselves by Indigenous status, sometimes marking their territory with sneakers on hydro wires (Buddle, 2011; Henry, 2019: 239).

If gangs did not target Jews they often went after immigrants or racialized minorities. As the media remind us, they can do serious harm, ranging from random beatings of the sort mentioned by Franca Iacovetta (1992: 107) to running drugs, extortion, and murder. But, although powerful and prominent, criminal gangs are the exception. Typically, male gangs indulge in minor thefts and property damage. The Burnside gang drank (illegally), shoplifted, and occasionally committed break-and-entries (Porteous, 1973). In *Boy Wonders*, Cathal Kelly (2018) ruefully describes how, for a short time, he and his pals entertained themselves by stealing hood ornaments from neighbourhood cars. They were caught and Cathal mended his ways. Hugh Garner (1971: 32) writes about gangs in Cabbagetown in the 1930s, but without suggesting that they were a cause for concern. Three decades later, Stuart Crysdale (1970: 90) implies much the same for Toronto's Riverdale, observing that they simply offered "alternative sets of meaning, interpretation and social cohesion." For boys, especially, gangs offered a form of community.

Especially in middle-class areas, most boys did not form gangs or intimidate outsiders. Their rebellions and infractions have usually been less organized and more modest. In the 1960s, a large survey of middle-class high-school boys in grades 9–13 gives us an interesting snapshot. Like adolescents everywhere, they were inclined to rebel: compelled to make a choice, most said they would prefer to incur the disapproval of their parents than break with a best friend (Vaz, 1965: 55). They pushed limits. Two-thirds had taken "little things" that did not belong to them and, in some way, half had damaged public or private property. But only 5 per cent had committed break-and-entry, intending to steal (64). Street-level violence has usually been confined to the rougher districts. In Toronto in 2006, 52 per cent of youth in trouble with the law lived in the city's 13 "priority neighbourhoods"; the remaining 48 per cent were distributed across 127 (Bania, 2009). In selected neighbourhoods, troublemakers fine-tune their activities to the character and daily rhythms of specific blocks (Weisburd, Groff, and Yang, 2012: 23–4; Boivin and de Melo, 2019).

Women, and the Norms of Gender

Mostly, however, whether anglophone, francophone, or allophone, in the past or the present, city or suburb, it is not young males that regulate local space but rather women, and especially mothers. Christine Lamarre (2010: 1014–15) reports that "chez les lexicographes les femmes sont les piliers de la sociabilité de quartier." When Polly Plant and her daughter (see chapter 2) needed help it was local women who rallied round. And when the local redevelopment proposal on our block was submitted to the Hamilton's Committee of Adjustment it was two women who organized and led the appeal. Like the old and the young, women have traditionally cared a lot about the very local places where they live.

That remains true, even though gender norms and roles have changed greatly since the late nineteenth century. Two truths remain: in families, women do more than men in and around the home, while those who work for pay are usually concentrated in specific occupations and earn less (Rose, 2015). The phrase "double ghetto," coined by Armstrong ([1978] 2010), overstates the point, but not by much. For decades, women's job and career options were constrained while the social expectations of marriage and child-raising were powerful. They are still non-trivial. Women marry later now, sometimes with other women, and more of them not at all. Single women, especially in professional occupations, have a variety of residential options that do not entail local connections. But, for substantial periods of their lives, most women have perforce cared about and engaged with neighbours. Often earning low wages, single parents are dependent on childcare, frequently informal and nearby. Where possible, and because fewer have access to a car, they prefer walkable neighbourhoods that are well-served by transit (Rose, 2015).

But even those who are better-off, typically with partners, have been constrained geographically by domestic responsibilities, adapting their lives to suit the needs of their children. A recent American study examined the social networks that grow up from chance encounters between neighbours. It turned out that, overwhelmingly, these stemmed from connections children made with one another, affecting their parents, and above all mothers (Grannis, 2009: 13, 94). James Lorimer confessed "I'm not good at getting to know people but Mif is" (Lorimer and Phillips, 1971: 5). It was she who made the first contacts when they moved onto Minster Lane and, although the couple had no children of their own, the ones who hung out on the street started the process: "we got to know their kids well, and then they [the adult neighbours] became acquainted with us through their kids" (16). No wonder, then, that in the homes and high-rises of Toronto Bill Michelson (1977: 142–53) found the wives were more attuned than their husbands to the social character of the areas in which they lived.

Class, of course, has always affected the picture. Writing about Montreal around 1900, Bettina Bradbury (1993: 38) declares that "the city that women inhabited was not the same as that of men." At that time, many working-class women worked from home. As Cynthia Comacchio (1999: 38) explains, the idea was "to avoid the costs of childcare and to take advantage of the existence of older children." Having large families helped, as did the presence of others in the same situation nearby. By the 1940s "proximity facilitated exchange of clothing, shoes and services, including the use of washing machines and telephones"; grown-up sisters who lived near each other "might exchange chores," one doing all the washing and the other the ironing (129).

In poorer working-class and poverty enclaves this pattern persisted after the dirty thirties. In her semi-autobiographical account of wartime life on an east end Hamilton working-class street,

Sylvia Fraser (1972: 33) reports that "the men of Oriental Avenue do not socialize beyond 'is it hot/cold enough for you Bill?' Oriental is a street of women." On The Flat in Prince Albert in the 1950s and 1960s, Robert Boschman (2021: 39) hung out with "kids whose dads worked at blue-collar jobs and whose moms stayed at home." The same was true in suburban Calgary. As one man recalls, "when I think about going to my friends' places, there was always a Mom there" (Onusko, 2021: 118). In the 1970s, Lorimer and Phillips (1971: 34, 40) found that "family roles are more sharply defined East of Parliament than amongst many middle-class people we know," with mothers being "primarily responsible for control of kids, partly because it is their function and partly because they see much more of their offspring." The same was apparent in public housing projects such as Lawrence Heights in Toronto, where some fathers were regularly absent. W.R. Delegran (1970: 89) reported that half of all residents talked to a neighbour at least once a day but that "the interactions between adult males are more likely to be terse and formalistic than those among their female counterparts." Today, after a transformation which saw British Canadians replaced by Jamaicans and Somalis, a similar contrast exists. Luca Berardi (2021b) noted that caregivers were mostly the local women. Mutual aid made their lives easier, while offering social and psychological advantages. As for working-class Montreal during the Depression, Baillargeon (1999: 116) reckons that "neighbourhood bonds ... contributed to the intense life of the district and helped to keep its housewives from the isolation that was frequently deplored in the period after the war" (116). To this day, working-class and poor women know that neighbours matter.

Of course there has always been a downside. Typically, Jovette Marchessault's (1975) and Kathy Dobson's (2011) portraits of women's lives in Montreal neighbourhoods contain social tensions. Drawing on the unusual records of the Halifax Relief Commission, which rehoused families after the catastrophic harbour explosion of 1917, Suzanne Morton (1995: 40) found that "conflict and strife between neighbours seems to have been a normal aspect of life in Richmond Heights," the new district that the commission developed. She found reports of verbal and physical abuse between women who "spent a great deal of time around their homes" (40). Bettina Bradbury's picture of women's lives in 1890s working-class Montreal probably gets the balance about right: "When the wage-earners had departed, women and children took to the street, heading for the market, exchanging goods and services, seeking household supplies from passing pedlars and hucksters, gossiping, arguing, or at times fighting." They sometimes fought, but encountered more strife with their menfolk at home (Strange, 1995: 77). Life, including family and neighbours, could be intense.

Women's isolation became more of an issue after 1945. Emerging prosperity enabled millions, and not only those of the middle class, to move to the suburbs. The new neighbourhoods lacked transit and convenience stores. Few families could afford two cars: in the mid-1950s, even upper-middle class Karen in *The Torontonians* did without most of time. The results were described by sociologist S.D. Clark (1966: 141) for a variety of subdivisions scattered around Toronto. "Women in particular," he writes, "were likely to feel the loneliness of suburban life ... [in] [a]lmost complete isolation" (ibid.). A speaker for a construction company put it bluntly: "A woman is there all the time, she lives there. A man just boards there: he gets his meals there" (Strong-Boag, 1991: 489). Fictional Karen hated that life, and eventually she and her husband moved back into the city. Many women, however, accepted and even welcomed their new, all-encompassing world. Clark (1966: 153) reports that "the number of housewives [interviewed] who expressed a dislike for the suburban way of life was an exceedingly small minority." And they did not change their minds. Thirty years later, Veronica Strong-Boag (1991: 502) interviewed women who pioneered the postwar suburbs; she summarized their retrospective assessment: "while not without flaws, suburbs were a good deal better than alternatives."

In those suburbs, as in older neighbourhoods, women created networks and ran the show. Not always, of course. In Vancouver's Cedar Cottage neighbourhood after World War II, a neighbourhood house evolved when "a group of fathers" worked "to get their boys off the street" (Levitan and Miller, 1986: 21). They persuaded the school board to give them a disused building, went door-to-door to raise funds, and organized woodworking lessons, dances, boxing, and gymnastics. But it was usually women who organized babysitting, while lobbying local government for better services, including sewers, libraries, and garbage disposal. Strong-Boag (1991: 495, 496) quotes a contemporary's judgment that "for most of the day while the men were away at work the women run the community," concluding that "women were transforming suburbs into good neighbourhoods." In urban settings they were almost as dominant. In Vancouver's Strathcona area from the 1950s, a residents' association formed to oppose redevelopment. Chinese-Canadian women mobilized through "block-based, multi-lingual, neighbour-to-neighbour communication systems," "informal networks" that leveraged *guanxi,* reciprocal gift-giving (Lee, 2007: 395–6). It seems the same is true in First Nations communities. After speaking with 26 activists in Winnipeg's inner city, Jim Silver, Parvin Ghorayshi, and colleagues (2006: 29) report that "it is aboriginal women who are, for the most part, the leaders in putting into practice an Aboriginal form of community development." In many respects Canadian women's neighbouring practices have mirrored those of their American cousins. In the only study of its kind, Amanda Seligman examined everything she could find about Chicago's block clubs throughout the twentieth century. She describes these as "dark matter ... hard to see, but their effects can be observed," in that they tackled the sorts of mundane issues that Strong-Boag noted in Canada's postwar suburbs (Seligman, 2016: 15). And, Seligman (2016: 7) notes, these suburbs were dominated by women.

Indeed, it is likely that many men were pulled in by their partners, just as Mif Phillips introduced James Lorimer to his neighbours on Minster Lane. One woman who spoke with Strong-Boag (1991: 495) said that "the fathers get to know their neighbours through their own ubiquitous wives." Men acknowledged this. Barry Wellman (1992: 12) surveyed men's social networks in Toronto in the late 1970s, reporting that while 33 per cent of women in the area relied on neighbours for companionship the same was true of only one-third as many men. One man commented that the "only relation I have with my neighbours is directly due to the conduct of my wife" (Wellman, 1992: 7). Men who become part of neighbourhood networks, social and political, often do so through their wives.

Of course, especially since 1945, some things have changed. Women have entered the labour force in unprecedented numbers (Armstrong and Armstrong, 2010). Many have done so eagerly, looking for challenges and rewards beyond the domestic sphere. Initially, as gender norms were slow to change, they took part-time work as a compromise, and, as incomes rose, two-car households became a suburban norm. But then, and notably from the early 2000s as house prices boomed, two full-time incomes became a necessary condition for homeownership. But, although more men are now actively involved in everyday parenting, even those women with full-time work still bear the brunt of domestic labour (Guppy, Sakumoto, and Wilkes, 2019; Michelson, 1985). Although many have less time to do so, it is still the women who form and sustain most local connections.

By a process of elimination, we are led to the conclusion that on average neighbourhood matters least to white, native-born, middle-class, middle-aged men. But "least" does not mean "little." There are very few of us who are indifferent to our immediate surroundings. When changing our address, we all weigh residential options, especially if we have children and are buying a home. To be sure, location matters: being within commuting distance to work, and to stores and other facilities, is important. But so is place, that difficult-to-define territory that feels like home turf. We look for neighbourhoods with parks, recreational facilities, good schools, and perhaps decent transit. We may prefer the aesthetic of older homes, or the open layout of newer ones. And, in so far as we can judge, using clues that include rents or house prices, and usually with the help of real estate agents, we look for neighbours with whom we might feel comfortable.

In this and the previous chapter I have treated separately a series of identities held by various groups – as well as their needs, preferences, cultures, and self-image – but these are always overlapping and intertwined. "Intersectionality" is currently the preferred term for this simple but important truth. Children are born into a class and ethnicity; women and men of all classes and ages have ethnic affiliations. People have their own idiosyncrasies, and their responses to neighbours are accordingly complex. No person, and certainly no block, can be typed in just one way. Westmount's streets are both home to an elite and a WASP counterpoint to francophone Outremont. Hugh Garner saw a working-class slum, Cabbagetown, as distinctively Anglo-Saxon; his contemporaries meanwhile saw another slum, the Ward, as "foreign." Perceptions and intersectionality make it dangerous to generalize: there are many variations on a theme. But the themes are there.

So far the focus has been on the people who live next door, across the road, or just down the street. These are the locals who have always mattered most. But when we choose where to live we consider much else besides: the people living on the streets beyond the block and who we (and our children) might run into; the stores located within walking distance; the schools, parks, services, and other facilities nearby. All of these features, and more, make up what we call "the neighbourhood," and to make sense of this amorphous entity – as if on satellite view in Google Maps – we need to zoom out a little.

chapter four

Neighbourhoods: How Varied?

Most people first pick an area in which to live in and then look for a home there.

Friedman and Krawitz, *Peeking through the Keyhole*

Apart from your immediate neighbours, who or what counts as part of your neighbourhood? The most obvious, I would guess, are physical features: the houses or high-rises, streets and side-walks, lawns, gardens and trees, and maybe a park, hill, creek, or even a waterfront. If you are lucky, a library and a public pool. If you live in a densely populated area there will be stores, a restaurant, a café or two. There are schools and a church, synagogue, or mosque. If more than a century old, there may be an old factory or warehouse building, probably now repurposed. Less visible, but perhaps more important, there are people. Some friends, maybe family, who you meet on the street or at a religious service, who use local stores and restaurants, whose children your own meet at the park playground or hang out with at school. And there will be other things besides. There is a lot more to your neighbourhoods than who lives on the block (Galster, 2001; Keller, 1968).

Neighbourhoods are diverse, physically and socially, which usually produces complicated and fuzzy boundaries. However difficult to pin down, such ambiguities are worth thinking about because neighbourhoods affect our lives, and those of our children. They may bring people together in various associations, sometimes for mutual aid but more often to articulate common needs or interests. Needs may be neglected, and interests challenged, by outsiders, including the municipality, whose actions depend on how the neighbourhood is perceived. Ignorance, stereotypes, and how those stereotypes are communicated, come into play. And so the existence of fuzzy boundaries raises questions about the effects of neighbourhoods, about their organization, and about their place within the larger urban setting. But first we need to think about how varied neighbourhoods are.

The Physical Landscape

The short answer is: a lot. The most obvious differences are physical, what you would see on an exploratory walk, or a digital ramble on Google Street View. The built environment can tell us a lot about culture, and its variants (Ford, 1994; Lewis, 1979). Areas are as different as the homes that define them: the nineteenth-century rowhouses or plexes, the 1960s side-splits on 70-foot lots, the present-day condo towers (Ward, 1999). In dense settings, buildings enclose public space, creating streetscapes; lower-density suburbs leave the impression of a more open landscape. They *feel* different. Types and sizes of homes tell us about the income and taste of their residents, while façades can offer other clues. Balustrated balconies, concreted, arcaded verandahs with wrought iron railings, and arches over windows are signature elements of a unique type of architectural bricolage once favoured by Italian-Canadian homeowners in Toronto, according to Luisa Del Guidice (1993). Southern Europeans in general have used some of these elements to leave their mark (Olson and Kobayashi, 1993: 147). Less obviously, and less frequently, a yin-yang symbol over a door might flag those of Chinese heritage. Local stores, like that depicted in the CBC sitcom *Kim's Convenience*, can say a lot. Cars are another clue, whether workaday trucks, upscale SUVs, or small, second-hand sedans. Run-down buildings, garbage, graffiti, or neglected front yards and public infrastructure point to poverty, or worse. All of this sends messages to the first-time visitor, repeated daily to local residents. The environment can give positive reinforcement; it can reaffirm one's status or identity, or, in the case of greenery, uplift one's mood. But it can easily do the reverse. An American study found that how well residents liked their neighbourhood depended on how well their neighbours maintained their homes (Lansing and Marans, 1969).

Street layout, property maintenance, and density help define the physical character of residential areas. Until World War II, except for some middle-class and elite areas, streets were gridded (Harris, 2004). Since then, loops-and-lollipops have reigned, these being enclosed within a generous grid of arterials. The thinking behind the deployment of curves and culs-de-sac is to create visual interest and ensure children's safety by reducing through traffic, with enclosures to encourage community. Those features affect social ties, but what matters more is the existence of destinations, such as a park or café (Small and Adler, 2019). Cafés are the classic "third space," for hanging out, meeting people, reading, scrolling, or just watching life go by. Parks are important that way too, and also encourage physical activity, provide therapeutic greenery, and in general improve public health (Wolc, Byrne, and Newell, 2014). All of this is especially important for people living on very low incomes, as a recent study in Ottawa underlined (Plane and Klodawski, 2013). Unfortunately, parks and greenery tend to be most scarce in the neighbourhoods, and to the people, who need them most. In general, the impact of design is hard to figure out, often being bound up with density. Regardless, we use it to distinguish two categories of neighbourhoods: urban and suburban. In their design, neighbourhoods can vary greatly, and in ways that allow us to read social signals (Ford, 1994).

But there are many design features that we cannot easily read. Many façades, notably those of apartment buildings and condominiums, are inscrutable, hiding subtle and not-so-subtle social disparities. To see how neighbourhoods differ, we need other sorts of information, statistical in nature, including those contained in city directories, property assessment records, and, since 1951, the Canadian census. These do not tell us everything we want to know, but they do reveal a lot.

Differences in Class Composition

The simple truth is that the social character of neighbourhoods varies in terms of class. It always has, and no wonder. In a private, capitalist system, aided by developers and landlords, different classes of people have sorted themselves, or been sorted. To what extent? The answer depends on what scale we are talking about. I have suggested that the block is crucial because immediate neighbours, and their actions, are unavoidable. It makes sense, then, that segregation – used in a neutral, descriptive way to refer to the way people are distributed unevenly across urban space – would be most apparent at that scale. That is indeed the case. But to establish this fact, and explore its implications, we need to consider some statistics.

A Necessary Statistical Digression

Oddly, the best information about segregation at various scales is historical. That is because Statistics Canada does not release current census information at the block scale, while the information for specific dwellings that was once reported in property tax records and, sometimes, city directories is no longer collected. The Jewish community in Toronto's Spadina area provides an interesting, and also extreme, example. Ethnically, this was as close to a ghetto as any large residential district in Canadian urban history could be. Here "ghetto" implies not only a great concentration of a particular group but also external constraints. In 1931, Dan Hiebert (1993) shows that this area contained 48,550 people, of whom 19,578, or 40 per cent, were Jewish. On Kensington Avenue, however, its commercial heart, fully 78 out of 82 households were Jewish. Indeed, at the finest scale, clustering got tighter still. On Huron Street north of Cecil Street, Jews were concentrated on the east side of the street (Hiebert, 1995: 66). South of Cecil, the west side of Huron was Jewish but there were none on the east side, a pattern which Hiebert carefully maps.

Contemporaries knew that the street and block scale was important and so sometimes went to the trouble of showing it on maps. Just before World War I, the Presbyterian Bryce Stewart (1913) prepared a series of block maps of immigrant districts in Sydney, Nova Scotia, as well as Hamilton, Fort William, and Port Arthur (now Thunder Bay) in Ontario. Typically, even on blocks dominated by one ethnic group, as on Tupper and Laurier Streets in Sydney, there was jumble of groups (figure 11). In 1897, in his ambitious pioneering sociological study, Herbert Ames had examined poverty patterns in the whole of Montreal's "city below the hill." This area was consistently working-class but, by contemporary standards, not especially poor. His house-to-house survey enabled him to report results at the square-block scale. In this densely occupied area of 38,000 people, poverty levels among families ranged from 26 per cent in block 24 to 5 per cent in block 17, even though these faced each other across William Street (Ames, 1972: 71). Block 24 was part of Griffintown, adjacent to the "Swamp." These two poor subareas were each three square-blocks in extent, underlining the varied micro-scales at which segregation has occurred.

Hiebert's Spadina and Kensington Avenue areas were unusual, while Ames's Montreal district accounted for less than a sixth of the city. More telling, and indicative of broader patterns, is the evidence for Hamilton in 1936 reported by Michael Doucet and John Weaver. They used the segregation index (SI), the most popular statistic because it is easily interpreted. When comparing one group with the rest of the population, it can vary between 0 and 1: "0" indicates that the group in question is evenly distributed across the urban area, "1" that it is completely segregated. All SI values can be interpreted as percentages. And so a value of 0.33 for a particular group suggests that one-third of its members would need to move in order to be physically integrated. Using information from property tax records, Doucet and Weaver divided occupational groups into four categories: professionals and proprietors, white-collar employees, skilled and semi-skilled workers, and labourers. In each case for 1936, they reported higher index values at the block scale than at the ward level. Those for white-collar workers, for example, were 0.34 and 0.15, respectively (Doucet and Weaver, 1991: 451). A similar contrast was apparent for other years.

Earlier evidence about the importance of scale comes from the extraordinary, systematic study of Montreal in 1881 undertaken by Jason Gilliland and Sherry Olson. I will be mentioning this study again, but for the present emphasize their analysis of scale effects. Using property records, they examined segregation at four scales: the 9 wards (average population 19,444), 67 census divisions (2,612 people), 370 block faces or "street segments" (473 people), and 942 blocks (186 people).[1] Using rent/property value to indicate income, they reported segregation across four categories, high to low. For the top group, the index value jumped from 0.60 at the ward scale – already high – to 0.76 for census divisions and then up to 0.87 for street segments and city blocks (Gilliland and Olson, 2010: 41). The authors found a similar scale difference for segregation by ethnicity and by occupational status: the smaller the area the higher the index.

This might seem to prove that it is the block scale that people care about most. In some ways it does, but the story is complex. For a start, large areas simply cannot be more segregated than small ones. If blocks are mixed, then neighbourhoods will be, too, but the reverse is not necessarily the case, because neighbourhoods can, and always do, contain blocks that are socially different. That was the case with the Spadina area in 1931: Kensington Avenue was almost entirely Jewish but other streets contained a relatively higher proportion of gentiles.

The same principle applies everywhere, including in Kingston, Ontario, in 1970 or Halifax in 2006. Kingston was – and remains – a city divided. After visiting in 1962, journalist Peter C. Newman declared that "few cities in Canada have such a clearly marked division between the right and wrong side of the tracks," where downtown Princess Street served as the tracks (Harris, 1988: 59). The North End was working-class, the West End was middle-class, and the South End included students, along with staff and professors at Queen's University. Altogether, these districts contained

---

1 Gilliland and Olson report the number of wards, census subdivisions, street segments and city blocks. I estimated the average population at each scale by dividing the city's total population by the relevant number of geographical units.

15 census tracts, each averaging about five thousand people. (This meant that, again on average, the population of each of the three districts was about half that of the Spadina area in 1931.) Owners and managers who lived in the city preferred the West End, where they made up 7 per cent of household "heads" with identifiable occupations (62). (Until well into the twentieth century, "heads" of households were deemed to be male, unless only women were present.) They had a special preference for tract 15, which included Alwington Place, where their proportion reached 15 per cent. Meanwhile, those in poorly paid and unskilled work concentrated in the North End, where they accounted for 16 per cent of employed household heads. In tract 9, however, the proportion reached 28 per cent. Accordingly, the poverty rate for families in that tract (24.4 per cent) was higher than in the North End as a whole (14.6 per cent), or indeed the West End (6.7 per cent), while in tract 15 (3.2 per cent) it was negligible – except to those affected (65).

Halifax in 2006 had comparable divisions. In its most affluent tract, households had an average income of $72,476, but when Victoria Prouse and colleagues (2014: 70) homed in on the smaller "dissemination areas" (DAs) of about five hundred people, extending three or four blocks, they found an elite area with an average household income of $122,987. Similar contrasts were apparent for the proportions of visible minorities at tract and DA scales: 32 per cent and 86 per cent respectively. No one who knew Kingston in 1970 or Halifax in 2006 would have been surprised by these statistics. Parts of Kingston's North End were visibly poor, and areas of Halifax were strikingly different. The point is that these examples illustrate a general truth. No matter the group in question, smaller areas are bound to be more differentiated than larger ones.

And we know that larger scales count for a lot. Vital social activities – schooling, shopping, and attending religious services – happen beyond the block. Indeed, Richard Wilkinson (2005: 126–30) argues that social inequality matters more between residential areas than it does in smaller or larger settings. Montreal again provides useful examples, historical and contemporary. Until 1914, the *paroisse* was "au coeur des relations sociales," providing social, economic, and spiritual sustenance, centred on the local church (Ferretti, 2001: 227; Olson and Robert, 2001). The Public Charities Act of 1921 enabled municipalities to play a larger role in providing social and health services (Bherer and Collin, 2013). The significance of parishes eroded through the interwar years, and then more rapidly (Germain, 2013). From the 1960s, with the declining influence of the Catholic Church and a rising wave of immigrants, *quartiers* such as Mile End and Parc Extension became hubs with stores, community centres, and religious institutions for multi-ethnic communities (Germain, 1999). But particular groups, including those comprising specific occupational or minority groups, rarely dominate large areas. They may not have enough members; the housing stock may be varied in ways that make clustering unaffordable, or undesirable; the location of jobs may compel wider dispersal. And so a lower level of segregation does not necessarily mean that congregation matters less at larger scales.

Certainly, larger-scale concentrations exist, and in ways with which we are all familiar: the rich and the poor are the most segregated, with the most distinctive neighbourhoods. For Montreal in 1881, Gilliland and Olson (2010: 41) found that, at all four scales, the most segregated rental groups were those at the bottom and, and even more noticeably, the top of the rent distribution. The pattern of SI values at the ward scale, running from high to low, was 60, 35, 11, 42: the highest and lowest rent groups were most segregated (ibid.). The same was true at the scales of census subdivisions and street segments. This is typical, regardless of how rent, property values, income, or social class is measured. Otis and Beverly Duncan, who invented the segregation index, first used it to describe the segregation of occupation groups in Chicago. The most segregated were, at one end of the spectrum, owners, managers, and professionals and, at the other, unskilled labourers. Dan Hiebert found the same in Winnipeg in 1921: the SI for owners of large businesses (0.50) and for managers and professionals (0.36) were both high, as it was for labourers (0.31), with those for other groups much lower (Hiebert, 1991: 70). The SI for skilled workers, for example, was only 0.11. The same pattern existed in Toronto in 1931, and for Kingston in 1970 (Hiebert, 1995: 62; Harris, 1988: 60). Indeed, many studies in Canada, and hundreds in other countries, have shown that this U-shaped patterning is ubiquitous. It shows what we all know: the richest and poorest occupy the most distinctive areas, in one case by choice, in the other of necessity. In Winnipeg in 1921, for example, two-fifths of household heads in wards 1 and 2 were owners, managers, or in professional occupations, more than double their proportion in the city as whole (Hiebert, 1991: 70). In counterpoint, in the North End wards 4 and 5, a quarter were labourers, more than double their presence citywide. Fully half were more skilled workers, firmly establishing the class character of the North End.

So how can we show the unique character of an area like that? The simplest way is to use another statistic, the location quotient (LQ). Let's use Hamilton to illustrate. There, in 1936, the segregation of white-collar workers meant that over a quarter of all blocks were unusually white-collar in character. The LQ enables us to put a number on "unusually." A value of "1" means that the proportion of a social group within the area in question is the same as its proportion in the city as whole. Values greater than "1" indicate an overrepresentation. There is no value above which overrepresentation *always* becomes significant. There are indeed thresholds, or tipping points, beyond which the social character of an area is likely to change (Galster, 2019: 139–45; Quercia and Galster, 2000). But these are hard to pin down, and vary according to time, place, and the group in question. We do not know what tipping point mattered in Depression-era Hamilton, but Doucet and Weaver (1991: 451) imply we should pay attention to any LQ above 1.66. On that basis, they found that just over a quarter of all blocks in Hamilton in 1936 were unusually white-collar, and that two-thirds of all white-collar workers lived on those blocks. Assuming those workers fared relatively well in that desperate decade, they helped their blocks resist the prevailing pressures of decline.

This patterning, of class segregation and neighbourhood differentiation, has defined all major Canadian urban centres for the past century, and of older cities for even longer. As described in chapter 2, in late nineteenth-century Montreal, the elite created the Golden Square Mile, with its landscape of "larger dwellings, fashion-conscious architecture, wider streets, and more reliable municipal services, as well as topographic advantages of drainage, winter sunshine, summer breezes, view, and visibility" (Gilliland and Olson, 2010. 36). Meanwhile, working-poor families lived in areas like those in Herbert Ames's *The City below the Hill* ([1897] 1972). In between, residential areas were more socially mixed, differing in more subtle ways. The same was true in Winnipeg in 1921, Toronto in 1931, Hamilton in 1936, and Kingston in 1970. It is surely true where you live today. Those middling, more finely differentiated, neighbourhoods are where most urban Canadians have lived. At least since they became possible.

The Complications of Size

The complication is that the range of possibilities increases with the size of the place. The point is clearest if we start in the smallest type of community, the village. Gerald Pocius (1991) has provided a nice example in the form of Calvert, a Newfoundland outport of under five hundred people at the time he studied it in the 1970s and 1980s. Even in such a small place, residents spoke about geographical subareas: broadly, north and south, and then breaking the south down into smaller clusters (figure 12). These were distinguished primarily by families, rather than by ethnicity, income, class, or, for the most part, even occupation. In the north, the Sullivans were known as fine fishermen, but that is because everyone else was in a good position to judge. In one way, the whole of Calvert was a single neighbourhood, although in this case "community" is the word that feels right.

With a population of 7,535 in 1891, rising to 11,852 by 1915, Galt, Ontario, provides an example of a place that could create distinct neighbourhoods, but only in a modest way (table 4.1). Even if it had sheltered a large minority group – which it did not – it could not have harboured an ethnic enclave the size of the Spadina district, or even an upper-middle-income area like Kingston's tract 15. Segregation could only exist at a smaller scale, as Florence Dickson knew. Together with her brother, Florence, a rich spinster, laid out a subdivision in stages from 1884 onwards. "Dickson's Hill" was meant to become *the* place to live in Galt. But she knew that, even if every doctor, lawyer, merchant, and employer wanted to live there, there were too few of them to fill the whole area. And so she divided it in two, with an affluent north for supervisors, managers, and professionals (64 per cent by 1919) and a pedestrian south, overwhelmingly (72 per cent) working class, with Gladstone Avenue serving as the tracks (Hagopian, 1999: 33). The social contrast would have been apparent to any visitor. The sizes of homes were different, as were the building materials, brick in the north and wood frame in the south. Altogether, when fully built-up, the two sections had a tract-sized population, but an average income statistic would have been unhelpful, disguising the neighbourhoods that residents cared about. City size mattered.

Up to now, in Canada, there has been no size beyond which it has ceased to matter. In postwar Canada, metropolitan areas have grown many times larger than Galt in the 1890s, or even Toronto in the 1930s, but segregation is still greatest in the largest cities. Consider levels of income segregation. Using a Gini coefficient – which in this context can be interpreted like a segregation index – Jill Grant and colleagues (2020) found that among seven metropolitan areas in 2015 the most segregated was the largest (c.f. Bourne and Hulchanski, 2020: 15). The average coefficient for Toronto (0.23) turned out to be the highest and Halifax's (0.13) the lowest. Switching perspectives, the authors then used location quotients to show how varied the census tracts in each could be. The largest cities contained the most extremely affluent tracts. In the Montreal area, one tract – probably in Westmount – earned a quotient of 9.56 while Toronto boasted one at 8.31. The best that Halifax (2.36) and Hamilton (2.13) could offer were several times lower.

The least predictable places are the small towns, which can vary in the extent to which small neighbourhoods differ. Joy Parr (1990: 6) sketches the extremes: in the first half of the twentieth century, Paris, Ontario, "was a community of separate precincts," with "Quality Hill" overlooking the low, occasionally flooded, flats. By contrast, in Hanover, which also had a population of about four thousand, "the dwellings of the factory owners and the town's merchants were ... intermingled with the plainer residences" (6). The Paris scenario has surely been more common. The historians of Alberta's town life observe that, before World War I, "towns were too small to have residential areas divided solely by class. Yet there was always a residential area where the wealthiest tended to live" (Wetherell and Kmet, 1995: 329n17). This was especially true in company towns such as Asbestos, Quebec, which grew up around the mine that Johns-Manville purchased in 1916. Historian Jessica van Horssen (2016: 45) notes that "Most JM officials lived on a knoll far from the Jeffrey Mine … Workers, on the other hand, usually walked to the mine, and many rented affordable housing from JM close to the edges of the pit." But segregation is also the norm in more diversified service centres, such as Minden, Ontario (pop. 6,000), which straddles the Gull River. In the older subdivision on the east side, the homes are modest, some built by owners and others in a subdivision of manufactured homes; to the west are typical suburban-style homes in McKayville and along Knob Hill Crescent, clustered near the new hospital and elementary school. Meanwhile, the most wealthy are scattered along the shores of nearby lakes.

In small centres such as these the particular character of class relations and physical topography can determine neighbourhood diversity. That is less true in larger centres, which are more diversified and where more residential possibilities are available. To be sure, in a city like Vancouver, squeezed between ocean and mountains, physical geography determines the pattern of settlement; elites have preferred, taken, and kept the sites with the best views. But in places that large, even a marked topography has little effect on the overall degree to which neighbourhoods differ. Like Toronto and Winnipeg – places with less striking topography – Vancouver has areas of concentrated poverty, along with elite and ethnic enclaves. Unlike Hanover, Ontario, c. 1930, no large urban centre is defined by residential areas that are socially mixed.

City Size in History

We have to bear in mind these city-size effects when comparing past and present. Montreal in 1881, with its elite and poverty-stricken districts, might seem like a metropolis, but its population was similar to that of modern-day Thunder Bay. Winnipeg's population in 1881 is roughly that of Paris, Ontario, in 2021. For all intents and purposes, Vancouver, Calgary, Regina, and Saskatoon did not exist (table 4.1). The scope, and even the possibility, of segregation was far less.

Slowly, things changed. Quebec City is a prime example, the earliest in Canada, where micro-segregation grew into the sorts of neighbourhoods we know today. It is one of Canada's oldest urban settlements and was, until the 1820s, the largest. Then, as Marc Lafrance and David-Thierry Rudell (1982: 162) have shown, "differences in street and housing conditions *within* the same area in the eighteenth century were transformed in the second quarter of the [nineteenth] century into differences *between* urban districts." The Lower Town became working-class, and by the 1840s Saint-Roch, with a population of 10,000 in an urban area of 32,000, "can be identified as one of the country's first ethnic and working class areas." "Ethnic," referred to the "cholera-stricken Irish," while the district's working-class character was unambiguous (159). In 1818, 13 per cent of household heads were labourers or domestics; by 1842 the proportion had reached 33 per cent, comparable to that of Kingston's tract 9 in 1970 (170).

Segregation on this scale emerged later in other cities. Montreal soon led the way as it grew from 9,000 in 1801 to 90,000 by 1861. By then, segregation existed at something like the tract scale. Almost half of the merchants and manufacturers would have had to move in order to be physically integrated (Lewis, 1991: 139). Of course, they had no intention of doing so. Wards, indeed whole districts, were markedly different. The working class made up just over a tenth of the population in east Saint-Antoine, four-fifths in south Saint-Antoine and Saint-Jacques, and over nine-tenths in Sainte-Anne and Sainte-Marie. As Gilliland and Olson (2011) have shown in a follow-up to their study of 1881, by 1901, with its mixture of French and Irish Catholics as well as Anglo-Protestants, Montreal had developed a broad and complex pattern of ethno-class districts. The main ethnic groups remained segregated, while the greater size of the city had enabled the growth of class segregation within each of them (figure 13). That was especially true for

Irish Catholics, a relatively small group. In 1881, the highest status households were not very segregated from others in this group, as an index value of 16 indicates. Within 20 years their degree of separation had almost doubled.

In Toronto and Hamilton differences emerged later. As late as 1861 in Toronto, with its population of 45,000, there was still an intermingling of residences and commercial activity, while "the pattern of segregation appears to have been finely detailed" (Goheen, 1970: 85). Class, and to some extent ethnicity and family status, distinguished small areas. Then, "in the three decades after 1870 the whole character of Toronto changed": social distinctions were still reflected spatially, but "the territorial frame has expanded" (220). In Hamilton, too. In 1851, with a population of 14,000, the city could not support large-scale segregation. The most distinctive areas were south-central, including what is now known as Durand and Corktown. South-central contained almost three times its share of the wealthiest 5 per cent while Corktown had more than twice its share of labourers (Davey and Doucet, 1975: 336). But the population of these areas was only about 800 and 1,450, respectively.[2] And, interestingly, they sat side by side. A modern census tract could comfortably have accommodated both, and other areas besides. It was only as the city continued to grow, exceeding 36,000 by the early 1880s, that larger-scale differences began to emerge, and by 1911 "the social ecology of the city had been transformed" (Sager, 2014: 444; Doucet, 1976).

That is about the time when the transformation began in St. John and Winnipeg. St. John was unusual because, at 41,000, its population barely held steady in the 1880s and 1890s. Nonetheless, its "residential districts ... became more homogeneous in terms of their occupations, religion and ethnicity" (Northrup, 1979: 3). Rapidly expanding, Winnipeg was more typical. It developed "a series of distinctive [residential] environments" by the time its population had reached 48,000 at the turn of the century (Artibise, 1975:

---

2 Davey and Doucet report the number of employed men in each subarea. Counts of total subarea populations were estimated by multiplying the subarea's share of employed men by the city's population.

151; Hiebert, 1991). The greatest contrast was between its immigrant North End and exclusive Armstrong's Point, symbolically distinguished by three gates. Vancouver's transformation happened at about the same time. By 1901, with a population of 29,400, its neighbourhoods were already growing apart: in the West End, one-quarter of household heads were the elite bourgeoisie, while another fifth were proprietors or managers (Galois, 1979: 318). In contrast, three-quarters of Yaletown was working class. The process of differentiation continued as the city boomed prior to World War I, incidentally separating residential from commercial and industrial districts (MacDonald, 1973: 34–5). This was an era of rapid urban growth everywhere. In 1901, Calgary (pop. 4,200) was no more of a city than Paris or Hanover, Ontario, but by 1914, with the introduction of streetcar lines, "differentiated residential districts emerged with astonishing rapidity" (Foran, 1979: 305). By then, most cities that are today of significant size had developed neighbourhoods on a modern scale, and they invariably varied in terms of class.

Ethnic and Racialized Minorities

But what about ethnic and racialized minorities, and the distinct character they lend to certain neighbourhoods? Indeed, as we walk around observing the stores and faces, we see features that are often more striking than those of class. However, it is difficult to generalize about how minorities shape neighbourhoods because they have varied so much in numbers and character, across time and space. But one thing is clear, scale and city size matter here as well.

Scale and City Size Effects Revisited

The scale effect for ethnic groups is shown in Gilliland and Olson's study of Montreal in 1881. With an SI value of 0.52, the segregation of French Canadians from Protestants and Irish Catholics was notable at the ward scale. It was successively more apparent at smaller scales, from census divisions (0.58), through street segments (0.67), to city blocks (0.69) (Gilliland and Olson, 2010: 41). Although lower, the sequence of index values for Anglo-Protestants, mostly of British origin, was similar: 0.42, 0.52, 0.60, and 0.63. Of course, the situation of French and Anglo Canadians in Montreal was fundamentally different from that of other ethnic groups in any city, then or in later years. They comprised the two European groups that had colonized, and were still remaking, the territory that had recently become Canada. Numerically they dominated the city that, in economic terms, the Anglos controlled. But scale defined their settlement in the same way as for later immigrant minorities.

The experience of those minorities has also depended on city size. That is why, in Quebec City, urban growth enabled recent Irish immigrants to give a distinctive character to – without dominating – a whole district. Fuller national comparisons become possible after World War II, when the Dominion Bureau of Statistics first identified census tracts. Boundaries have never corresponded perfectly, or sometimes at all, to those identified by local residents (Germain and Gagnon, 1999). That said, they were not arbitrary, being originally defined to maximize their internal social homogeneity, while acknowledging natural or artificial barriers. In Toronto, for example, the bureau's statisticians received the "valuable assistance of a special committee representative of organizations interested in social and economic research" (Toronto City Planning Board, 1949: 11). The first decennial census to report tract-scale data was published in 1951, but not for all major urban areas. The first complete national view, then, is for 1961. Considering 11 ethnic groups in that year, the sociologist T.R. Balakrishnan (1976) calculated average levels of ethnic segregation in 16 census metropolitan areas. He excluded Quebec City, because it lacked enough ethnic diversity for tract-scale calculations to be meaningful. Among the other centres, the most ethnically segregated, with an average index of segregation of 0.5, was Montreal. Not surprising, given that this was still the country's largest urban area and largely populated by the "two solitudes." Significantly, Montreal was followed by Toronto (0.4), while ranked at the bottom were smaller centres, including Regina (0.2) and Victoria (0.22). In part these city differences reflected differing proportions of immigrants in each place: Montreal and Toronto were attracting the most immigrants while Regina and Victoria – not to mention Quebec City – were bringing in fewer. But city size evidently counted.

Averages are useful up to a point but they hide variations in the experience of particular groups. That is true regardless of how ethnicity is defined, whether by religion, birthplace, first language, immigrant status, or self-identity. With all of these, segregation is apparent, and varies in similar ways (Darroch and Marston, 1969). Regardless, it is difficult to generalize. Warren Kalbach (1990) has shown that, in Toronto from 1871 to 1971, groups defined by country of origin varied enormously in their degree of segregation. The SI index for the Scots and Irish hovered at 0.1 while that for Jews was always 0.5 or more. Values for Italians and people of African origin varied, but lay in between.

Fluctuations over time reflect changes in immigrant numbers. The index for Italians peaked at 0.57 in 1901 but had fallen to 0.25 by 1931 because the immigrant stream had dried up. After 1945, a large wave reversed the trend, producing an SI value of 0.57 by 1971. Then, as the wave declined, the index slipped back to 0.50 in 1981 (Kalbach, 1990: 98). Immigrants and their children moved, dispersing somewhat in the process (Stanger-Ross, 2009). This experience is nicely illustrated by Luisa Del Giudice and her family. Her parents brought her in the 1950s from Terracina on the Tyrrhenian coast to Toronto, where "our entire extended family of aunts and uncles settled in the Rogers and Dufferin area … we lived on two or three streets only" (Del Guidice, 1993: 59). Initially, that block scale was vital. When the extended family moved further into the suburbs they still clustered, but now in ways that were "no longer communicable on foot." Women became "marooned at a substantial distance from [co-ethnic] neighbours and family" because they "frequently" did not drive (72). Cultural assimilation depended on physical integration.

Versions of this experience have been common. The Chinese in Vancouver, so clustered in the 1910s, soon began to disperse. By 1931, barely a third of Chinatown's residents were Chinese (Barman, 1986: 113). In postwar Toronto, just as the Italian community had suburbanized, so did the Portuguese: sons and daughters who grew up in downtown Little Portugal moved to suburban

Mississauga (Teixeira, 2006). Using co-ethnic agents, they looked for larger homes on larger lots in enclaves, where they favoured Portuguese restaurants and grocery stores. But the new enclaves were more spread out. The SI for Portuguese Canadians, too, slipped during the 1970s, from 0.68 to 0.62. Like the Italians that Jordan Stanger-Ross (2009) has written about, they might return to "the neighbourhood" for old times' sake. But they had not recreated Little Portugal in the suburbs.

These trends fit in with the "melting pot" model of assimilation and, even in Canada's self-styled mosaic, they express a truth. But not *the* truth. After all, those second-generation Italians and Portuguese still sought some proximity. Indeed, the sons and daughters of immigrants are often more self-conscious about their ethnic background than their parents, which is why "second-generation" is a meaningful term. The same pattern holds for South Asians, who have arrived in even larger numbers since the 1980s. Faced with a gentrifying city, many immigrants to Ontario settled in suburban Brampton. Acknowledging that detached suburban homes express "traditional ideas of socio-economic success," their adult children also value family ties and ethnic connections (Kataure and Walton-Roberts, 2013). They have bought homes near their parents, maintaining the enclave. Other recent immigrants, notably Chinese Canadians, have done the same. Second-generation immigrants have experienced social mobility, expressing this by moving to slightly more distant suburbs, but they still act to define neighbourhoods. That is what T.R. Balakrishnan (1982) reported for the 1960s, and what Valerie Preston and Brian Ray (2020) found recently, when the diversity of origins has grown. In light of this, and recognizing that Toronto and Vancouver are the prime destinations, Dan Hiebert (2012; Vertovec and Hiebert, 2021) predicts that second-generation enclaves are the foreseeable future. In that context, the experience of visible minorities, especially, is distinctive.

Racialization

Although races do not actually exist in a biological sense, throughout history many Canadians have acted as though specific groups have intrinsic, unalterable qualities, usually negative. A hundred years ago, Anglo-Protestants viewed Irish and French-Canadian Catholics as inferior, as they did Jews and the Chinese, not to mention "Indians." In the early postwar decades Italians and Portuguese were also stereotyped. Reflecting more recent immigration patterns, Black people, South Asians, and Chinese have been subject to discrimination, the latter again partly because of a pandemic originating in Wuhan. Based on general outward appearance, immigrants and the native-born are often lumped into the same category; meanwhile cultural differences between, for instance, Chinese from Hong Kong and Hangzhou, Indians from Karachi and Kolkata, and Black people from Kano and Kingston, Jamaica, are disregarded. Given that erroneous inferences are common, a catch-all term "visible minority" has come into use for various official, and unofficial, purposes.

Visible minorities are fairly segregated in Canada. For example, Chinese Canadians are quite numerous in most CMAs. In 2001, predictably, they were most segregated in the largest cities, Toronto (0.58) and Montreal (0.57), and least in the smaller ones, including Kingston (0.40), Saskatoon (0.38) and Abbotsford, BC (0.30) (Walks and Bourne, 2006: 283). Much the same was true of South Asians, except that Vancouver (0.56) edged Toronto (0.54). Although their proportions vary from place to place, even First Nations followed this pattern, except that the cities where they tended to live did rank high. Their segregation in Saskatoon (0.33) and Winnipeg (0.35) was markedly higher than in other centres of comparable size. As Evelyn Peters (2005: 356–9), the leading geographer of First Nations settlement, has observed, this is not a group – in fact, a diverse group of people – who are typically ghettoized. To be able to congregate in significant numbers at the tract scale (or larger), an ethnic or racialized group needs to number in the thousands.

With sufficient numbers, in principle a visible minority can dominate a neighbourhood. In fact, this rarely happens. True, parts of Montreal were for many years – and in some cases are still – overwhelmingly anglophone or francophone. In 1901, for example, whole swaths of the western portions of the city, including the Golden Square Mile and Westmount, were more than 90 per cent anglo-Protestant, while even larger areas, east of Boulevard Saint-Laurent, were French-Canadian to the same degree (Gilliland, Olson, and Gauvreau, 2011: 478). Other groups, smaller in numbers, have clustered tightly, as with the first Chinatowns in larger cities. But in general, even the tract-scale neighbourhoods that were publicly associated with particular groups were in fact diverse. In 1911, Toronto's Department of Health surveyed six inner city areas. Three, referred to as "districts," were deemed of special concern and residents were sampled to discover their "nationalities." In the Niagara district, enumerators counted 38 Poles, 15 "Hebrews," and 11 Swedes; in the Eastern district, 42 Macedonians, 24 Hebrews, and 20 French. Even in the Central district, which overlapped with the Ward, 1,207 "Hebrews," 180 Italians, and 32 Poles were counted (City of Toronto, 1911). As John Lorinc (2015: 13) has observed, in the Ward at that time there was not a community, but a collection of communities.

The same is true today. In his marvellous *Vancouver: A Visual History*, Bruce Macdonald (1992: 73) includes a map of the city in 1981 that shows just how unusual it has been for any group to dominate a neighbourhood. For each of the major ethnic-origin groups, the map shows the tract in which the group is most common. Exceptionally, Chinese Canadians make up 59 per cent of the traditional Chinatown and adjacent territory, the majority of whom are probably Canadian-born, and some of whom might be third or higher generation. Leaving aside those of British heritage, the peak figures for all other ethnic groups are much lower, ranging from highs of 27 per cent for Indo-Pakistanis, 24 per cent for Italians, and 20 per cent for Jews. Most other groups do not come close to dominating any neighbourhood: in the most Portuguese tract only 8 per cent were of Portuguese heritage; the highest proportions for Greeks and First Nations (7 per cent), as well as Japanese (5 per cent), were lower. As cross-sections show, the typical neighbourhood is ethnically mixed.

A comprehensive, nationwide study of visible minorities was undertaken by Alan Walks and Larry Bourne (2006). Looking at all census metropolitan areas in the 1990s, they calculated the frequency of six types of neighbourhoods. At one extreme were "isolated host communities," in which minorities accounted for less than 20 per cent of the population. Towards the other are "polarized enclaves," where minorities make up at least 70 per cent of the population, with one being dominant. Here is where it might be appropriate to apply a label like "Little Italy' "but only two metro areas, Toronto and Vancouver, contained any tracts that met those two criteria. Even in those cities, they were rare. Moreover, no tracts in any Canadian city fit the most extreme category, "ghetto," meaning that 60 per cent of the population were from one minority group, with at least 30 per cent of all the group's members living in such ghetto-like neighbourhoods. Many areas in American cities would fit those criteria, notably Black ghettoes, but none in Canada.

In fact, the most common type was the "isolated," being the most numerous in every metro area except Toronto and Vancouver. Elsewhere, including Montreal, Calgary, and Hamilton, they were dominant; sometimes, as in Quebec City and Kingston, overwhelmingly so. In seven places, including St. John's, Sherbrooke, and Sudbury, they were the only type (Walks and Bourne, 2006: 284). Perhaps the most distinctively Canadian type, however, was the "non-isolated," containing between 20 and 50 per cent minorities. These were the second most common type overall, and the most numerous in Toronto and Vancouver. Here, at least in a numerical sense, diversity of the sort that Canadians are supposed to welcome was apparent.

Although ethnic and/or racialized minorities have commonly been more segregated than social classes, they have rarely been as effective in defining whole neighbourhoods. In Winnipeg in 1921, segregation indexes indicate that French Canadians (0.68), East Europeans (0.63), and Jews (0.60) were all more segregated than even the mostly highly concentrated occupational group, the owners of large businesses (0.50) (Hiebert, 1991: 70). The same was true for Jews in Toronto in 1931 (0.65) (Hiebert, 1995: 64). But it was above all class that defined the social character of most areas. Three-quarters of Winnipeg's North End, comprising Wards 4 and 5, was working-class, while only a fifth was East European; a large majority were of British origin. To be sure, in Ward 5 alone, Eastern Europeans were by far the largest ethnic group but, at 42 per cent, their population share did not match the level of its working-class character. As Craig Heron (2015: 53) says of Hamilton's enclaves in the early 1900s, "the immigrant quarter was too small to allow for complete segregation by nationality." A small exception was "Chirperville," in Crown Point, where English immigrants played football (soccer) and cricket (54). In the massive boom that led up to World War I, there were other districts where the wave of British immigrants inundated whole neighbourhoods, Toronto's Earlscourt being a case in point (Harris, 1996). But these exceptions prove the rule. Even at the time, what defined the ring of suburbs, of which Earlscourt was a part, was above all its working-class character. Embodied in small lots and modest homes, that character helps to define that ring to this day, long after the British, and indeed the following wave of Italian immigrants, have moved on.

Other Orientations

In some ways, it is the homegrown minorities whose geography has been most difficult to pinpoint. As Anne-Marie Bouthillette (1997: 214) has observed, "much like turn-of-the-century ethnic minorities, gay male and lesbian communities have developed … identifiable neighbourhoods in the inner cities of large metropolitan areas." Identifiable eventually, but for many decades less visible. Until at least the 1970s, homosexuals were widely stigmatized, criminalized, and sometimes attacked. No wonder they kept a low profile, that is, when they indeed acknowledged to themselves their sexual orientation. In that regard, gays and lesbians were treated no better than the Chinese were a century ago. Unlike the Chinese and other immigrants, however, their identity was not recorded in official sources, and for many decades most would have publicly disavowed their status. That said, there have long been commercial spaces, and then areas, where gays and/or lesbians congregated, at least at night. Bars and other meeting places existed, even during the interwar years, as Valerie Korinek (2018) has shown for prairie cities. Julie Podmore (2006) suggests that Montreal harboured a lesbian bar culture from the 1950s. A commercial enclave developed by the 1980s, becoming a "gay village"

(Germain and Rose, 2000: 207; Remiggi, 1998). Toronto's gay district around Church and Wellesley also emerged in the late 1960s and had become well established by the 1980s (Nash, 2006).

There has been a separation between gay and lesbian areas, associated partly with class. Church and Wellesley was for men. Similarly, in Vancouver the best-known "gaybourhood" was in the West End, populated by middle-class men, many living in high-rise apartments close to the bar scene (Bouthillette, 1997). Less visible was a concentration of less-prosperous lesbians on The Drive, the Commercial Drive section of Grandview-Woodlands (Podmore, 2006). Some had children and occupied cheaper housing, sometimes in collectives and often in converted homes. Both areas were, to use a term coined for ethnic minorities, "institutionally complete," with bars, clubs, bathhouses, community centres, and gay-owned stores that provided validation, community, and an element of protection, while catering to everyday needs. Up to a point, such areas reflected cultural needs in the same way that ethnic areas had always done. In fact, many activists were initially sceptical or even hostile to the creation of such areas, preferring alternative, lower-profile, spaces (Nash, 2006). But as the original defensive communities came to be accepted and turned into "villages" – as much by local municipalities as by their residents – the gay community came to accept them more wholeheartedly.

Or at least the older generation has. Lately, across North America, many LGBTQ2S+ communities have fragmented and partially dispersed (Ghaziani, 2014). In many ways they have joined the mainstream and there is less need to form defensive huddles. Meanwhile, the proportion of people who identify as gay or lesbian has risen, while some publicly identify as bi- or trans-sexual, or non-binary. As the community has grown and diversified, so has its pattern of settlement, nowhere more obviously than in Montreal and Toronto (Podmore, 2006; Nash and Gorman-Murray, 2014). Added to the complications of class and ethnicity there were generational considerations, with aging members tied to the original districts of gay and lesbian life. Like second- and third-generation immigrants, the young disperse to, or remain in, the suburbs. Getting together can be difficult, especially for those without cars. Ann Verma (2017) speaks about queer organizing in Mississauga, where a monthly meetup was launched to overcome "the small everyday traumas of being in a place where reaching across the expanse of the city, and of the region, is hard." What sorts of queer spaces they have created remains unclear (Podmore and Bain, 2021). Some small enclaves, such as Columbia Street in downtown New Westminster, with its hotel, bathhouse, and gay-friendly pub, are tight-knit (Bain and Podmore, 2021). But there, and in other suburbs, activists in different subgroups have varied goals. Like some members of the queer community of an earlier generation, albeit in a changed context, the Rainbow Event Network wants nothing to do with enclaves. Alison Bain and Julie Podmore (2021: 1510) quote one activist's positive spin: "That's what's unique about New Westminster. We don't have a village. The whole community is our inclusive village."

Coming Closer to Reality

It is useful to think separately about the class and minority aspects of neighbourhoods, but in practice they are interwoven in complex ways. Rare indeed is the sort of situation that existed recently in Winnipeg: a cohort of refugees lacking skills and income all ended up in the inner North End, along with other minorities in similar situations (Carter, Polevychuk, and Osborne, 2009). To be sure, historically, most immigrants were poor, but they always exhibited some class diversity, or soon created it by starting businesses and by getting some training that led to better-paid work. In recent decades, many have arrived with skills and savings that would classify them as middle or upper class. Although some areas of concentrated poverty are associated with visible minorities, immigrants today are, in general, heterogeneous and upwardly mobile (Ley and Smith, 2000).

However, the Black population has not necessarily followed this pattern. In Canada, perhaps the best-known ethnic enclave – in truth, a ghetto – has been Africville (figure 14). Settled by Black people in the mid-nineteenth century, it sat on Halifax's fringe until it was cleared during the urban renewal of the 1960s (Clairmont and

Magill, 1999). Stigma and isolation created a tight-knit community. No wonder: on the eve of the clearance of 394 people, it was much like a village where everyone knew each other's business. After the clearance, 56 per cent of the evictees believed that their previous neighbours had been "very trustworthy" (239). This was important because most of the homeowners lacked legal deeds to their property, and so the person responsible for compensating displaced owners "had to rely on local knowledge of neighbours and long-time residents" (Loo, 2010: 38). An even higher proportion, 80 per cent, recalled that their neighbours had been "very friendly," and fully 87 per cent felt that they had "really belonged" in the community (239). Even allowing for hindsight nostalgia this is impressive, especially since Africville lost some of its "close-knit" quality after 1945 (52; c.f. Africville Genealogy Society, 2010).

There were other early enclaves: Cokeville, in Sydney, Nova Scotia, and a two-block section of Main Street in Dresden, Ontario (Reynolds and Robson, 2016: 54–5). Not much is known about most of these places, but Carol Talbot takes us into one of these, a district – she mentions no name – of a few blocks in Windsor, Ontario, where she grew up in the 1940s (figure 15). She has sketched a mental map of the area which, according to her dad, contained "95 percent of all black children in Windsor" (Talbot, 1984: 15). It's doubtful that a census would have confirmed this, but it expresses what Carol herself felt, that this community "was an extension of home." Another community, larger and better-known, was Little Burgundy in Montreal, settled in the early twentieth century by train porters (High, 2019; Williams, 1997). It was large enough to support the Colored Women's Club, the Negro Community Centre, and a branch of the Universal Negro Improvement Society. But such neighbourhoods were rare. Harry Gairey, a Jamaican who emigrated to Canada permanently in 1917 and became a Black leader, reckons that "in the 1920s there wasn't any West Indian settlement as such in Toronto" (Hill, 1981).

The number and scale of Black districts grew when immigration from the Caribbean picked up in the 1960s; there are now substantial districts in several cities. Many households are disadvantaged, and sometimes in ways different from other visible minorities. In Toronto in 2001, for example, Black people's average household income ($40,000) was higher than that of Bangladeshi immigrants ($34,000), so that the proportion spending over half of their income on housing was higher among Bangladeshis (30 per cent) than Black people (22 per cent) (Murdie and Ghosh, 2010). But other factors worked against the Black community, including gang activity, job discrimination, racial profiling, and weaker family structures. Only 4 per cent of Bangladeshi households in Toronto were headed by single parents, but among Black people the proportion was 44 per cent. Thus many more Black families have qualified for public housing, a mixed blessing: it concentrated their poverty and vulnerability to violent crime in places like Lawrence Heights, creating as many problems as it solved.

How do we make sense, and generalize, from this complexity? Let's take Montreal. In some respects it was a mosaic: the Scots, English, and Irish viewed themselves as distinct, with their own churches, cultural institutions, and economic niches (Germain, 2016). But, as is well-known, there was a broader binary, as Jason Gilliland, Sherry Olson, and Danielle Gauvreau (2011) have shown. An anglophone elite controlled the city and its trade while French Canadians were largely relegated to menial work. But the association was never perfect, and in the latter two decades of the nineteenth century, while ethnic segregation remained high, such that many streets in effect belonged to one group or the other, class differences grew within each group, and within those large areas. And so it is possible to identify what Gilliland and colleagues call "ethclass" districts, constituted by class and also by a culture that comprised language, religion, and memory. Having grown up near the intersection of Keele Street and Eglinton Avenue in Toronto, Stephen Muzzatti (2012) writes of a local, personal experience that was, in equal measure, Italian and working class. Many writers have portrayed this kind of intersectional experience, whether Jewish, French-Canadian, or Métis.

Currently, an interactive website devised by Steven Vertovec and Dan Hiebert (2021) enables anyone to explore the intersection of class and ethnicity. For any area in Vancouver, Toronto, Montreal, Calgary, or Edmonton, we can picture its position in terms of income, education, or ethnicity.[3] It allows us to see the degree of income or ethnic mix and, ingeniously, by combining these two, a type of "ethclass" ranking. It shows how the degree of income and ethnic diversity in Vancouver rises from the northwest to the southeast. Most usefully, it shows all of this at a meaningfully small scale: census dissemination areas, with on average 250 households or, in an urban area, about four square blocks.

But class and ethnicity do not exhaust the possibilities. Residential areas have also varied in their age composition; the suburbs, for instance, have attracted more families with children than have inner cities. Since at least the 1960s, large numbers of young adults have defined particular areas. In the 1960s, many baby boomers populated high-rise apartments created through inner-city redevelopment. Since the 2000s, attracted by the urban lifestyle and constrained by high real estate prices, millennials have fuelled the tsunami of high-rise condos that has reshaped downtown Toronto, a process that some have, rather clumsily, called "youthification" (Moos, 2006). Recently, a brave attempt has been made by Ivan Townshend and Robert Murdie (2020) to track this phenomenon in eight metropolitan areas. The authors brought the complex statistical method of factor analysis to bear on a wide range of census tract data for 2006. There was almost no limit to the number of neighbourhood types they might have identified. Initially they characterized fifteen, later honing these to six. These categories signal neighbourhoods that combine class, ethnicity, family status, and other elements including housing tenure, in different ways. All were present in every city, but in varying proportions. The results are meaningful. With a youthful population, Calgary stood out as having a high proportion of tracts of young, single, mobile renters. Smaller and slower-growing metropolitan areas, including Winnipeg, Hamilton, Ottawa, and Halifax, had many examples of two types of neighbourhoods: older, working-class areas, mostly lying within the inner city, and stable, postwar homeowner suburbs. Vancouver and Toronto, both cities with many prosperous

3 As of May 2021. The authors plan to add Montreal and New York City.

immigrants, stood out for their middle-class "family ethnoburbs," such as Richmond, BC, and Brampton and Markham, Ontario(see also Wang and Zhong, 2013). Montreal lacked these "ethnoburbs" but, with Toronto, it had by far the highest proportion of "disadvantaged" areas, mostly in terms of household income. This is as good a capsule summary of neighbourhood social diversity, Canada-wide, as it is possible to imagine.

At least at the tract scale, and assuming that those boundaries are meaningful. But often they are not, and indeed boundaries of all sorts are questionable. That is not a major issue if we want to make broad generalizations. But it becomes non-trivial in looking at particular places, like Halifax, where tracts obscure important micro-scale variations (Grant and Ramos, 2020). This is not just an academic matter. As I was writing an early draft of this book, the residents of a housing co-op in False Creek South, Vancouver, were upset by a misleading planning report (Gold, 2021). The area was scheduled for partial redevelopment and the report implied that the resistance of residents reflected their privilege and affluence. But its author had used data for the whole census tract, which obscures the fact that False Creek South is relatively poor. Clearly, tract boundaries can be a problem, and the problems are not confined to tracts.

Boundaries

Up to now, I have implied that neighbourhoods have clear boundaries. Indeed most people speak about them as such (Lee and Campbell, 1997: 926). And up to a point, that makes sense. A block, certainly, has a beginning and an end. But at larger scales, things usually get complicated.

Of course, there are a few places with clear boundaries. That's true of the individual high-rise and of isolated high-rise clusters, such as Toronto's Thorncliffe Park. Boundaries are most apparent with public housing projects, which are typically set apart, both physically and socially. Their municipal developer-landlord, of course, is different from the norm, as are the qualifications for residence. Stigmatization is made easier by boundaries while incidentally reinforcing them (c.f. August, 2014; Delegran, 1970). This discourages people from stepping across them. In Lawrence Heights in the 1960s, "the daily routine of the [numerically] dominant members, that is, the women, is confined to the point of highlighting the locality" (Delegran, 1970: 94). Half a century later, Luca Berardi (2021a) reports that some younger residents had never ventured outside it. Appropriately, he describes it as "an island tucked away behind a plaza" (Berardi, 2021b).

Another type of neighbourhood with distinctive ownership and borders is the common interest development (CID). Most of these are not gated, but symbolic entrances signal a unique public-private territory: stone posts with name boards and a "private road" sign (Grant, Greene, and Maxwell, 2004; Grant, 2005). These markers, too, inhibit boundary crossing. As for Calgary, which has proportionately more of these developments than any other major Canadian city, Ivan Townshend (2006) suggests that neighbouring is affected (figure 16). Elite CIDs, or conventional packaged suburbs, have had some of these elements. Suburban Westmount put signs on its playgrounds to exclude non-residents (van Nus, 1998). Commonly, developers post subdivision names. Depending on their size, subdivisions may be bordered on two or more sides by arterial streets, establishing "clear-cut boundaries" (Clark, 1966: 12). These, as much as proximity and a destination park or landmark, can affect social connections and whether residents agree on a neighbourhood's name and edges (Guest and Lee, 1983; Small and Adler, 2019). It may produce what Clark (1966: 12) calls a "distinctive personality," or the illusion of one. Certainly, that is what developers want buyers to think. But, as Clark also noted, after a couple of years the physical boundary may carry less social weight.

And then, more ambiguously, there are so-called urban villages. Writers, sociologists, and historians have a tradition of speaking about working-class inner-city neighbourhoods as if each has had a clear identity. This is a type of romanticization, discussed in the previous chapter, that is often bestowed upon immigrant districts, such as Boston's Italian-American West End (Gans, 1962; Fried, 1973) and Toronto's Cabbagetown. Hugh Garner (1971: vii), for example, insists on its boundaries in the 1930s: Parliament,

Gerrard, Queen, and the Don River. He even personifies it: "Cabbagetown went on its way, not caring whether the stock market crashed or not" (36). Writing later, but about an earlier phase in the area's development, Maurice Careless (1985: 27, 41) claims it was "a neighbourhood. A locality with perceived limits and identity, with internal bonding, a built environment and a history of its own," although even he conceded that "only the eastern boundary," the Don, "seems beyond much dispute."

Montreal before World War II provides other examples. As Harold Bérubé (2019: 76) points out, writers and historians have often treated Montreal's neighbourhoods as villages, viewing each as both a unique space and "une affaire de sentiment." An example is Gilles Lauzon's (2014) account of Pointe-Saint-Charles, a multiethnic area. Each ethnic group had its own institutions, churches, and schools, and although the area was divided by the tracks there was a strong sense of collective identity and limits. The effect of this way of thinking is to make the whole city a "community of communities," each well-defined (Bérubé, 2019:76; 2016). Unfortunately, the historical evidence is fragmentary. Using his daily diary, Mary Poutanen and Jason Gilliland (2017) mapped a rabbi's visits and appointments in downtown Montreal between 1907 and 1918; all lay within a 15-minute walk, a modern planning ideal for the walkable city (figure 17). But few people had a job that enabled them to remain so close to home. Places like the Spadina district in the 1910s have been rare indeed. There, Jews could live, work, shop, and attend synagogue without leaving the area (Hiebert, 1993). That is why Stephen Speisman (1985: 115) claimed that "Jewish society in St. John's Ward was virtually self-contained." Less visible were patterns of investment. Using nineteenth-century tax records for Montreal – a city where most households were renters – Robert Sweeney (2020: 115) found that a high proportion of landlords only owned properties in the ward where they lived. This may help to explain the ties of reciprocity that existed between many of Montreal's landlords and tenants decades later (Krohn, Fleming, and Manzer, 1977: 125). So there have been, and continue to be, places and times where boundaries carried weight.

But, because they are unusual and distinctive, such districts have attracted a disproportionate amount of attention. Boundaries are usually ambiguous, complicated. Robert Harney (1985: 17), for example, speaks about "sub-neighbourhoods" in Toronto a century ago, each a small ethnic cluster. James Lorimer (Lorimer and Phillips, 1971: 9) recognized that his "immediate neighbourhood," Minster Lane, lay within a "mini" or "local" neighbourhood, while this itself fitted into a larger area which local residents called East of Parliament. Histories of Halifax talk about subareas within the North and South Ends (Erickson, 2004; McGuigan, 2007). They still exist. In 2011, Waye Mason, a local councillor, crowd-sourced them using emails and message board posts. He found that North, South, and West Ends each contained smaller, locally recognized communities (Mason, 2011). The West End, for example, included Armcrest, Ardmore, and Westmount. Similarly, in Kingston, Ontario, in the 1960s, everyone talked about the North End, but the existence of several distinct areas, including Kingscourt, the Near North, and Rideau Heights, was also acknowledged (Harris, 1988: 61–3).

Difficulties arise, however, when two or more groups living within the same area define their neighbourhood in different ways. A good illustration is provided by Jacklyn Hwang (2015) for a sub-area of South Philadelphia, a predominantly Black area. White newcomers defined the neighbourhood narrowly, in part because they were in the minority but also because they preferred to exclude a housing project, whose presence did not align with their particular vision of what the neighbourhood could and should be. Gentrification is common in Canadian cities as well, and comparable struggles have existed, and still do, in areas undergoing transition.

More challenging still is the fact that to some extent everyone defines their neighbourhood in a unique, egocentric manner. In my neighbourhood, there are major streets at the north end of my block and four blocks to the east, and a less busy one four blocks west. I regard those streets as three of the four physical boundaries, the fourth being an escarpment. But someone living three blocks east or west, but still within "my" neighbourhood, would probably see things differently. Even the couple living across the road might, if only because of their daily commute – they cannot work from home – takes them in a particular direction. It turns out that most people give egocentric definitions, using their residence as the point of reference (Lee and Campbell, 1997). Some years ago, a study in Glasgow nicely illustrated the point. Michael Pacione (1984) asked residents of several adjacent areas to draw their neighbourhood boundaries. He then combined these to create "consensual" maps. It turned out that personal and consensual maps never corresponded perfectly, the proportion ranging from 93 per cent concurrence in an area with fairly clear physical limits to 54 per cent where these were fuzzy. Not surprisingly, apps that have tried to define neighbourhoods by using social media come up with varied and conflicting results (Capps, 2015). Moreover, although many people identify their neighbourhood as being roughly similar in population to that of a census tract – about 5,000 people – the lines they draw on a map are not those of statistical agencies (Coulton et al., 2001).

The situation is complicated further by various administrative boundaries (Downs, 1981; Hunter, 1983). Defined top-down rather than bottom-up, these are doubly distinctive (Kuper, 1951: 238). First, their boundaries are unambiguous. This makes them functional for planners but can give arbitrary weight to specific lines on a map (Wellman and Leighton, 1979). Second, they are comprehensive, covering the whole city. Many urban residents might be persuaded to define some sort of home neighbourhood, but in some areas residents apparently are unable, do not have the time, or don't think it worthwhile, to do so. In 2009, the *Toronto Star* posted an online map of Toronto neighbourhoods and for six months curated readers' suggestions (Kidd, 2009). The outcome differed from the original map, and markedly from planners' larger districts. A comparable project mounted in Halifax found much the same (Mason, 2011). None of this was surprising. In Glasgow, Pacione (1984: 384) found that none of the consensual districts coincided closely with planning areas although, significantly, they did fit well with those used by real estate agents. What was most intriguing about the *Star*'s project was that, judging from gaps in response, some areas appeared to lack any sort of identity. Four years later, the journalist Edward Keenan was able to confirm this when he talked to people in various inner suburbs of Toronto. He found that, in the area planners called Woburn, "residents say a lack of any sense of 'community' or 'neighbourhood' is one of its defining characteristics" (Keenan, 2013: 116).

For anyone with children in the public (or in some provinces separate) school system, vital boundaries are those of catchment areas. The catchments of the three schools – primary, junior high, and high – that my children attended ignored major streets, and of course differed. Different again is the planning area in which my block sits. The City calls this Kirkendall South. Through "consultations," they invented the name, along with 108 other planning areas, in the 1960s. Not surprisingly, a reporter commented that "few Hamiltonians know which of the city's 109 neighbourhoods they live in" (Smith, 1969). In many areas that remained true for a long time. In 2006, speaking about one of the poorest areas, a reporter observed that "there is no Beasley neighbourhood. Beasley is a construct, a 'neighbourhood' created by city planners" (Dunphy, 2006). But in time many locals learned the lingo. After all, Beasley has a city-funded community centre. Kirkendall does not, but it does have a neighbourhood association which responds to the proposals of developers or planners. Larger still is the city ward in which residents can vote.

Noting all of this, many observers have suggested that there are three or four types of neighbourhoods: the face block, the residential area, the institutional (or administrative) unit, and then the "regional" district, equivalent to a small suburb or city ward (Chaskin, 1998: 22; Downs, 1981: 13–14). They are not necessarily nested. In the 1960s, a study sponsored by the United Community Services of Greater Vancouver showed how different and incompatible the boundaries of tracts, wards, community centre areas, and school districts were (Mayhew, 1967). Trying to figure out how it should draw boundaries, the City's Planning Department later threw up its hands: "neighbourhood is a term which can mean anything from an area of several blocks to a local area (several miles within a large city) depending on the context" (City of Vancouver Planning Department, 1978: 25). Of course, it proceeded to define planning areas, the ones that Larry Beasley (2019: 107), a Vancouver planner, later reckoned to be "well-established." In his dreams.

One response would be to follow George Galster's (2001) advice. He argues that we should worry less about neighbourhood boundaries and instead consider them as complex commodities. Property values vary continuously across space. Occasionally, a distinct subdivision or administrative boundary – a catchment area, for example – may create discontinuity, but that is not the norm. There is a logic to that argument, and it does shape the way residents – or at least property owners – think and act. Why else would prices vary in the way that Galster suggests, and as real estate agents know, including, it seems, those in 1980s Glasgow? But that does not capture the way that, at the neighbourhood scale as in other contexts, we like to draw lines between us and them.

chapter five

The Effects of Neighbourhoods

Most of us make three important decisions in our lives: where to live, what to do, and with whom to do it.

Daniel Gilbert, *Stumbling on Happiness*

We know that where we live affects us, and so we're careful about where we move. Our home territory affects us, and our children, daily and over the decades. The challenge is figuring out how, and how much. The first half of this chapter discusses that challenge.

Once settled, people usually enjoy, accept, or at least tolerate their circumstances. Most try to make connections with some neighbours, and a rare few set out to build and maintain local community on a bigger scale. Usually, however, neighbourhoods organize when they feel neglected by the municipality, or when faced with an external threat, such a proposed development. All of these actions – building, connecting, and organizing – respond to local conditions, and so count as indirect neighbourhood effects. Whether and how a municipality responds to pressure depends in part on the image of the area held by planners and politicians, an image that traditional and now social media promote. As I will discuss later in the chapter, then, how neighbourhood effects play out is not a purely local matter.

All types of residential areas have effects, but observers – including social scientists, planners, novelists, and politicians – have been most interested in figuring out their impact on the disadvantaged (Leventhal and Brooks-Gunn, 2000; Pebley and Shastry, 2004; Oreopoulos, 2008; Galster, 2012; Kwan, 2018). This makes sense, because public and private agencies are often concerned with ameliorating, if not eliminating, poverty. One way is to help people in need, wherever they live, as income support programs do. Another is to target areas of concentrated disadvantage, as with the "hot spot" approach that Toronto adopted for vaccinations in the spring of 2021 (Grant, Keller, and Perreaux, 2021). The geographical approach makes sense if neighbourhoods exert an independent influence on people's lives. To justify such actions, then, we need to understand *how much* neighbourhoods matter and also *how* (Harris, 2020a; Porteous, 1977: 68–89; Turok, 2004).

How Neighbourhoods Matter

The recent pandemic has shown us how difficult it is to answer those questions. Early on, journalists reported that rates of infection were highest in low-income areas, often with concentrations of minorities (e.g., Gee, 2020; c.f. Grant, 2021). Echoing public health experts, they speculated about causes. It was obvious that localized peaks reflected the characteristics of the residents: frontline workers who couldn't work from home and were exposed to greater risks, and whose low rates of vaccination might reflect the wariness, and language challenges, of minorities. Once vaccines became widely available the take-up rate turned out to be low in those areas, in part because those with unpredictable shifts found it difficult to make appointments to get a jab.

But neighbourhood factors were also discussed. Many residents lived with extended families in crowded conditions, while those in apartments faced risks when sharing semi-public spaces such as elevators and hallways. In *No New Land*, M.G. Vassanji (1991: 2–3) dramatized the crush that could develop at the end of the day as people waited for elevators in an older high-rise, especially when one was out of order. At a larger scale, those without cars relied on public transit which, despite precautions, itself offers dangers. By the spring of 2021, the impact of such factors was being confirmed in the research (e.g., Sundaram et al., 2021). And then it became apparent that take-up rates were also low because lower-income areas had fewer pharmacies or centres where vaccines were available (Ouellet and MacMillan, 2021). Within this complex mix of influences, there is a catch-22: living and shopping in an area where many neighbours are sceptical of vaccines, or are infected, itself increases the probability of infection.

Broadly, then, these influences fall into two categories. Some reflect personal characteristics and circumstances – type of work, income, car ownership, attitude to vaccines – while others are neighbourhood-wide – residential density, the availability of vaccines, and the characteristics of neighbours, who also rely on transit. Figuring out the relative importance of each is tricky. Much of the information we need is lacking, and then there would be the question of how to make sense of it all. One approach is to compare areas that are similar in all important ways except one. But influences are too varied and neighbourhoods too few in number to make this viable. And so some researchers resort to statistically sophisticated methods that can, at best, yield only rough estimates (Galster, 2019: 173–208).

As if this were not frustrating enough, there is the problem of selection bias, mentioned earlier. Take a hypothetical, middle-class neighbourhood with schools whose pupils perform well on standardized tests. Statistical analyses show that test results reflect a mix of influences, both personal (educated parents who join parent-teacher associations and employ tutors) and environmental (motivated peers, well-resourced school). That's useful to know. But such analyses ignore the process by which the neighbourhood is sustained: once its character had been established it attracted ambitious parents who looked for the "right kind of people," who could afford local house prices, and who thereby strengthened its character. The situation is not purely hypothetical. It exactly describes Toronto's Forest Hill in the 1950s, and doubtless beyond. Seeley, Sim, and Loosley (1956: 224, 234) report that that the area was "literally, built around its schools" and that "the reputation of the school has been, and still remains, the magnet drawing residents to the area." Where we draw the line between personal and environmental influences, or cause and effect, depends on how far back in time we want to go.

The same principle applies to low-income areas that had high rates of COVID-19. Presumably, no one chose to live in such areas out of personal preference, any more than they head for high-crime areas. But many had no choice. They took an affordable basement suite or high-rise apartment near a transit route, which, with a connection or two, gets them to work. In so doing, they perpetuated the area's health problems by crowding into hallways and buses, rubbing shoulders with others who have no choice. Meanwhile, those who could afford something better went elsewhere. Here too, then, a historical dynamic muddied personal and neighbourhood influences.

Arguably, "neighbourhood sorting," the process by which people choose neighbourhoods and reinforce their character, is part of the neighbourhood effect (Hedman and van Ham, 2012: 97; Sampson, 2012: 64; van Ham et al., 2013: 1–21). A few have tried to estimate its significance. In a case study of Denver, Colorado, Lina Hedman and George Galster (2013) found that selection and neighbourhood effects are both important and that both are underestimated if viewed separately. A Dutch study confirmed this (van Ham, Boschman and Vogel, 2018). This is worth knowing because social and neighbourhood inequality is lower in the Netherlands than in Canada, suggesting that neighbourhood effects are greater here than there.

Neighbourhood sorting accounts for much of the distinction that we make between urban and suburban areas. As mentioned in chapter 2, Sandrine Jean (2016) reported that residents of suburban Laval found that large yards enabled private lifestyles, but this was hardly an unlooked-for environmental influence: it was exactly what households had wanted when they moved in (Jean, 2014). It is a familiar dynamic. Comparing two areas in Toronto, one gentrifying and the other stable lower-middle-class, Alan Walks (2006) found that gentrifiers wanted an area that would enable an urbane way of life while expressing an appropriate self-image. From an earlier, extensive comparison of city and suburban movers, William Michelson (1977: 182) concluded that people "choose different forms of environment to represent different needs and wants." In so doing, they are creating what Gerald Suttles (1972: 4) has called "self-fulfilling prophecies." Where, then, does the neighbourhood effect end?

Some would say: not there (Slater, 2013). It is the job of builders and developers to understand what attracts different people to different areas. They target homes to market segments defined by income and – even beyond the 1940s, when legal covenants were declared illegal – sometimes on the basis of race and ethnicity (Harris, 2004; Walker, 1997). Over the past century, as their scales of operation grew, more and more developers acknowledged the importance of neighbourhood by building and marketing whole subdivisions geared to particular kinds of buyers. The lenders who provide credit to these entrepreneurs, and to homebuyers, understand the deal. This marketing mindset is applied in many older residential districts, at least those occupied by the middle classes. Residents lobby municipalities and sometimes work to create a public image that attracts the right outsiders. Real estate agents are vital parts of the matching process. In her study of the Highgate agency in Montreal, Edith Martindale (1977: 49) found that agents were choosy about clients; meanwhile "the potential purchaser resents it if the agent does not choose suitable houses." And so residents, realtors, developers, lenders, and sales agents are part of the sorting process by which neighbourhood effects are put to use.

So what are these effects? To put it simply, there are two types. Some affect our daily lives, such as crime and social capital. These effects are related to the social connections that enable people to thrive, or at least to manage their lives. And then there are longer-term effects, including those involving the prospects of our children. The distinction between the two types of effects, the everyday and long-term, is not always clear-cut. Exposure to a virus happens in seconds, and its consequences depend on the response of individuals and the health care system. But our vulnerability to infection, and to its consequences, reflects a lifetime of environmental exposures, as well as overall nutrition and levels of stress. Even so, for convenience, it is useful to consider these two types of effects separately.

Short-Term Patterns and Effects

Short-term effects are the more obvious. Our experience of COVID-19 mirrored that of those who lived through the Spanish Flu a century ago, down to geographical patterning. In both periods, neighbours helped out, albeit cautiously. During the earlier pandemic, a Montreal girl was sent to deliver hot soup to her music teacher, who lived in a rooming house, but in a socially distanced manner: she was told to place the bowl over the door sill, but not to enter (Pettigrew, 1983: 88). Amid disaster and social chaos, social connections were reinforced: "people who worked hard to help their neighbours remember with pleasure the friends they made" (132). Volunteer organizations geared up: the Toronto Neighbourhood Workers' Association received thanks from the wife of a veteran – for broth: "'a returned man is thankful for a service rendered him … I am grateful from the bottom of my heart'" (103). Then, as now, some areas suffered more. Across Hamilton's eight civic wards, flu mortality varied from 1.8 to 4.0/1000 (Herring and Korol, 2012). The middle-class wards fared best while the highest rates of infection occurred in four of the five closest to industry. A similar pattern developed in Winnipeg. The crisis peaked between October 1918 and January 1919, when mortality rates ranged from 4.0/1000 south of the Assiniboine to 8.4/1000 north of Portage, south of the tracks (Jones, 2007: 59). Scale mattered. Infant mortality was high in a middle-class area, but this contained a "pocket" of rooming houses and crowded rental accommodation (62–3). Predictably, the North End fared poorly. There were no vaccines, and public health experts were still flummoxed by a disease that, unlike cholera and typhus, was not waterborne. But they could see the geography, and knew it mattered (Koch, 2017; Vaughan, 2018).

There is no question that residents of low-income areas have poorer health. The variations are infinite because causal influences vary both geographically and historically (Sharkey, 2017; Kwan, 2018). Although they are intertwined, in principle there are two types of environmental influences, physical and social. The power of the one is clearer, but not necessarily more important.

Physical Influences

Flu viruses are mostly airborne, but other pathogens, notably those causing cholera and typhoid, are water-borne. Clearly, location matters for transmission of communicable, and other, diseases, and working-class areas have suffered most. In late nineteenth-century Montreal, child mortality was highest where crowded spaces bred infection, which meant that working-class districts were most affected (Thornton and Olson, 2001; Olson and Thornton, 2011: 119) (figure 18). It didn't help that some areas, notably St-Gabriel and Ste-Anne, were subject to flooding (Boone, 2011). That remained true into the twentieth century. In 1922, infant mortality rates – mostly due to gastrointestinal diseases – were four times higher in St-Henri (213/1000) and Ste-Marie (214/1000) than in Outremont (57/1000) and Westmount (55/1000) (Copp, 1974: 94). Jews fared better, possibly because more mothers breastfed (96). Similar contrasts existed elsewhere. In Ottawa in 1885, mortality rates in low-income wards were double those in the rest of the city (Taylor, 1986). A generation later, the typhoid epidemic of 1911/12 was triggered by leaking outhouses in working-class Hintonburg. Effluent emptied into Cave Creek, flowing into the Ottawa River before being drawn into one of the city's water intakes. The Lower Town had the worst mortality rates (Lloyd, 1979). Characteristically for the era, the city government blamed residents for poor hygiene and did little as long as the problem was confined to the poorer districts.

The geographies of air pollution and poverty coincide. Except in extreme instances, the health effects of air pollution were ignored for many decades, but noxious odours were not, and so areas downwind from industry came to be reserved for workers. Given prevailing westerly winds in northern mid-latitudes, east ends typically became working class. An ingenious study of British nineteenth-century cities found that this pattern only became true after coal-powered industry became common (Heblich, Trew, and Zylberberg, 2021). It shaped the geography of Montreal, Toronto, and most obviously Hamilton, Canada's most industrial city. Its geography of mortality during the pandemic of 1918/19 amplified, but didn't change, that of previous decades: the northeast fared worst (Gagan, 1989). Significantly, however, accidents aside, the greatest ward-scale variations in morbidity were related to cancer and respiratory illnesses, often directly related to poor air and water quality. Both types of pollution grew worse into the interwar decades as Hamilton industry specialized in steel production. Emissions from Stelco, one of the two main steel companies, ballooned: dumps of cyanide, ammonia, oils, and sulphuric acid increased 20- and 30-fold between 1912 and 1957 (Cruikshank and Bouchier, 2004: 483). Largely oblivious to the effects of these products, workers and their families were exposed to all sorts of toxins. This was nowhere more true than in the Brightside area, so close to lakefront industry that it was annexed when Stelco expanded in the 1960s. Paul Palango (1983: 3), who grew up in the area, recalled that "playgrounds were where you made them, like at the dump or incinerator at the foot of Plymouth." In time, a local resident, Dr. Cecilioni, documented the effects of pollution on cancer rates in the area (Bouchier and Cruikshank, 2020: 57).

If health effects were at first unrecognized, smells were very apparent. The northeastern districts were already working-class in character, and became more so (Doucet and Weaver, 1991: 453). In 1911, 17 per cent of residents in the northeastern districts were proprietors or white-collar workers, similar to their citywide proportion, 25 per cent. By 1945, however, the contrast – 16 versus 40 per cent – had grown (Cruikshank and Bouchier, 2004: 472, 480). This was the outcome of several generations of selection bias. Well into the 1990s, the area suffered more air pollution, in line with the city's social geography (Jerrett et al., 2001). Deindustrialization has cleared the air to a large extent, but the built environment and social prejudices have remained. The British study found that the patterns of poverty created in late nineteenth-century cities have persisted; here, too, the east ends of Hamilton and Montreal are still the poorest (Heblich, Trew, and Zylberberg, 2021). Vancouver's east end is also poorer, although the western sector of the city also has the advantage of the best views and access to the water.

The physical geography of cities still matters. Today traffic emissions are the main source of air pollution, with the impact being more widely distributed. Nonetheless, a recent study found that levels of nitrogen dioxide were greatest in areas with high levels of social deprivation, tenancy, and immigrant populations (Pinault et al., 2016). And now, with climate change, new risks emerge. In Vancouver's summer heat wave of 2021, surface temperatures were 4°C cooler in Kitsilano than the Downtown Eastside, a contrast that was surely attributable to the geography of air-conditioning (Baum and McClearn, 2021). Plus ça change.

Environmental impacts on lungs and digestive tracts are fairly well-documented, but what of the social and psychological effects? There are all sorts of correlations. Hamilton, again a prime example, is the subject of the most comprehensive study of the geography of health in a modern Canadian city. The Code Red project, a collaboration between the *Hamilton Spectator* and experts from McMaster University, yielded striking evidence (DeLuca, Buist, and Johnson, 2012). At the tract scale, the average age at death ranged from 66 to 86; the use of acute-care hospital beds ranged from 46/1,000 to 729/1,000; annual emergency room visits per thousand varied from 97 to 1,291. The pattern for mental health was similar: psychiatric emergency visits ranged from 2.6 to 88/1,000. And, to anyone familiar with the city, the geography was predictable: 25 of the 27 least-healthy tracts were in the lower city, especially northeast of downtown. Perceptions of physical and mental health, along with social capital, followed the same pattern (Kitchen, Williams, and Simone, 2012; Wakefield and McMullan, 2005). Lower-income residents sensed the overall precariousness of their situation (Williams and Kitchen, 2012). The worst areas of the city were those that were physically challenged, then, but they were also the poorest and most socially deprived. Social influences too were at work.

Social Influences

It makes sense that, leaving aside physical influences, living in an area of concentrated poverty – in what used to be called a slum – poses risks and harms our health. That might seem to be doubly true for children. But intuitions can be wrong. Perhaps such patterns simply reflect the fact that residents have stressful work and home settings; perhaps a middle-class person in such an area would be immune to such harms. Fortunately, a number of researchers, in Canada and beyond, have explored that possibility.

Although areas of concentrated poverty are not as marked, or severe, in their effects, in Canada as in the United States, they do affect mental health and crime (Oreopoulos, 2008; c.f. Galster, 2012: 43). In Toronto, homicide rates are high in such areas, including some immigrant areas (Thompson and Gartner, 2014). In counterpoint, lower-income households fare better in middle-class neighbourhoods (Hou and Myles, 2005). A comparison of residents in Mount Pleasant with those in Sunset in Vancouver has shown that living in a poor area harms health and well-being (Dunn and Hayes, 2000). The effect is not confined to Mount Pleasant, or Vancouver. A national survey revealed that material deprivation, both personal and area-wide, causes anxiety and depression, and follows a distinct geographical pattern (Matheson et al., 2006). This is especially true for women, because they spend more time in and around the home (Matheson et al., 2010).

The effect extends to children. Abundant American research has shown that, although family background is crucial for children's health, happiness, and educational prospects, neighbourhoods also matter a lot (Pebley and Shastry, 2004). Children who grow up in a poor district are more prone to stress, physical and mental health issues, behavioural problems, substance abuse, and delinquency (Pebley and Shastry, 2004). Good schools and recreational facilities make a difference, but what matters most are the norms, and pressures, of peers (Leventhal and Brooks-Gunn, 2000: 322). Canadian research is scanty, but it confirms our expectations. A study of kindergartners in Saskatoon in the 2000s found that those in poor areas had worse physical health, weaker communication skills and general knowledge (Cushon et al., 2011). A Canada-wide study produced similar findings, adding "behavioural problems" to the mix (Kohen et al., 2002).

In light of this, we can plausibly interpret the geography of health problems in Hamilton. Much is attributable to personal circumstances: poverty, unhealthy and precarious employment, and long-standing mental health problems (DeLuca and Kanaroglou,

2015). The latter is especially significant in Hamilton, where the clustering of people with mental illnesses around downtown stems from deinstitutionalization policies, first implemented in the 1960s (Dear and Wolch, 1987). These are people with diverse challenges, including weak social networks. Their concentration reflected, and then influenced, the location of group homes and service agencies, creating a social service ghetto. Concentration also helped to make them targets of crime while, as David Baillie (2015) has portrayed, growing up poor here has pushed young adults in socially dysfunctional directions. Generally, the city's current geography of public health is testimony to its long and continuing history of air, water, and soil pollution and, for those reliant on transit, the challenges of finding green space to relax in or in simply obtaining fresh produce (Latham and Moffatt, 2007).

Although the class character of a neighbourhood plays the largest role in shaping its social effects, its ethnic composition matters too, although in less predictable ways. In the nineteenth century, cities were filthy places. Back then, in much of Montreal people relied on box privies, which contributed to sewage overflows, while ubiquitous horse dung attracted flies, which could contaminate unprotected food, "as well as cups and spoons, rattles and pacifiers, kitchen surfaces, clothing, and hands" (Olson and Thornton, 2011: 121). The problem was compounded by crowding in homes and back alleys. Maintaining high standards of cleanliness was vital. Jews, in following kosher laws, were careful about this, and Protestants to a lesser extent, but the more fatalistic French Canadians often came up short (Thornton and Olson, 2001; c.f. Copp, 1974: 88–105). Children suffered. As late as the 1890s, 38 per cent of children in francophone families died before their fifth birthday (Olson and Thornton, 2011: 125). The equivalent proportions for the children of Irish Catholics (25 per cent) and anglophone Protestants (20 per cent) families were still bad, but appreciably lower. Location as well as family mattered. Mortality rates were lower among the minority of francophones who lived in Protestant districts. Here was a neighbourhood effect rooted partly in ethnic culture.

Ethnic influences have varied not only by group but also by immigrant status. A recent national study by Lu Wang (2014) compared the native-born with South Asian, Chinese, Italian, and Portuguese immigrants. She found overall variations in health and also in specific ailments – the Portuguese and Italians, for example, were more susceptible to arthritis. Some of this could be explained by income: the neighbourhood effects associated with poverty were most apparent for the native-born, suggesting that the mutual supports available to immigrants mitigated their deprivation. Being congregated has all sorts of material and psychic benefits for immigrants (Breton et. al, 1990; Driedger and Church, 1974; Saunders, 2010). Even Port Arthur, Ontario, had a Finn Town with therapeutic saunas, clubs, and churches, which doubtless provided comfort and reduced stress (Zucchi, 2007). Carried too far, enclaves may slow integration while encouraging an overreliance on ethnic networks and nepotism (Fong and Berry, 2017: 56–7). But in general their effects on health are good. Is it class, diet, genes, or their resistance to dispersal that has made the Chinese among the healthiest of immigrant communities (Mendez, 20089; Wang, 2014)?

Enclaves illustrate a general point. Residents with connections to many people in their neighbourhood have better health (Carpiano and Hystad, 2012; c.f. Ellen and Turner, 1997). A massive Canadian study of half a million people distinguished the independent effects of material and social deprivation, the latter being indicated by the number of single parents and people living alone. If anything, social deprivation mattered more, but with limited impact in immigrant areas (Ross, Oliver, and Villeneuve, 2013). And, confirming one of the themes of this book, a study of three neighbourhoods in Quebec City suggests that all of this is true on various scales (Pampalon et al, 2007). Especially, but not only, in immigrant districts, social networks are good for health.

Another factor related to population health is the availability of resources. In the nineteenth century, drains, sewers, and effective scavenging – the collection of night soil from privies – counted as being the most important. As Thornton and Olson (2001) note, although French-Canadian parenting practices at that time were partially responsible for high mortality among their children, it did not help that the Anglo-dominated municipality provided more resources to the Protestant districts. Since then, urban residents everywhere have come to expect basic services, but some neighbourhoods are ill-served in terms of other amenities. A recent study of Mississauga found significant variations in access to health care, especially for those whose English was poor (Bissonette et al., 2012). These usually correlate with limited access to parks, other green spaces, and playgrounds. Remarkably, Hamilton's Bernie Custis High School, recently built in one of the poorest neighbourhoods, lacks a playing field. All of which underlines the important role that the state can play in shaping neighbourhood effects, an issue to be discussed more fully in the next chapter.

The health effects of neighbourhoods depend on prevailing levels of material and social deprivation. They also help to make that deprivation worse. A study based on national survey data used a sociological distinction between "bonding" and "bridging" capital. "Bonding" arises from established associations, for example with neighbours or co-workers; in contrast, as the name suggests, "bridging" capital refers to weaker connections across these spheres. Both can be useful, but it turns out that bonding capital, for example knowing and getting help from neighbours, improved people's economic well-being (Weaver, McMurphy, and Habibov, 2013). At the extreme, it can help them get a job, or a better job. Building such social capital in a deprived area is a challenge. That may explain why, as the authors of a study using tax data for residents of Canada's three largest cities found, the longer someone lives in a low-income area the less likely they are to leave, another type of selection bias (Frenette, Picot, and Sceviour, 2004), resulting in a cumulative drag on social and neighbourhood mobility.

Longer-Term Effects

If, over time, deprived neighbourhoods hold adults back, they have an even greater impact on children. Here again, Canadian research is limited, but abundant American evidence consistently supports this premise (Leventhal and Brooks-Gunn, 2000; Neal and Neal, 2012; Pebley and Shastry, 2004). Children who grow up in materially and socially deprived areas have lower IQs, more limited verbal and reading ability, do worse in school, are more likely to drop out of high school, and are less likely to go on to higher education. They are more likely to exhibit behavioural problems, become delinquent, get drawn into criminal activity, and, in the case of girls, run the risks of teenage pregnancy (Crane, 1991). Robert Putnam (2015), the best-known political scientist in the English-speaking world, lends his weight to the argument that income inequality, as reflected in space and manifested in schools, has placed the American Dream in crisis. The author of another survey agrees that "to raise cognitive skills, it is even more important to avoid concentrations of poverty than concentrations of minority students" (Kahlenberg, 2019). This message easily crosses the border (Kohen et al, 2002).

As Putnam points out, access to various public amenities, including libraries, parks, and athletic facilities, affect children's lives, but the most important factor for child well-being and success is the quality of the local public school, including its physical facilities and the competence of its teachers. As such, it is a major object of reform and public policy. But what matters at least as much are local social networks (Galster and Killen, 1995). These include family, friends, and adult neighbours, but school and neighbourhood peers are especially important (Leventhal, Duprée, and Shuey, 2015). That is true from the earliest age: in the playground, toddlers learn how to cooperate – or not. But it becomes especially true in adolescence, as young people push parents away while searching for their own identity (Ellen and Turner, 1997). That is why the effect of neighbourhoods on school achievement is at least as apparent among the social elite as for those who live in areas of concentrated poverty (Howell, 2019b). The elite's children have everything: school and neighbourhood resources, together with peers who reinforce the educational and social expectations of their parents. Those children go on to higher education, better jobs, and socially advantageous marriages, a process known as assortative mating (Hou and Myles, 2008; Putnam, 2015). This in turn hinders social mobility across generations (Heisz, 2015: 96). An American study found that the average income in the neighbourhood where a child grows up has almost as much effect on lifetime earnings as the income of their parents (Rothwell and Massey, 2014).

Parents know this intuitively. That is why those who care do what they can to "set the stage" for their children's success, as in Forest Hill in the 1950s (Robinson and Harris, 2014: 218; Seeley, Sim, and Loosley, 1956:224). That is also why, when that setting is threatened, they organize. The issue was dramatized recently in Kingston, with the planned closure of KCVI (Kingston Collegiate and Vocational Institute), Ontario's oldest high school and alma mater of Sir John A. Macdonald. With the highest test scores in the city, its threatened closure aroused heavy opposition (Collins, All-man, and Irwin, 2019). But these days, concern about closures is not confined to the elite and the middle class. In Montreal, schools in Little Burgundy and Pointe-Saint-Charles have been ranked at or near the bottom in standardized test results. Occasionally parents were able to evade the problem. One family used a fake address on the north side of Saint-Antoine so that their children could go to a much better school in Westmount (Williams, 1997: 70). But this approach was risky and rarely possible. Instead, many local residents protested when deindustrialization and depopulation led to threatened school closures in the area (High, 2022: 193–7).

In Montreal, as in Kingston, the opposition was unsuccessful and closures went through. But here, in microcosm, we see the options that residents have when confronted with something they dislike, whether it pertains to schools, urban redevelopment, crime, or pollution: move, or complain (c.f. Hirschman, 1970). Either way, their response is indirectly part of the effect that neighbourhoods have.

How Residents Respond

Residents usually organize for one of three reasons: to build community, to attract attention, or to resist change. Occasionally there is a fourth: because they were asked to. These are not mutually exclusive, and most organizations combine at least two purposes. After all, building community makes the other two main goals more attainable.

Given that some organizations are long-lived, the mix of motives shifts over time (Ley, 1993: 226–7). In Toronto one example is the Annex Ratepayers' Association, which organized in 1923 to prevent construction of a hospital, and which then persisted partly through inertia and as a way to maintaining community. It faltered during the 1930s and early 1940s, when the City backed away from enforcing bylaws against rooming houses; property owners found conversions profitable, and there was a growing demand (Moore, 1982: 31). But with postwar prosperity it briefly restructured as the West and East Annex Neighbourhood Associations to lobby for neighbourhood improvements. In time, they amalgamated, reviving when asked by the Toronto Planning Board to gather information for planning purposes, and then roared into top gear in the 1960s to resist piecemeal redevelopment of multi-storey apartments and then the proposed Spadina Expressway (Jacobs, 1971).

Organizing to build community is probably the most common purpose of such organizations. On my street, for example, an annual block party is organized; in recent years, a group has gone carolling; in the summer of 2020, a different group organized a series of front-porch events where young musicians played to socially distanced audiences dispersed across the sidewalk and road. The main purpose of such gatherings is to build or reaffirm community, and most neighbourhoods can provide examples of this sort. It is the other types of activity, however, that attract outside attention.

Pay Attention to Us

Structured organizations either seek or resist attention from outsiders (Fisher, 1982; Hasson and Ley, 1994). Many are started by property owners who want better services. In the 1880s, before Toronto's Annex had a formal association, indeed before it was fully developed, landowners were lobbying for annexation by the City to get piped water, sewers, and paved roads (Moore, 1983). In Nepean (pop. 10,000), a suburb of Ottawa, groups coalesced to form the Britannia Line Citizens' League, struggling to raise taxes (true!) and to acquire services, against the resistance of frugal farmers (Elliott, 1991: 221–2). John Weaver (1979) provides similar examples from early twentieth-century Vancouver. In Eburne, South Hastings, and Kitsilano ratepayer associations lobbied for sewers and tram (streetcar) service. They also demanded protection of the single-family homes. Versions of this were happening everywhere. In piecemeal fashion from the 1920s, Nepean enacted various building and land use controls in different subareas (Elliott, 1991: 207–9; 244–6). And, as described in the next chapter, in Toronto de facto zoning crept in piecemeal over a period of decades in response to local petitions (Moore, 1979). All this to satisfy homeowners. Other associations include, and may even be dominated by, tenants, Vancouver's Downtown Eastside Residents' Association being a case in point (Ley, 1994). There the main concern is with social circumstances. Elsewhere, physical conditions are the issue. In Hamilton in the early 1990s, residents northeast of the downtown were concerned about air pollution. Sceptical of official statistics, they successfully pressured for more reliable, and geographically specific, information (Elliott et al., 1999).

Although less frequently, lower-income households and tenants also organize to bring attention to local needs. For many decades, most local efforts were organized by non-profit and charitable agencies, more about social work than community organizing. Change came in the 1960s, partly inspired by Saul Alinsky's work in Chicago and then by the New Left (Fisher, 1982; Hasson and Ley, 1994; Shragge, 2013). In Kingston, organizers formed the Association for Tenants' Action, Kingston (ATAK), which successfully lobbied for tenants' rights, hence better living conditions, in the city's North End (Harris, 1988: 106–21). Many organizers, and members, of such organizations saw their goals as being both local and national, or even international: they were part of a larger movement. Some got carried away. In *The Power to Make It Happen*, based on his experience with community organizing in Toronto's Riverdale neighbourhood, Donald Keating (1975: viii) declared "even a superficial glance around the world shows that one way or another there is going to come about a vast redistribution of power." He was right, of course, except that he got the direction of change wrong.

Organizations composed of tenants, and of the disadvantaged, are more likely to see themselves as part of a wider movement than those of property owners. Owners care about what affects their property values, and those influences are mainly local. The exception came during the Depression when homeowners' organizations lobbied provinces to enact mortgage moratoria. Tenants, too, care about the actions, or inaction, of their landlords. But in general landlord and tenant rights are determined by provincial legislation. Their frustrations are more likely to become a springboard for wider action.

This can be true for all sorts of disadvantaged persons and groups. Activists, reformers, and militants grow up in particular neighbourhoods, and their views and actions are often shaped by that experience, and in relation to that of the more fortunate. Movies, television, and now the internet have shown us how the rich and famous live, but there is nothing quite like witnessing it in person. As a delivery boy for the Winnipeg *Telegram* in the 1920s, James Gray (1970: 119–20) was astonished to discover the landscapes and lifestyles of the rich in Armstrong's Point. He reckoned that, knowing only the North End, his parents were only dimly aware of the prevailing degree of inequality. Half a century later, growing up in Pointe-Saint-Charles, Kathy Dobson (2011: 47) had used textbooks that were "ripped up … with missing pages and extra shit written inside." When transferred to a school in Westmount, she was impressed by the higher quality of her new teachers and facilities. What struck her most forcibly was the language of her peers. She learned that "there was a wrong way to swear … Nobody in the Point tries to say fuck on purpose. It's just a word that's part of our sentences" (173). Not so in Westmount.

How people dress can be just as telling. When the young protagonist of David Baillie's (2015: 27)*What We Salvage* visits a new acquaintance in middle-class Westdale, he is told "look, this isn't downtown or the north end. If you're going to walk around dressed like that, I'd prefer it if you used my back door." Just a mile or two away from home, like Kathy and James, he was discovering a new world.

Experiencing such disparities can turn someone into a radical. A striking, quintessentially Canadian, example is Pierre Vallières. The French-English divide has been a defining feature of the Canadian experience, and certainly of its national politics. Emerging from the Quiet Revolution in Quebec came a jarring event that drew international attention: The October Crisis of 1970. Rivalling the Winnipeg General Strike as the most dramatic moment in twentieth-century Canadian politics, it involved the abduction of two men and the killing of one, in the name of Quebec separatism. Pierre Vallières, the person whose ideas inspired the abductions, and Paul Rose, the killer, both grew up in Ville Jacques-Cartier, on Montreal's South Shore. In his autobiography and political statement, Vallières reports that his radicalism was fuelled by youthful experiences. Until age 7, he lived in Frontenac Park in the "tough" East End, where he joined a gang at age 5 (Vallières, 1971: 89, 90). The family then moved across the river to a poorly serviced semi-rural slum, an area controlled by "gangsters" (Vallières, 1971: 102). He became embittered because just "two thousand feet" away were "modern cottages being built for the petty-bourgeois families who wanted to live in the suburbs," and also by the way that in high school the children from those houses viewed him as an "outsider," making him "more and more ashamed of my milieu" (Vallières, 1971: 107, 113). In Winnipeg in 1918, this might have made him a labour radical: what rankled here was the class difference. He reports that residents of the adjacent neighbourhood built a fence to protect themselves from the "dirty people," and "treated us with unbelievable contempt," but then he adds (in parentheses) "and yet they were not English" (Vallières, 1971: 113). If Ville Jacques-Cartier helped make him into a radical, it was his later experience in Montreal that turned him into a *séparatiste*. In a recent documentary about Paul Rose, his son reports that Rose grew up on a similar dirt road in the same area, set apart from nearby Saint-Lambert and its fine public swimming pool (F. Rose, 2020). Both Rose and Vallières, like the workers who organized the Winnipeg General Strike, were highly influenced by their neighbourhood experiences.

In some cases, neighbourhood groups are funded and commissioned – some might say "co-opted" – to provide and/or lobby for local services. In Vancouver in 1995, for example, the Collingwood Neighbourhood House was built to help immigrants adjust to their new lives in Canada. It differed from a conventional community centre in that it was run by local residents, not the city

(Sandercock and Attli, 2009). Many variations on this theme are possible. Nationally, the most significant was the federal Neighbourhood Improvement Program (NIP), 1974–8. Funds were made available to cities for the improvement of designated neighbourhoods. Some cities, for example London, Ontario, were more active than others; Hamilton showed an almost complete lack of interest (Patterson, 1993: 327). Participating cities were active in different ways, but everywhere local residents were given a say in how NIP money was spent (Filion, 1988). Sometimes, local organizations were organized by municipalities, eager for federal money; in others, pre-existing residents' associations lobbied for public money and attention. Here, as in other respects, local associations embodied diverse goals.

Leave Us Alone

And then there are situations where residents resist change. Today, the most familiar are "not in my backyard" (NIMBY) groups that oppose forms of intensification and integration in their neighbourhoods, but there has been a long history of such opposition. One consequence of Toronto's first apartment boom in the 1900s was a movement to prohibit apartment buildings on specific streets. Apartment buildings were considered to be at odds with the image of Toronto as a "city of homes" and so were confined to major streets (Dennis, 1994: 312). The trajectory of successive Annex associations in Toronto was mirrored by those in Kitsilano. Established in 1906, a Kitsilano improvement association lobbied for services and utilities, but then switched to fighting upzoning that threatened its quiet, family-oriented character (Carr, 1980: 184). Resistance comes and goes in waves, reflecting new redevelopment pressures. The apartment boom of the 1960s was especially productive. In Hamilton in 1972, using a name recently coined by planners, the Durand Neighbourhood Citizens' Association formed to prevent further high-rise redevelopment (Elman, 2001: 11–12). Much the same was happening across the country, and planners were wringing their hands. In a statement that could have been written anywhere by any planner or developer in the past half-century, the Toronto-based consultant Frank Lewinburg (1983: 178) noted that, faced with proposed redevelopments, "they [residents] feel that their life styles, their children and their property values are threatened." The result? "The neighbours turn out in force and say 'no'" (178).

Occasionally, residents have been open to the idea of change. Adriane Carr (1980: 189) suggests that, by the late 1950s, some of the first residents of Kitsilano, now aging, welcomed the chance to sell at a profit to the developers of mid-rise apartment buildings. Even more striking is More Neighbours Toronto (MNTO), a recently formed organization that seeks to promote densification by declaring "yes in my backyard." But these are exceptions by and large. In Vancouver, residents have usually resisted low-rise co-ops, even after learning that they did not depress property values (City of Vancouver, 1986). Ann McAfee (McAfee et al., 1989), the City's chief planner, puzzled over what to do. The City worked hard to sell the idea through community "engagement," with initial success, but once redevelopment began so did the opposition. As McAfee (2016: 229) wryly comments, "communities shifted from enraged (1992) to engaged (2004) and returned to enraged (2007)." What Poulton (1995) calls the "incumbent's club" can be a powerfully conservative force.

Despite, or perhaps because of, the pressures for intensification, and the efforts of groups such as MNTO, there are no signs that NIMBY resistance will abate. In a high-profile confrontation in midtown Toronto in 2015, residents organized against a modest mid-rise condominium development, exactly the sort of project that planners most favour (Krishnan, 2015). Such resistance has always crossed political boundaries. In their study of redevelopment proposals in downtown Toronto in the 1950s, Paul Hess and Robert Lewis (2019: 281) report that the Wellesley-Bloor Ratepayers Association "displayed a mixture of progressive and conservative values." We see similar situations today. In 2017, Margaret Atwood supported her neighbours' opposition to a mid-rise project, although it was on a major street (Bozikovic, 2017). Here, in the resistance to piecemeal redevelopment, we have the most enduring and ubiquitous form of neighbourhood organization.

There is even stronger opposition to the infrequent, larger-scale projects. These are usually different in character as well as scale. The target is typically an older, run-down district occupied by tenants and lower-income homeowners. These are often stigmatized areas where residents feel resentful, sceptical of assistance, and primed to feel threatened. Glimpses of this are provided in works of fiction by Louisa Onomé and Kathy Dobson. In her young adult novel, Onomé draws on her own experience of growing up in a lower-income district in Mississauga to dramatize the resentment that residents feel at how outsiders view their neighbourhood. Nelo, the 15-year old central character, insists "Ginger East may be a bit scary but it's not trashy" (Onomé, 2021: 25). For some, stigmatization breeds scepticism, so that even the motives of those with good intentions are doubted. I sensed this among residents in my study of ATAK in Kingston, where students helped to organize the group and others worked to get the city's NIP area designated (Harris, 1988: 136). One type of response is illustrated by Kathy Dobson, for Montreal's Pointe-Saint-Charles. Her mother became a community organizer, working with well-intentioned middle-class outsiders. But Dobson recalls her father saying, "all your new little friends are just playing at being poor. They think it makes them more real or authentic or something but trust me, as soon as they get tired of eating bologna … of stepping on roaches … all they have to do is walk away" (Dobson, 2011: 83–4). Faced with a proposed redevelopment of their whole area, many respond with resignation. But a minority, incensed, act.

Large urban renewal projects became common from the late 1950s, and so did organized resistance. This happened everywhere, with varying success. In Hamilton, whose city council embraced renewal more aggressively than that of any other major city, opposition proved ineffectual (Robick, 2011). Much the same happened in Halifax's Africville and Montreal's Little Burgundy (Loo, 2019; High, 2022). In Vancouver's Chinatown and Strathcona areas, however, opposition worked, as "block-based, multilingual, neighbour-to-neighbour communication systems" using "informal networks" proved effective (Lee, 2007: 395; Loo, 2019). More typically, in Toronto there was mixed success. The Spadina

Expressway was stopped, but other achievements were modest. With help from middle-class organizers, the residents of Trefann Court, adjacent to Cabbagetown, halted the bulldozer, but only temporarily and after most of the general area had been redeveloped (Fraser, 1972). Donald Keating's optimistic radicalization happened only after the Don Mount public housing project had been carried out. Sometimes there were compromises. In Montreal, the residents of Saint-Michel, an immigrant area, had long complained about the local Miron quarry. As a consolation, they were given the TOHU, a large multi-purpose community space (Trudelle et al., 2016).

Recently, low- and moderate-income areas have organized against more subtle forms of change that can be cumulatively significant. Since the 1960s, many inner-city neighbourhoods have gentrified: average incomes have risen as educated, middle-class people have moved in, fixed up homes, and attracted new stores. This has brought benefits to run-down areas, and in Toronto's Little Portugal many existing residents accepted, or embraced, the change. Some homeowners welcomed upgrades to the neighbourhood while others capitalized on rising real estate values and moved out (McGirr, Skaburskis, and Danegani, 2015; Murdie and Teixeira, 2011). But most observers have expressed concerns for the people displaced, as Steven High (2022) has done so eloquently in his account of deindustrialization in Pointe-Saint-Charles and Little Burgundy. And many residents have resisted gentrification. In Toronto's South Riverdale, some homeowners and businesses sought its designation as a Heritage Conservation Area, a step which would have underlined its attractiveness to certain types of middle-class households (Valverde, 2016). In response, a grassroots organization calling itself Planning South Riverdale arranged meetings in local venues where lower-income residents would usually gather, including a health centre and shelter. Their goal was to develop an alternative vision, one which might include affordable coffee shops and so on (204). Occasionally, resistance becomes violent. In March 2018, an extreme example occurred in Hamilton – also somewhat surprising, given the city's relative lack of gentrification. A local anarchist collective led 30 masked vandals on an evening rampage, smashing windows on Locke Street, an event that attracted national coverage (Harris, 2020b: 166). But such resistance is rare and usually piecemeal, indeed personally motivated.

It is not only the slow, insidious character of gentrification that inhibits resistance, but a complex brew of class, immigrant status, and housing tenure. Those displaced are typically low-income homeowners and tenants, people with little time, resources, or inclination to organize. The middle class are at an advantage: those with education, a decent regular income, and the belief that their voice will be heard, speak up. In Vancouver in the mid-1970s, the more affluent west-siders were more likely to stand up against proposed redevelopment than the east-enders (Ley and Mercer, 1980). Again, class and tenure matter. Tenants have been less likely than property owners to become involved in local politics and residents' associations. A striking example was Toronto's Bloor-Carleton Ratepayers' Association. In the 1950s, 95 per cent of the residents of this area were tenants, but the main issue raised by the association concerned property rights (Hess and Lewis, 2019: 286). The most systematic research on this topic is American. For example, in a study of Columbus, Ohio, Kevin Cox (1982) found that homeownership had a significant effect on neighbourhood activism, especially when families had children. An exhaustive study of planning board meetings across Massachusetts found that participants were disproportionately white, male, long-time residents and … homeowners (Einstein, Palmer, and Glick, 2019; Einstein, Glick, and Palmer, 2019). Homeowners become active because they have more at stake, have lived in the area for more years, and have developed personal and psychic attachments. For those reasons, William Fischel (2001: 6) put forward the influential "homevoter hypothesis," arguing that homeowners support municipal actions and policies that "maximise … the value of their primary assets." It is more truism than hypothesis, and not just in the United States.

In Canada, homeowners have always moved infrequently, in part because of the higher costs. In mid-nineteenth century Hamilton, tenants moved four times as often as homeowners (Katz,

1975: 131). Much the same was true in late nineteenth-century Montreal (Gilliland, 1998: 33); today the ratio is about 5:1. The effects are apparent. A recent study of Winnipeg found that neighbouring, and a sense of community, were strongly associated with homeownership and length of residence (Farrell, Aubry, and Coulombe, 2004). Where planners have sought local input they have invariably found homeowners to be more receptive. Consultations that planners carried out under the NIP program were dominated by the concerns of property owners, one reason why municipal strategies in Toronto and Montreal diverged (Filion, 1988). Forty years later, and despite the efforts of city-funded organizers, workers and minorities rarely participated in the Neighbourhood Action Strategy that Hamilton developed for 11 underprivileged districts (Pothier et al., 2019). Barriers included language, literacy, inadequate childcare, and physical access to meeting places. The same applies to resident-initiated organizations. In Toronto's Mount Dennis area, low-income residents were concerned that the city's plans to revitalize the area would neglect their needs, but it was white, middle-class homeowners who ended up dominating the Ratepayers' Association – the name itself tells the message (Rankin and McLean, 2015). And, where tenants do organize, they may find themselves at odds with local property owners. In the battle for Trefann Court, they ended up forming their own, weaker, association (Fraser, 1972). Planners have sometimes worked hard to engage tenants. Recently, for example, the City of Seattle cut funding to 13 Neighbourhood District Councils and replaced them with a city-wide commission that would be more inclusive (Rojc, 2017). But it is an uphill struggle.

Occasionally, an issue unites everyone. Airbnb is a recent example. In tourist areas, some property owners have seen Airbnb as a profitable option. Only a minority of clients are noisy, but this is unpredictable. Some owners fret about the effect on property values while tenants worry that, by removing units from the rental stock, Airbnb will push rents up. In affected areas everyone mobilizes. A prime example was when a socially inclusive organization, the Friends of Kensington Market, lobbied the City of Toronto to take action. (Mintz, 2017).

There, as in Don Mount, Strathcona, Africville, and dozens of other neighbourhoods, the way municipalities act depend somewhat on the facts: the actual conditions. But what matters just as much are their perceptions of the neighbourhood. Bureaucrats' mental images, often stereotypes, govern how, and indeed whether, municipalities respond to cries for attention, or choose to promote redevelopment. At some remove, these perceptions too may be thought of as parts of the neighbourhood effect.

How Neighbourhoods Are Perceived

Most people know only a handful of neighbourhoods in their city: the places where they, their friends, or family live. For the rest, they will have some sort of impression, often only of those areas that make the news. This means areas known for crime and drugs, or perhaps celebrities and public figures. At both extremes, there is likely to be a gap between popular perception and on-the-ground reality.

People are essentially ignorant about the parts of their city that they never visit. Low-income households have limited reason or capacity to explore. In *The Fat Woman Next Door Is Pregnant,* speaking about adjacent districts, Michel Tremblay (1981: 20) comments that "no one had ever gone to Saint-Henri and no one from Saint-Henri had ever come to the Plateau Mont-Royal"; at best, "they met halfway, in the aisles of Eaton's." Not surprisingly, a recent study found that, for shopping, work, and recreation, people living in disadvantaged areas in Montreal circulated in limited, similar areas (Shareck, Kestens, and Frohlich, 2014). Less obviously, privileged people had a similarly restricted – although very different – geography. A study of Los Angeles showed that both groups were socially isolated, one through constraint, the other by choice (Krivo et al., 2013). This is generally true. Erna Paris recalls that lives in 1950s Forest Hill were "insular." In *The Heart Laid Bare,* by Michel Tremblay (1989: 39), Johanne declares "we hardly ever leave Outremont. I don't know why we'd go to St. Denis when we have everything we need right at our doorstep."

And Mordecai Richler (1969: 60) recalls that 1950s Outremont was "where the emergent middle class and the rich lived," comprising "an almost self-contained world." No world was more self-contained, or constrained, than Canada's residential schools. Most were located in rural areas, but even those in urban areas in effect remained secret. The Assiniboia Indian Residential School sat in River Heights, Winnipeg, an affluent area. Gary Robson, who grew up just round the corner, later recalled that for many years he was simply unaware of its existence ("Survivors …" 2021: 152–3). More surprising, although it is their job to be well-informed, many planners have inaccurate perceptions of neighbourhoods, and assess them differently than local residents (Lansing and Marans, 1969). Planners know the relevant statistics but are subject to the same prejudices and blinkers as other members of the middle classes.

Stereotypes can change when neighbourhoods do, but remain inaccurate. In its early years, Toronto's Parkdale was idealized as a decorous middle-class suburb. In her history of the area, Carolyn Whitzman (2009: 63) quotes an extravagant newspaper article: "She became austere, proud and chaste. Ostracized the saloon-keepers, frowned on negro minstrels, erected several churches … and became pious." In fact, from the outset the area was home to a number of workers, as well as the Magdalen Asylum for Fallen Women. After half a century of decline, however, it acquired the reputation of being a slum, an image that it has lately begun to shrug off, declaring itself an urban village. None of these labels has ever captured its true, always socially complex, character.

Given their lack of direct knowledge, it's no wonder people harbour inaccurate impressions of both rich and poor areas. Elite enclaves may seem homogeneous. In Hamilton, the Durand area has always had that image but there is, and always has been, a social mix. Historically there were servants' quarters in the attics of mansions and in small houses, not all along back alleys. Today there are apartments and group homes. No doubt much the same is true of Outremont, and even Westmount.

More common are the inaccuracies pertaining to lower-income areas. Some can be inconsequential. Speaking about his home turf, between Main and Park in Montreal, Mordecai Richler (1969: 29) comments that "to the middle class stranger, the five streets would have seemed interchangeable. On each corner a cigar store, a grocery, and a fruit man ... An endless repetition of ... peeling balconies and waste lots ... But, as we boys knew, each street ... represented subtle differences in income." Here, misperceptions did not matter much. The same can be true of ethnic neighbourhoods. Toronto's "Little Italy" never contained a majority of Italian Canadians, and that has been true of most ethnic "villages." Sometimes the label can be very wide off the mark. One of Stacey Zembrzycki's (2007: 89) interviewees recalled that Sudbury's "Polack Town" contained few Poles but many Ukrainians. Another commented, "they [Anglos] didn't understand the difference," while another shrugged, "you just got used to it" (89). Such ignorance is common. It drives the plot of Louisa Onomé's story about how outsiders judged a smashed store window in fictional Ginger East. But insults alone have few consequences.

Ignorance matters when it affects how outsiders, including planners and politicians, act. That is why some writers, reporters, reformers, and academics have worked to inform their contemporaries. In nineteenth-century Britain, "social explorers" published reports about working-class districts in industrial cities (Keating, 1976). The photographer Jacob Riis's *How the Other Half Lives* (1890) exposed readers to the realities of New York tenement life. The closest equivalent in Canada was Arthur Goss who – in Canadian fashion – was made Toronto's first official photographer; among other things, he depicted the homes of the poor (Greenhill and Birrell, 1979: 146). The impulse to reveal hidden worlds also inspired Herbert Ames (1897) to document conditions in Montreal's *City below the Hill*, while, a decade later, J.S. Woodsworth (1909; [1911] 1972) told Canadians about the immigrant "strangers" who were becoming their metaphorical, if not necessarily literal, neighbours. On the effects of segregation, Woodsworth (65) quoted another reformer, J.J. Kelso, founder of the Children's Aid Society: "the better class of citizens do not know what is going on, so these wretched social conditions are allowed to grow."

Later writers have perceived a need for such commentary and reportage. In the 1960s, Michael Harrington, in *The Other America*, exposed the reality of poverty in the United States, in the process shocking President Kennedy when he read it on Air Force One. Channelling Harrington, the Canadian economist Harvey Lithwick (1971: xx) attempted something similar in an influential study of urban poverty, declaring that "the poor are not seen and, being out of sight, are out of mind." Toronto sociologist W.E. Mann (1970: 11; 1961: 40) uncovered the city's "soft underbelly," "a hidden 'no-man's land' to most respectable Torontonians." Lately, academics, reporters, and social agencies have done much the same, through statistics and narratives. Both can be effective. David Hulchanski's (2010) analysis of the "three cities within Toronto" drew the attention of planners and social agencies to the decline of inner suburban neighbourhoods, as did Shawn Micallef's (2017) account of his walking tour of those areas during the 2016 civic election. In different ways, many writers have brought deprived neighbourhoods to the attention of a wider public.

Other observers have explored unfamiliar territory simply to inform themselves. Dan Hiebert (1991: 61) reports that a "gentleman" who explored Winnipeg's North End was "so shocked that he later returned with enough food and fuel for several particularly needy families in the area." During the Spanish Flu epidemic, middle-class volunteers went into hard-hit areas and came out wiser (Fahrni, 2011: 78–9). Others explored with more selfish motives. In late nineteen-century London and Paris, and then in New York, "slumming" became a popular pastime among the well-off, especially for entertainment purposes; notably, Depression-era Harlem attracted many to its jazz clubs. In Montreal, Little Burgundy had taverns "on every corner" and drew visitors to clubs with live musicians, including Oscar Peterson (High, 2019: 27). In Canada, Quebec City's Old Town became a less racialized destination. In Roger Lemelin's (1948: 158) *The Town Below*, Denis Boucher, a clerk, observes a "sightseeing trolley" full of tourists from Montreal, Trois-Rivières, and the United States. Observing the "sloping-roofed houses" in the Lower Town, the slummers "clucked delightedly over this debris of the past" and threw pennies to the urchins who followed them (ibid.). The social stereotypes held by these tourists were probably reinforced by their brief sojourn – they were looking for entertainment, not enlightenment – but other explorers came back better-informed.

The same applies to some of the straight – cisgender, in today's language – tourists who ventured into the early gay bars and neighbourhoods. They enjoyed the frisson of the exotic. In time, as gays and lesbians became accepted, "gaybourhoods" became just another urban village, attracting consumers or tourists, although more by day than at night (Gorman-Murray and Nash, 2017; Remiggi, 1998). In the minds of planners and politicians such districts have become intrinsic to their city's cultural industry; in 2011 the City of Ottawa officially recognized Le/The Village (Lewis, 2013; c.f. Ghaziani, 2014). In this way, artistic and gay communities have sometimes been unwitting pioneers of gentrification.

Often it is the stigma attached to its residents that stereotypes the neighbourhood. This is obvious in the case of racialized districts, such as Vancouver's Chinatown and, after the attack on Pearl Harbor, its Japantown (Anderson, 1991; Olson and Kobayashi, 1993; Wideman and Masuda, 2018). In 1913, speaking about Toronto's Ward, a contributor to *Canadian Magazine* made passing comments about the "the canny shopkeeper" – a transparent reference to Jews – and "indolent Southerns" (Bell, 1913: 234, 242). Where no single group was dominant, a more general dismissal served. In early twentieth-century Toronto, "foreign quarters" were seen "as a form of urban blight" (Harney, 1985: 4). In Winnipeg, the *Telegram* generalized about immigrants "whose ignorance is impenetrable, whose customs are repulsive, whose civilization is primitive, and whose character and morals are justly condemned," in sum, "the scum of Europe" (quoted in Hiebert, 1991: 65). And part of Montreal's Little Burgundy was known as "Nigger Town" (High, 2019: 25). On other occasions, racialization has been implied, even when inappropriate. In the 1960s, Lawrence Heights, then occupied by Anglo-Canadians, was sometimes referred to as a "jungle" (Delegran, 1970: 84).

In other times and places, the culture of the poor was invoked. In the 1920s, according to the Canadian planner A.G. Dalzell (1927: iv), the chair of the Social Services Council of Canada had commented that "a slum population creates slum conditions." That was a commonly held view until at least the 1960s, a decade which saw a good deal of public housing built. When a project was proposed for a middle-class area of Kingston it was strongly resisted. One resident spoke for many when he damned it as "substandard houses for substandard people" (Harris, 1988: 98). Indeed, given the status associated with homeownership, renters in general are sometimes viewed sceptically, as a recent study of Calgary has shown (Rollwagen, 2015).

Reversing the direction of influence, once an area has acquired a negative image, any resident is marked. In the early twentieth century a version of environmental determinism held sway. As Maria Valverde (2008: 133) observes, reformers believed that a "deviant environment … naturally produces deviant people." Hence, those living in so-called slums become "stigmatized as slum-dwellers" (Mann, 1961: 45). Today, residents of the Jane-Finch area of Toronto know the routine. From her interviews with local residents, Mariam Zaami (2015: 81) reports that several, including a respondent identified as "Hanson," complained that, during job searches, "as soon as you tell potential employers of your postal code that's enough," to which "Austin" added, "we're all just scum to them" (81). Interestingly, several African immigrants blamed Jamaicans for giving the area a bad name.

Many of the residents that Zaami interviewed blamed the media for creating or reinforcing the area's stereotypes. Lately there has been much discussion about how the police treat visible minorities, especially Black people. As Bajan-Canadian Austin Clarke's (1986: 63) narrator in Nine Men Who Laughed says, "the Police does-be up in this St-Clair-Oakwood district like flies round a crocus-bag o' sugar at the drop of a cloth hat." Journalists follow. Zaami's "Serwan" observed that "an ordinary conflict or police arrest from this area gets too much unnecessary attention" (Zaami, 2015: 81). It is an old complaint. In the 1970s, Graham Fraser (1972) noted that "the people of Trefann Court had been constantly humiliated and condescended to by the press." In Vancouver, and indeed nationally, the Downtown Eastside has long been maligned for its poverty, drugs, ill-health, and crime: we are told about "people dying in the streets," where it is "too dangerous for undercover operations" in "Canada's poorest postal-code area" (Liu and Blomley, 2013: 124, 126). The media's influence is a major theme in Louisa Onomé's (2021) story set in fictional Ginger East. Media coverage matters, and often underlines stereotypes. In Toronto, April Lindgren's (2009) extensive survey of local newspapers confirms that they pay disproportionate attention to crime, especially in disadvantaged areas, while a more focused study by Chris Richardson (2014) makes the point for Jane-Finch.

But what should local reporters, editors, and TV stations do? In the spirit of Herbert Ames, J.S. Woodsworth, and David Hulchanski, they have tried to prick the consciences of the middle classes and elites, making the case that something should be done. Most people would praise them for that. But to accomplish that goal they must highlight the challenges faced by local residents, problems which are unpleasant, even repellent. That is what the *Hamilton Spectator* has done over a decade in its Code Red series, which yielded significant data on the geography of public health. This left it open to the criticism that it had homogenized and further stigmatized neighbourhoods in the lower city, in effect emphasizing the public costs of concentrated disadvantage and downplaying the agency of local residents (Cahuas, Malik, and Wakefield, 2016). As a resident of Hamilton (and subscriber to the *Spectator*) I believed the critique to be overdone. But then my wife, who taught at the downtown high school, told me that the students she taught hated the series. But here there is an important question, not just for the media but more generally: how should we speak about areas where many people are suffering?

For a start, we need to be careful about the names we use for residents, as well as their homes and neighbourhoods. During the Depression, Hugh Garner (1972: 145) noted that, while many people lived in "houses," those in Rosedale occupied "homes," and those in obituary columns had "residences." He could have extended the linguistic options with "tenements" and "mansions." Names matter most for lower-income areas because "language always supplies pejorative terms to describe places of poverty" (Topalov, 2017: 45; c.f. Topalov et al., 2010). There are generic terms, most commonly "slum" (Mayne, 2017). Its connotations have shifted since the nineteenth century, as moral condemnation of slum-dwellers made room for a new emphasis on environmental determinism (Ward, 1989). But both interpretations are negative. Contemporaries had no hesitation in describing and deploring Cabbagetown and the Ward as slums (c.f. Lorinc, 2015: 16). As physical conditions deteriorated during the Depression, "blight" was also used (Robick, 2011). When prosperity resumed, blighted slums became targets for redevelopment, in the belief that better homes would solve social problems. This affected cities of all sizes across the country, including St. John's, a city of barely 50,000 in 1951. Following Newfoundland's entry into Canada in 1949, it came under the influence of CMHC and federal programs, and the clearance option was soon implemented (Phyne, 2014). Some contemporaries recognized that the areas called slums contained some decent housing, along with residents who were not especially poor. For example, discussing a proposed project in Vancouver's Strathcona area, Leonard Marsh (1950: 23), also the author of the influential *Report on Social Security for Canada*, was careful to emphasize that the only crime *most* residents had committed was poverty. But there, and elsewhere, clearance went ahead nonetheless. Few of those displaced were rehoused. Part of the problem was a misguided faith in filtering. A classic case was Churchill Park in St. John's. Begun, unusually, in 1944 when Newfoundland was under British administration – hence its patriotic name – this garden suburb was supposed to relieve the housing shortage and improve the living conditions of the residents of inner-city slums. Typically, as Christopher Sharpe and A.J. Shawyer (2021) have shown, that never worked out.

"Slum" was an outsider's term. Residents disliked its connotations, especially when used to justify displacement. In their influential study of London's East End, Michael Young and Peter Wilmott (1957: 110) report that one woman was shocked that her home and neighbourhood might be described that way. Mostly, residents prefer a specific to a generic name. The question is, which one? Topalov (2017: 45) observes that "inhabitants immediately grasp the positive or negative overtones attached to the names of places or areas in the city." Any Londoner knew what the East End meant. Until the 1950s, "Cabbagetown" was perhaps accepted by most residents. At any rate, Hugh Garner felt free to use it for his home turf. But in the era of slum redevelopment people became wary. By the early 1970s, local residents preferred "East of Parliament," while planners, aware of sensitivities, invented the neutral "Don Vale" (Lorimer and Phillips, 1971: 13). Meanwhile, no resident of Rosedale, Westmount, or Shaughnessy challenged the way outsiders referred to their areas of residence.

But the middle class can be name-sensitive, especially where the character of a neighbourhood is unclear, or changing. Recognizing that "The Beach" (a.k.a. "The Beaches") was a desirable area, Toronto real estate agents and then residents began to speak about the adjacent "Upper Beach." In the Junction, a once-industrial working-class district, gentrifying residents came to prefer "Junction Triangle." A real estate agent reckoned that "that's the moment the neighbourhood turned the corner" (Lorinc, 2015). Planners can cooperate in the process, especially where they are keen to attract middle-class residents and new investment. In Montreal, following redevelopment in the 1970s, it was planners who gave part of Saint-Antoine ward the name Little Burgundy and helped to play up its musical past (High, 2022: 11, 16). As Albert Hunter (1987: 212) has observed, names "help to create the very reality they are attempting to describe." Indeed, once a neighbourhood has turned around, its old name might be recycled. In Toronto, along with its brick Victorian homes, Cabbagetown was sandblasted. Anyone growing vegetables would now be showing environmental awareness.

Some names, like "slum," rise and fall. Others, like "Cabbagetown," are reborn or, like "the Junction," renovated. But the most important name of all, the one we take for granted, and which I have used so far, also has a history. Its usage has flowed and ebbed and then flowed again, changing its connotations for over a century. It is time to look more closely at the word "neighbourhood" itself because, as the next two chapters show, its meaning has expressed, and in modest ways influenced, broad trends in Canadian society.

PART 2

Changed Meanings and Contexts

chapter six

Language, Meaning, and Governance

Unless qualified (e.g., "noisy"), "neighbour" has always had positive connotations, but the same has not been true of "neighbourhood." A century ago, it carried mixed messages, and other terms were used to refer to the residential parts of cities. It is reasonable to speak about neighbourhoods since the late nineteenth century; that's what I have been doing so far. But we need to recognize that such discourse can be anachronistic.

The profile and meaning of "neighbourhood" have changed, in part because of developments in the larger social and economic context. But again it makes sense to start at the local level. Relative to its closest synonyms, "residential district" and "residential area," the profile of "neighbourhood" has shifted because of the ways in which municipalities – and sometimes land developers – have regulated and shaped land use. Intriguingly, from a different starting point, we can see a similar trajectory for *quartier* in francophone Quebec, as indeed in France. The language has been different, but meanings have converged, along with the forms of urban governance that shaped them.

The Meaning of "Neighbourhood" and Its French Equivalents

The idea, and ideal, of neighbourhood has fluctuated and become more prominent over time. Today the word is widely used, with the municipal governments of Toronto, Montreal, and indeed of many other places around the world styling themselves as "cities of neighbourhoods."

We seem to be at a historical peak in the term's usage. Zane Miller (1981: 4), a prominent American urban historian, surveyed the changing role of neighbourhoods in American cities. He first suggests that popular usage of the concept appeared in the late nineteenth century, an intriguing observation which few have noted. The same may well have been true in England. Apparently, Charles Booth, who undertook exhaustive surveys of London's residential areas in the late nineteenth century, used a range of terms, including "district," "precinct," and "quarter," without attaching weight to "neighbourhood" (Reeder, 2010: 813). In the United States, "neighborhood" then "flourished" for a couple of decades, "all but disappeared" from the 1920s through the 1940s, and then revived, gaining momentum from the 1960s (Miller, 1981: 5; see also Mooney-Melvin, 1985: 361).

Numbers, specifically Google's Ngram Viewer, confirm Miller's impressions. Ngram Viewer is a software tool that enables keyword searches of a huge database of works published between 1500 and (currently) 2019. Usefully, it shows the relative, not the absolute, frequency with which words appear, therefore indicating their shifting profile. It is possible to make separate searches on "neighbourhood" and "neighborhood" but, although Canadians now use the British spelling, the American version is more relevant. It speaks to the North American experience as a whole because, as indicated in J.S. Woodsworth's epigram in chapter 2, for decades we used the American spelling. An Ngram search shows an early peak in the late nineteenth century, a deep interwar lull, and then waves reaching a higher peak today.

That pattern is borne out by the writings of Canadian urban experts and historians. The extended trough, for example, is striking. In the midst of the Depression, the influential Report of the Lieutenant-Governor's Committee on Housing Conditions in Toronto (Ontario, 1934) described conditions in various inner-city "districts" but never used the word "neighbourhood." At the speech in March 1934, that prompted the report, the Lieutenant-Governor, H.A. Bruce, contrasted "fine residential areas" with "slum districts," terms which the report itself uses, along with "parts of the city" (Ontario, 1934: 5, 22). Apparently, Bruce did not see "neighbourhood" as a useful generic term. The same was true, more surprisingly, of a study of Vancouver completed a decade later by a geographer, Donald Kerr. Kerr (1943: 112) spoke only about "all the residential districts in Vancouver," while as late as the mid-1960s Barry Mayhew's (1967) survey for Vancouver's United Community Services spoke of "local areas." None mentioned neighbourhoods.

Historians likewise. This is telling, given that they are supposed to be sensitive to past usages. In reality, we always see the past through the eyes of the present and historians may unconsciously use terms that contemporaries would not have. One way to minimize this problem is to consider the writings of those historians who did not think of themselves as "urban" historians, and who would not have imbibed the modern social science discourse about cities. The results are sobering to those, like me, who believe that spatial patterns matter. For example, two accounts of Vancouver written by general historians make only casual reference to residential areas, and none to "neighbourhoods" (McDonald, 1996; Morley, 1974).

Of course, some writers are more sensitive to the social geography of cities than others. Historian Terry Copp (1974) brought this kind of sensitivity to his study of poverty in Montreal before the Depression. No Canadian city was larger, or contained more striking socio-geographical contrasts (c.f. Olson and Thornton, 2011). Copp acknowledged this. Taking his cue from Herbert Ames, he spoke about the city's working-class "city below the hill." Early on, he refers in turn to "district," "quarters," "quartiers," "section," "ward," "locality," and "area" before first mentioning "neighbourhood," but even then only because he is quoting Ames (Copp, 1974: 15, 17, 18, 19, 21, 23). He then reverts to "section," "wards," "area," and "district," adding "parts of the city," all on one page (25). His uncertainty about what language to use echoes Charles Booth's. Copp was writing at a time when a neighbourhood movement was shaping urban politics in many Canadian cities, but he respected contemporary usage, or at least its anglophone version.

Certainly, in the depths of the Depression, when the Montreal Board of Trade's Civic Improvement League (1935) reported on housing conditions, it, too, ignored "neighbourhood."

Popular Usage in English

But it's popular usage that matters (Harris and Vorms, 2017). There is no way to know how people referred to neighbours in, say, 1922, but daily newspapers offer reasonable clues; they spoke to men and women on the street, or else they were out of business. Fortunately, at least one English-language daily, the Toronto *Globe* (to 1936) and then the *Globe and Mail* (1936–2017), starting from 1844, is available for full-text searches (see appendix). True, the *Globe* has never spoken to people on every street. In working-class Newtonbrook between the wars, Jean Bradley (1996: 35) recalls that "we cut the *Daily Star,* the *Telegram*, or the local weekly *Enterprise*, into small squares to hang on a nail in the outhouse" [but] "I never saw *The Globe and Mail* used this way ... presumably its subscribers had indoor plumbing – although they may have used it at summer cottages in the Muskoka Lakes." Point taken. But the historical depth of its database, coupled with a versatile search engine, makes it possible to explore trends in usage in Toronto with more subtlety than for any other place. It has always had a Toronto bias but, where comparison is possible, the trends it reveals are consistent with those in Ngram. Important patterns are surely valid across English Canada.

Leaving aside terms like "suburb" or "slum" that denote specific types of areas, the *Globe (and Mail)*'s text indicates that, apart from "neighbourhood," two terms have been widely used since the 1880s: "residential district" and "residential area." Adjusted counts, explained and discussed in the appendix, reveal notable shifts (bar chart 6.1; table 6.1). "Residential district" peaked early, in the 1900s, fluctuated, and went into decline after the 1950s. "Residential area" gathered momentum over the first half of the twentieth century, peaking in the 1960s before falling thereafter. In contrast, confirming the impression of Zane Miller and the evidence of Ngram, the pattern for "neighbourhood" has been bimodal, peaking in the late nineteenth and early twentieth centuries, falling through the 1940s, 1950s and 1960s, before coming back in force since the 1970s. Indeed, its recent dominance cannot be overstated. Because of the various ways in which the word can be used, "neighbourhood" was searched only in the title of newspaper articles, whereas counts for the other terms reflect their appearance in the full text. Even so, "neighbourhood" became dominant.

A selective reading of the *Globe*'s articles shows that the meaning of these three terms overlapped to only a limited degree, while that of "neighbourhood" has shifted. The origin and evolution in usage of each of these terms illustrates Christian Topalov's argument about the naming process. In the 2000s, Topalov led an international project to provide concise accounts of how the popular usage of a range of urban words in seven languages had evolved (Topalov et al., 2010). The 1,489-page result is only available in French, but usefully the author has summarized in English his conclusions about the historical process. A word will emerge, or be reconceptualized, he argues, when three conditions are met: there is a linguistic resource available to fill it, a linguistic niche has opened up that needs to be filled, and there are one or more interest groups to promote it (Topalov, 2017: 60–1). "Residential district" easily fits the bill. Responding to popular pressure, following a major fire, in 1904 a new bylaw consolidated and expanded the scope of land-use regulations (Fischler, 2007). It enabled residents to petition the City to exclude non-residential uses from defined areas, in effect zoning (Moore, 1979: 320–1). Residents organized, with effect, at scales ranging from the block to whole districts (figure 19). "Residential district" denoted the areas affected. Here then, combining common enough terms, was a phrase that responded to new conditions and which was seized upon by interested groups.

Within 20 years residential districts covered much of the city. In June 1905, for example, a deputation to the recently established Board of Control asked that the area between Yonge and Spadina, College to Bloor, should be designated, as a title declared, "a residential district" (1905). Residents had been goaded by a proposal to build stables in the area. Such lobbying, and usage, persisted until 1954, when a comprehensive zoning bylaw rendered it moot. By then, "residential district" had acquired social connotations.

In August of 1951, for example, observers were wringing their hands at the deterioration of the area that included Jarvis Street, once "Toronto's finest residential district" ("Blighted area," 1951). Something more than mere land use was implied here.

Some residents adopted a refinement of these connotations, "residential area," although the terms could also be used synonymously. For example, when the annexation of North Toronto to the City was impending, the *Globe* reported that the Toronto Civic Improvement Committee was asking to have it designated as "a residential district" ("North Toronto is a residential area," [1911]). Land-use restriction was clearly implied. In response to the city's first apartment boom, the municipality prohibited the building of such structures in most districts, and in 1921 it allowed the identification of areas for "private dwellings" only, these in turn being subdivided into "detached" and "other" (Dennis, 1989; Moore, 1979). "Area," then, as opposed to "district," came to imply greater exclusivity, with the social connotations. In 1925, for example, a journalist waxed eloquent about North Toronto as an "exclusive residential area to particular people" ("Distinctive houses ... ," 1925). But, perhaps because the language was so similar, the term remained ambiguous. By the mid-1950s it was displacing "residential district" to denote any area that was zoned residential in any way. The case of a frustrated music teacher illustrated this in 1954. On her behalf, and that of its other 1,199 members in Toronto, the Ontario Registered Music Teachers Association argued that since anyone could play their piano at home, this teacher should be allowed to move her business from a commercial into a "residential area" ("Ask to operate in home zones," 1954). Such usage, along with the social connotations, has persisted. In 2017, for example, it was reported that Maha's Fine Egyptian Cuisine on Greenwood Avenue was "tucked in a residential area" ("Ethnic food," 2017). Since the 1960s, however, this term has been overtaken, and lapped several times, by "neighbourhood."

Although "neighbourhood" implies a mainly residential area, usage in the *Globe (and Mail)* underlines it has always embraced other land uses, notably commercial operations such as restaurants ... or music studios. The wrinkle is that, in Toronto, its prominence and connotations have changed. In the late 1800s the term was rarely used and, even then, the word often implied "in the vicinity of." For example, "Business lively in the neighbourhood of the Limestone City" (1881) referred to the Kingston region. This is consistent with how Charles Booth gave it no special attention in his surveys of London (Reeder, 2010). In the early 1900s, however, the word began to be used more exclusively for residential areas, as in "New Methodist church for the Don neighborhood" (1922), American spelling and all. But it was mainly associated with poorer districts. From the 1890s to the 1940s, between 13 and 16 per cent of "neighbourhood" articles also contained references to immigrants or the poor (table 6.2).

An even stronger association was with people and organizations who helped those in need. Settlement houses had been established in lower-income areas, and became bases from which volunteers provided supports. In the 1910s, references to these "neighbourhood workers" occurred in 33 per cent of all articles with "neighbourhood" in the title, a proportion that rose to 56 per cent in the 1920s, when the number of such articles first peaked. The association with immigrants was strong. A typical early article noted that volunteers were "Making Canadians of little foreigners" (1916). In time the emphasis broadened somewhat, following the creation of a city-wide Neighbourhood Workers' Association (NWA) in 1914. By the mid-1920s, the NWA's annual presentation claimed that "all over the city a tremendous work is being carried on by volunteer assistants" ("Volunteer workers take leading part ...," 1925). This might suggest that "neighbourhood" was coming into general usage. Not so. The NWA received reports from several districts, but entire areas, notably North Toronto, were absent. Their focus was on needy households in "slum areas," whether or not occupied by "newcomers" (ibid.). The connection between "neighbourhood" and volunteer workers tightened still further during the Depression, before rapidly trailing off (table 6.2).[1] For the first

---

1 The slight resurgence of "worker" from the 1990s signals reports on the activities of "community workers" and the like. See, for example, Raveena Aulakh's (2006) report on Jane-Finch.

half of the twentieth century, in Toronto at any rate, "neighbour-hood" was associated with poverty. It had been seized on, initially by reformers and the press but then more generally, to denote a new type of urban settlement.

Since 1945 conditions have changed, and so this noun has come to denote all types of residential areas, including a range of land uses. Indeed, commercial activity serving local needs came to be seen as intrinsic, essential even. An article by Judith Knelman (1983) spelled this out. She praised Riverdale, with its Danforth Avenue, as "a neighbourhood of bustling markets and residential streets," along with parks and convenient transit service. Such an article could not have appeared a generation earlier, and would have been inconceivable before World War II. "Neighbourhood" had come to mean something more inclusive than "residential area" or "district," and it had overshadowed these terms. Today, with its connotations of warmth and inclusivity, "neighbourhood" is ubiquitous. It was not always that way.

If "neighbourhood" overtook other terms it also absorbed a vital connotation: the defence of residential space. The growth of munic-ipal planning after 1945 entailed the creation of planning districts; "neighbourhood" became associated with planning. The propor-tion of "neighbourhood" articles with "planner" or "planning" in the text rose slowly from the 1930s and then jumped in the 1960s (table 6.2). As discussed in the next chapter, this reflected a wave of grassroots activism. Coupled with the City Planning Board's deci-sion to seek citizen input by holding neighbourhood meetings, this nudged the proportion even higher in the 1970s (Webster, 1970). This surge settled back, but has lately revived as inner districts have come under pressure for redevelopment. There has always been a tension between accommodating local preferences and doing what is best for the city as a whole: planning is a political balancing act. For that reason, as Alex Bozikovic (2017) notes about neighbour-hood resistance to densification in the Annex – resistance similar to that against apartment buildings a century earlier – "the planning system is maddeningly ambiguous." Inevitably, "neighbourhood" has inherited the wrangles over land-use regulation that were once encompassed by "residential area" and "residential district."

The Francophone Tradition and Close Synonyms for "Neighbourhood"

As Terry Copp's language suggests, English usage in Montreal paralleled that in English Canada, but what of francophones in Quebec? The meanings of related terms in French are rooted in a different cultural tradition, sustained by distinctive institutions, notably the Catholic Church. But, although the starting points were different, lately there has been a little-recognized convergence.

Historically, the most important neighbourhood-scale entities were Catholic parishes. Colonial administrators had delegated many civil and territorial responsibilities to religious institutions (Fougères, 2018: 300). The Church loomed large, literally: Mark Twain supposedly claimed that it was impossible to throw a stone in Montreal without breaking a church window (Marsan, 1981: 69). Straddling the nineteenth and twentieth centuries, Lucia Ferretti's (1992) study of a poor Montreal parish underlines the pervasive role that it played in local life. Apart from spiritual succour, it provided community as well as social services. Often associated with the *paroisses* were local *caisses populaires* (Sweeney, 1995). Almost as sharply defined as modern local planning areas, *paroisses* once provided an all-encompassing frame to neighbourhood life. Lucia Ferretti (1992: 190) has suggested that, although historically the Catholic Church had deep rural roots, it also undertook a "creative and functional response" to new urban conditions, putting *paroisses* "au coeur des relations sociales." Their role waned slowly and steadily after 1914 as "organizations, receipts, participation in lay groups, everything, now, went into decline" (Ferretti, 2001: 227; 1992: 181). But they remained important into the early postwar years. In Quebec's Lower Town, for example, one parish maintained and took on a range of pastoral activities. For example, eight *vicaires* were each made responsible for a subarea, where they made calls to the sick (Falardeau, 1949). Indeed, Nancy Christie and Michael Gauvreau (2010: 182) have argued that, for a while in Quebec, the Church was again "highly effective" in responding to rural-urban migration, creating 491 new parishes between 1940 and 1968 and "reinvigorating the idea of the parish as an urban community centre."

Complementing them, at the smallest scale, has been the language and practice of *voisinage*, which speaks to the immediate environs of the home, and of social connections with immediate neighbours, *voisins* (Ferretti, 2021; Jean, 2015). The concept of being "good neighbours" exists in English Canada, and the idea behind Neighbourhood Watch groups is that keeping an eye on activity on particular blocks is important. But there is no exact equivalent to *voisinage* (c.f. Maltais, 2014).[2]

The closest French equivalent to the current meaning of "neighbourhood" has long been *quartier*, but the trajectory has differed. The word comes above all from Paris; other French cities used other terms (Lamarre, 2010: 1014). There, and then in Quebec, *quartier* has always been significant but at a larger scale, as shown by Dale Gilbert's (2015) study of Saint-Sauveur in Quebec City in the middle decades of the twentieth century. In the 1800s, *quartiers* were equivalent to wards: their residents elected politicians. Although they lost that power, they remained as administrative units and eventually became *quartiers de planification* (Lamarre, 2010). When the federal government introduced the Neighbourhood Improvement Program in 1972, in Quebec it was known as the Programme d'amélioration de quartier. But the term had social significance, too. In *Bonheur d'occasion*, set in Montreal's Saint-Henri ward, Gabrielle Roy (1945) uses the term 120 times, gesturing towards an administrative division but emphasizing its social meaning. (She makes 20 references to *paroisse* and 4 to *voisinage*.) Writing about Quebec City at the same time, Roger Lemelin (1942) also makes liberal use of *quartier*, and in much the same way.

Increasingly from the 1960s, and indeed in parallel to the experience in France, the social aspects have become more important (c.f. Lamarre, 2010: 1015). This has come to be acknowledged in planning documents. In 1992, Montreal's first Master Plan subdivided all boroughs into planning districts, emphasizing that each "are living environments ["milieu de vie"] with very distinct

---

2 I would like to take this opportunity to thank Alexandre Maltais for his assistance in surveying the French-language planning documents for Montreal since 1944.

character," whose character was "not limited to the quality of dwellings" (Ville de Montréal, 1992: 20; trans. Maltais). It recognized the role of parks, local services, and, in the same spirit as Judith Knelman's tribute to Riverdale, the dynamism of commercial streets. Indeed, later documents treat *quartier* and *milieu de vie* as synonymous. The revised plan insisted that these *milieux de ville* were "diversified and complete living environments" (Ville de Montréal, 2004; trans. Maltais). A transportation document suggested that "for a lot of Montréalers, the term *quartier* refers to the traditional notions of *paroisse* and *voisinage*, that we also call *milieu de vie* in reference to the lived space of the majority of the members of the community" (Ville de Montréal, 2013: 1.4; trans. Maltais).[3] The conflation of *quartier* with *paroisse* and *voisinage* is puzzling, but the spirit was summarized by Annick Germain (2013: 3): *quartier* "is not an institutional space but a social one."

That is going too far; it is both. And so there has been a convergence of meaning across linguistic solitudes. For France, Christine Lamarre (2010: 1015) has acknowledged that, in its social aspects *"quartier* se rapproche de *voisinage* ou du *neighbourhood* anglaise" but in Canada the coming together of French and English terminology seems to have been overlooked. Both *quartier* and "neighbourhood" have uneasily combined meanings, juggling rather than smoothly integrating the planning and social functions. But their ambiguity is part of their appeal. At any rate, in recent decades both have grown into unprecedented prominence. Underlying that trend are the same stakes, notably real estate and schools, which will be explored in the next chapter. But before considering them, we need to examine the actions of municipal governments, for they have been persuaded to play a large role in creating and defending those stakes.

---

3 "Pour beaucoup de Montréalais, le terme 'quartier' se rapport à la notion traditionelle de paroisse ou de voisinage, qu'on appelle aussie milieu de ville en reference à l'espace senti et vécu par la majorité des occupants d'une communauté."

The Rise of Municipal Governance

By nature, cities have always required, and in varying degrees received, both regulation and services. All those people and their activities create challenges: the need for clean water and safe disposal of waste, for infrastructure to help transport people and goods, for managing disputes between neighbours, and much else besides (Harris, 2021). Rapid urban growth in the late nineteenth century magnified these challenges, compelling municipalities, and some private agencies, to undertake a growing range of activities. Growth and change came in four phases: the urban reform era (1880–1920), the interwar period (1920–45), the early postwar decades (1945–70), and the modern era (1970–). Nothing magical happened on these dates, except 1945, but these periods are recognized as being distinctive. In each, neighbourhoods were understood, and treated, in distinctive ways.

The Reform Era, 1880s–c. 1920

Following close behind the United States, in the late nineteenth and early twentieth centuries Canada became urban. The proportion living in urban areas doubled from 23 per cent in 1881 to 47 per cent in 1921. At the same time, the total population was growing rapidly, which created challenges that municipalities were ill-prepared to handle. This mattered because, as Frank Underhill ([1910–11] 1974: 326) observed, "city government touches the individual more closely than any other kind of government." Reformers like Underhill sought to promote civic efficiency by limiting party politics and creating a larger, better-organized bureaucracy that could regulate utilities, development, and land use (Rutherford, 1974, 1977; Weaver, 1977). They also aimed to improve public morals (Valverde, 2008). This was a tall order.

It is easy to overstate the impact of urban reform in Canada. It never acquired the momentum that it had in the United States, partly because the challenges were smaller (Weaver, 1977). Paradoxically, it was especially weak in the largest city, Montreal, where the challenges were greatest. With reference to Montreal, Annick

Germain (1984) reports a contemporary saying: "le maire règne mais ne gouverne pas"; little changed. But, almost everywhere, two developments affected local areas. First, reformers believed that efficiency and morality could be improved by abolishing the ward system in civic elections. Ward politics had become corrupt, a tail wagging the civic dog. Reformers like Underhill ([1910–11] 1974: 328) thought that "the interests of the ward as compared with the broad interests of the whole community are comparatively petty," and should be subdued. As historian Paul Rutherford (1977: 378) puts it: "it was essential to subordinate the neighbourhoods to the city." Toronto accomplished this in 1904 by establishing a Board of Control, which took over many powers of ward councillors. Comparable initiatives were taken in other cities, notably out west. In Winnipeg, the abolition of wards became a priority because the growth of a large immigrant population threatened the power of elites who, supposedly, had wider civic interests at heart (Anderson, 1979). Rutherford flags the way civic reformers were chiefly concerned about the influence of the poorer districts, where immorality was arguably rampant and where diseases were nurtured. The influence of their residents was reined in (Weaver, 1977: 412).

Those residents were targeted in a different way. Care of the poor had been the responsibility of churches and private charities, whose volunteers worked in such areas (Rutherford, 1977: 371; Magnusson, 1981: 65). In Halifax, when the unemployed became destitute it was the "voluntary, non-profit-making agencies [that] did what they could" (Fingard, 1977). After 1900, as immigration revived, missions and settlement houses dealt with what the Reverend William Knox described as "the problem of the downtown districts and the foreign settlements in our great cities" (quoted in Irving, Parsons, and Bellamy, 1995: 68). The Canadian settlement movement, too, had a lower profile than its British and American models (James, 2001: 17). Even so, it had an impact. Inspired by J.S. Woodsworth, who had taken over Winnipeg's All People's Mission in 1907, Toronto became the most prominent, and now best-documented, centre of activity. By 1912 it boasted six settlements, including Central Neighbourhood House (James, 2001; Irving, Parsons, and Bellamy, 1995). Other projects were mounted where needed. Notably, in suburban Earlscourt the Reverend Peter Bryce made his Methodist church into a "neighbourhood hub" for British immigrants, the best-equipped of any in the country (Stubbs, 2018: 239, 253). Here, in contrast to the inner-city missions that aimed to "Canadianize" European immigrants, the goal was simply to support struggling families (ibid.). Bryce eventually became Moderator of the United Church and, as a first step in that direction, he took over as the president of Toronto's NWA in 1916 (Wills, 1995: 49).

Conditions in the poorer districts, combined with pressure from volunteers, pushed the municipality to become involved. The Neighbourhood Workers' Association had been established in 1914 at the City's request because it needed help screening relief applicants (Wills, 1995: 47). At the same time, the Toronto Playground Association and the Trades and Labour Council persuaded City Council to create a playground on Elizabeth Street, a first (Kidd, 2015). More significantly, in 1912 Council appointed a Social Service Commission, which introduced the idea of professional social work. The University of Toronto opened Canada's first school of social work in 1914, and by 1918 the NWA was employing qualified case workers (Pitsula, 1979: 40; Wills, 1995).

The City's most substantial role, however, was in the field of public health. It brought piped water and sewers to recently annexed outer districts like Earlscourt, while employing building and health inspectors in the inner neighbourhoods (Harris, 1996: 151–2). Directed by Charles Hastings, the City's Medical Officer of Health, Toronto became a leader (Bator, 1979). Hastings's approach was enlightened: a survey of six "slum" areas indicated that conditions did not reflect the choice or culture of residents (City of Toronto, 1911). Where Toronto led, other cities such as Winnipeg soon followed (City of Winnipeg, 1921). No longer through the stratagems of ward politics, municipalities were having a growing, positive impact on the poorer neighbourhoods.

But the second initiative that affected local areas put poorer districts at a relative disadvantage. When, after 1904, Toronto allowed local residents to petition to exclude non-residential land uses, a bias soon emerged. In principle, residents anywhere could have their say. Indeed, in 1909 the Earlscourt and District Ratepayers' Association asked the City to annex that working-class area; then, after annexation, it lobbied for a tax to pay for watermains, a fire hall, and improved transit ("Earlscourt ratepayers," 1909; "Ask many improvements," 1911). But Earlscourt was atypical. As Mary Clarke ([1917] 1974: 183) observed about the downtown Ward, such associations "do not exist in the district where the need for them is greatest." The key term was "ratepayer." Indirectly via their rent, tenants have always paid property taxes (rates), but this is still not widely acknowledged. Tenants could not vote in civic elections and "ratepayer" meant property owner. Indeed, it has been owner-occupiers, not landlords, who have formed and joined ratepayer associations because they see their lives as well as their investments being at stake. And so it was areas with a high proportion of homeowners, and of people who reckoned they had a right to be heard, that shaped the developing pattern of de facto zoning in Toronto. Class, as well as homeownership, was involved.

The dominance of ratepayer associations is shown in the *Globe*. They were effectively the only form of local residential organization in Toronto until the 1960s, when residents' and eventually neighbourhood associations replaced them (table 6.3). Some simply built community but their noteworthy activities were political. When not lobbying for services or infrastructure improvements they were agitating to prevent development. In Toronto in the 1890s, a city-wide association played a significant role in municipal politics. When it faded after 1900 and the Board of Control was established in 1904, local groups sprang up. Many organized at the ward scale; from the 1890s to the 1910s, these were the subject of two-thirds of all the articles about ratepayer associations. But the number of local groups jumped after 1904, once petitions were allowed. Thereafter, through the first half of the twentieth century, ratepayer associations like that in the Annex, defended the interests of homeowners in the more affluent parts of the city (Jacobs, 1971).

Toronto was typical, and not just in Ontario. In Berlin (soon renamed Kitchener), the Building Line Act of 1904 allowed residents to petition for the exclusion of non-residential land uses. As in Toronto, its purpose was to protect property values, and by 1920 it had been well used (Bloomfield, 1982). In Hamilton, Durand residents were active in submitting petitions (Elman, 2001: 10). And in the Montreal area, Westmount already had a comprehensive land-use bylaw (Fischler, 2014; 2016). The West kept pace. John Weaver (1979) has documented the demands for trams and sewers from local associations in Eburne, South Hastings, and Kitsilano, with tenants having little clout, but no area carried more influence than Shaughnessy (c.f. Hasson and Ley, 1994: 45). In time, Vancouver adopted ad hoc bylaws for "residential areas," systematizing these in 1926 (Bottomley, 1977: 234). Across the country, prospering property owners exercised control over their territory (Gunton, 1979: 163, 182). But there are few signs that they thought of their areas as neighbourhoods in the modern sense.

With growing interest in land-use regulation, the idea of urban planning began to take shape. Curiously, however, the two hardly connected (Moore, 1979). Today, planners are kept busy controlling (re)development at the neighbourhood scale, but in the 1910s this was barely on their radar. An interest in city planning had emerged in Germany and Britain around 1900 and soon spread to North America (Ward, 2002). In Canada, a federal Commission of Conservation was established in 1909, and in 1914 a prominent British planner, Thomas Adams, was brought over to become its town planning adviser (Hodge and Gordon, 2008: 87, 94–5). A small group of people began to style themselves as planners, and some provinces enacted legislation. In 1912, for example, Ontario passed a City and Suburban Planning Act (Hulchanski, 1981: 91–5). In this way, planning became a strand in the urban reform movement (Bottomley, 1977). Extravagantly, in 1913 Winnipeg's new Planning Commission claimed that "the vital role of improved housing and city planning schemes is everywhere acknowledged," but this was delusional (City of Winnipeg, 1913: 5). Like urban reform as a whole, Canada's planning movement "lagged far behind" its counterparts in Britain and the United States (Simpson, 1985: 110); in Montreal, it had little or no effect (Germain, 1984). Planners threw around big ideas – beautifying cities, shaping growth – but these were disconnected from the everyday petitions and bylaws that were actually determining land use. As a result, in Ontario, legislation typically had "little effect" (Hulchanski, 1981: 104). Planners were not thinking about residential districts, still less about neighbourhoods. One of Adams's achievements was the design of Lindenlea, Ottawa, a prototype suburb laid out on a sort of neighbourhood principle (Delaney, 1991). But when, as he was leaving Canada, he talked about the five appropriate scales for planning, neighbourhoods were not on the list (Adams, [1922] 1974).

The Interwar Years

After the trauma of war and the Spanish Flu, prosperity resumed in the 1920s and a new understanding of neighbourhoods began to take shape, initially in the suburbs. Then, when the shock of the Depression drew attention back to the inner city, a more encompassing view of housing, neighbourhoods, and neighbourhood change emerged. A great deal did not happen until after 1945, but some seeds were sown.

There was of course some continuity with the prewar years. Earlier trends took varying trajectories, some fading while others grew. Perhaps because civic administration had been reorganized before 1914, the obvious change was that "the public zeal for urban reform dissipated" (Rutherford, 1974: xxi). This included interest in planning. By mid-decade A.G. Dalzell (1926), a rare person who still called himself a planner, reckoned the movement was at a "standstill." As a past assistant to Thomas Adams, a founding member of the Town Planning Institute of Canada in 1919, and future president of the institute, he was in a good position to know, and also to see a glass half full. In fact, the movement was in retreat. Ottawa's only town planner, Noulon Cauchon, devised big schemes but by the late 1920s even the capital's Town Planning Commission was faltering (Taylor, 1986: 148). By 1929, the movement was effectively dead. Fading, it still showed no interest in neighbourhoods. Dalzell (1927) himself ignored the subject in reports he wrote on housing and land development. Even Vancouver, the city where planning ideas endured the longest, did not address the issue in a sustained way (Bottomley, 1977). At the end of the decade, it did commission a report from an influential American consultant. This discussed a range of planning issues, including transit, harbour development, zoning, public recreation, and civic art, but mentioned neighbourhoods only in passing (Harland Bartholemew, 1929). Thereafter, urban planning went into hibernation for a decade and a half.

Other prewar initiatives gained momentum. The profile of neighbourhood work rose through the 1920s and into the 1930s, helping to fix the association of "neighbourhood" with lower-income districts (table 6.2). Meanwhile, thanks to growing activity by ratepayer associations, residential districts grew in number, refined their manner of exclusion, and consolidated their influence at city hall. In the *Globe*, references to ratepayer associations ballooned sixfold between the 1900s and the 1920s (table 6.3). As ever, their concerns were with services and land use. For example, inspired by Westmount, North Toronto threatened to secede if the city did not improve its infrastructure ("North Toronto going strong for secession," 1920). In sum, the complementary growth of neighbourhood work and of ratepayer activity underlined existing contrasts between the territory of the haves and the have-nots.

The 1920s also saw the emergence of what became a significant trend, one that revived and reshaped planning. Surveying Hamilton's land registry records, Michael Doucet and John Weaver (1991: 100–1) found that as early as the 1850s some developers were imposing building restrictions, but this practice only picked up during the property boom after 1906. The same happened in Toronto (Harris, 1996: 171–4; Paterson, 1985). The new ratepayer groups were making it clear that buyers cared a lot about what was built next door. As a result, entrepreneurs not only supported land-use regulation but often pioneered it, in Canada as in the United States (Harris, 2004: 85–8, 136–41; Weaver, 1979; Weiss, 1987; c.f. van Nus, 1979: 238). Some also imposed legal covenants that defined who was not allowed to buy. The Chinese were a target on the West Coast, but elsewhere "restricted" was often code for "no Jews" (Harris, 2004: 87–9; Pask, 1989: 76; van Nus, 1998; Weaver, 1979). The most restricted subdivisions, of course, were those for the social elite. Across the country, these included Tuxedo

(1904–) in Winnipeg, the Uplands (1907–) in Victoria, Point Grey (1907–8) and Shaughnessy Heights (1908–9) in Vancouver, Lawrence Park (1907–) and Kingsway Park (1912–) in Toronto, Forest Lawn (1912–) and Mountainview (c.1912–) in Hamilton, as well as Mount Royal/Mont-Royal (1910) and Hampstead (1911) in Montreal (Doucet and Weaver, 1991: 102; Forward, 1973; Germain and Rose, 2000: 61–2; McCann, 1996, 2017; Pask, 1981; Paterson, 1985; 1989; van Nus, 1998). Notice the frequent reference to attractive physical features such as parks and heights with views. In the Uplands, the most restricted, high-value lots were clustered, with many facing the ocean (Forward, 1973; McCann, 2017). In Mount Royal (today the town of Mont-Royal), codes specified building materials, and designs were reviewed by an architectural commission (McCann, 1996: 281–3). In Westmount, houses could not cover more than 60 per cent of the lot. In Hamilton's Mountainview, no apartments, tenements, stores, or mercantile businesses were allowed (Doucet and Weaver, 1991: 102). Everywhere, single detached homes had priority.

When the economy revived after 1922, regulations became more sophisticated and – a crucial shift – planned subdivisions emerged. It is unclear how much Canadian developers were influenced by American initiatives. There, in 1908 various agents in the development industry had formed a national organization, renamed the National Association of Real Estate Boards (NAREB) in 1916. Developers came into prominence, including J.S. Nicholls, famous as the developer of the Country Club District (1906–) in Kansas City (Weiss, 1987; Worley, 1993). Speaking to the group in 1912, Nicholls observed that "in the early time (1906–08) I was afraid to suggest building restrictions; now I cannot sell a lot without them" (quoted in Weiss, 1987: 69). But what he and other American developers came to see was that the most marketable subdivisions were those that included more than just houses. Jeffrey Hornstein (2005: 62), historian of NAREB, observes that "the trade press" favoured "those who advocated planning and zoning and designed subdivisions with an eye towards permanency rather than quick profit." That was a false opposition: buyers were willing to pay more for homes in planned subdivisions, making these profitable indeed.

By the 1920s, NAREB's members thought of themselves as "community builders." Their projects included parks, schools, churches, community centres, and sometimes commercial hubs and strips. This was land-use planning, and at the neighbourhood scale.

Based on his experience as a resident of Forest Hills Gardens in New York, Clarence Perry formalized these ideas and practices in his ideal of the "neighbourhood unit" (Perry, 1929; Rohe, 2009: 210–12; Talen, 2018: 38–45). Ideally, a neighbourhood could support an elementary school, which, along with other institutions, should be situated in the centre. Small parks and playgrounds might be scattered. Shops should be at the periphery, perhaps on boundary arterials. Meanwhile, interior streets should be designed to discourage through traffic. This appeared to make sense at the time, and to this day still does for many observers. But it implied that physical form was very important, and at the same scale for everyone, young or old, single or married, rich or poor (Patricios, 2002). Since it was typically advocated at a specific scale, it also assumed a certain fertility rate and level of attendance at religious instututions.

But the idea caught on. In 1932, incoming President Roosevelt established a conference on housing issues and no less than four of its subcommittees endorsed Perry's idea (Gillette, 1983: 417). The Committee on City Planning and Zoning began its report by discussing which aspects it deemed to be "vitally important" because "the stabilization of the neighborhood through planning has the effect of stabilizing the individual home" (President's Conference on Home Building and Home Ownership, 1932: 6). In the United States, then, the development industry made neighbourhood planning part of the urban agenda. Indeed, encouraged by the industry, it was soon supported by national policy. Under the National Housing Act of 1934, NAREB members were hired into many senior positions at the new Federal Housing Administration (Weiss, 1987). Then, "because FHA could refuse to insure mortgages on properties ... in neighborhoods that were too poorly planned and therefore too 'high risk,'" reputable subdividers soon conform to FHA standards (Weiss, 1987: 148). Between 1935 and 1940, these were codified in four documents which, in Marc

Weiss's estimation, framed "the entire development pattern of modern American suburbia" (149).

Developers and the federal government in the United States were moving faster than in Canada. Here, comparable subdivision guidelines were not framed until after the Central (later Canada) Mortgage and Housing Corporation was established in 1946. But the same interwar trend was apparent. Developers began to plan as well as restrict. The best-documented example is Westdale (1920–) in Hamilton, designed by Robert Pope, a landscape architect from New York (Weaver, 1979: 414). It was laid out around a central shopping district that eventually included a movie theatre. Sub-areas were subject to varied building restrictions with ravine sites being – no surprise – the most expensive. Land was set aside for community institutions. A primary school (1927) and high school (1930) opened and, encouraged by the development syndicate, McMaster University was enticed to move from Toronto. As John Weaver (1988: 427), Westdale's historian, has commented, "with the arrival of McMaster University, a child could progress from kindergarten through university within a mile radius." This might not always have appealed to the children, but parents liked the idea and the syndicate reckoned, correctly, that this addition would be a selling point. They also aimed for cultural exclusion. Covenants prohibited the sale of properties to various ethnic minorities and only Protestant churches were encouraged. Westdale was unusual in how thoroughly it was conceived but it signalled a wider trend. Across Canada, the 1920s saw the tentative emergence of planned suburban neighbourhoods.

Another trend began to emerge in the following decade. Builders and developers had rarely erected cheap dwellings, and poorer households had always had to make their homes in buildings that had been subdivided and/or allowed to deteriorate. The economics of low-income house-building grew worse during the 1920s with increasing regulation and rising costs of construction, and then the Depression hit (Harris, 2013). The deterioration of older neighbourhoods became a public issue. This was not the first time. In the 1910s, Toronto's Bureau of Municipal Research (1918: 6) had published a study of the Ward in which they documented dire living conditions, hoping "to bring home to the citizens [i.e., property taxpayers] the real meaning of 'the Ward,' its cost in money and lost civic efficiency and the necessity of preventing the spread of such conditions." It was not until the Depression, however, that that message began to resonate. Henry Whipple Green, a shrewd practical statistician in Cleveland, made the case that slums imposed public costs while undercutting tax revenues (Navin and Associates, 1934). It was an argument designed to appeal to municipalities, and it resonated in Canada, first in Montreal and later in Vancouver (Montreal Board of Trade, 1935: 5, 22). In reference to Vancouver, Leonard Marsh (1950: ix, 23) argued that "social ills cost money,", with slums being "deficit areas." There was an economic as well as a social logic behind slum clearance and redevelopment, one which might point to the need for replanning at the neighbourhood scale.

More important in the long run, the decline of many inner-city areas nurtured a new, all-encompassing view of neighbourhood change in cities *and* suburbs. Deterioration and decline had been around for years, but a new term, "filtering," signalled its sudden prominence, while implying that all parts of the city were connected (Harris, 2012b). Decline was seen as inevitable, threatening all areas. Filtering became *the* dynamic element in the urban "housing market," a term effectively coined in the 1930s at the FHA (Harris, 2012b), although it did not appear in the *Globe and Mail* until 1948.[4] It followed that residential areas could be placed on three or four ranks, ranging from "stable," through "threatened" and "blighted," to "slum." This thinking took some years to gel, but its elements were emerging by the mid-1930s. For example, in his chapter on housing in *Social Planning for Canada* for the League for Social Reconstruction (1935), Humphrey Carver, trained in Britain as a landscape architect, threw around the new concept of "blighted areas," implying that these either needed help or, if too far gone, redevelopment. Three years later, the term first appeared in the *Globe and Mail*. W.S.B. Armstrong, a city councillor and

---

4 The term appeared once before, in 1938, in an Associated Press article referring to an American initiative.

long-time activist for housing and planning for a quarter century, noted the expansion of blight across a swath of territory. The solution, he reckoned, was "zoning and comprehensive planning," which would "give stability to development" (Armstrong, 1938).

This new thinking embraced the suburbs and the policy for new construction ushered in by the Dominion Housing Act (DHA) of 1935 (Bacher, 1993). Canada's mortgage system had weathered the Depression better than that of the United States, and the federal government was too conservative to adopt measures approaching those of Roosevelt's New Deal. Accordingly, the DHA was limited in its effect. The first DHA-financed house, in Westmount, was built of stone and occupied by an NHL star (Hulchanski, 1986: 34). This was probably a publicity stunt, but proved prophetic: the act's provisions helped finance expensive homes in nice neighbourhoods (Belec, 1997). With cooperation from developers, and influenced by the FHA's interpretation of mortgage risk, Canadian policy favoured well-regulated and planned subdivisions. The hope was that such areas might remain stable for decades, resisting, even if not preventing, eventual decline.

The view of suburban projects that emerged in the 1920s, and the new thinking about inner-city slums in the 1930s, were tentative but complementary. Linked, they pointed to the existence, in every city, of an integrated housing market that connected all neighbourhoods. They also underlined the point that regulating land use alone was not enough. There was public value in organizing land use, not only in new subdivisions but in inner-city districts that required redevelopment. In other words, a need for neighbourhood planning.

The Early Postwar Decades, 1945–1970

The new understanding of housing markets, neighbourhoods, and planning coalesced in the latter part of the war before acquiring weight after 1945. In Canada, as in the United States and Britain, wartime disruption showed the importance of central leadership, offering scope for new thinking about how to manage the peacetime economy and society. Reframed, new ideas once seen as fanciful rapidly became the new normal. Planning, increasingly linked with neighbourhoods, was one.

As David Hulchanski (1981: 243) has observed, "what was unique about the WWII period was how planning suddenly emerged as a very significant political issue long after the original town planning movement had disappeared." In 1944, as allied victory became likely, a subcommittee of the Advisory Committee on Reconstruction prepared a report entitled *Housing and Community Planning*. Known as the Curtis Report after its chair, it argued the need for a master plan in each city that would include a "neighbourhood unit plan," which, using clipped language, "indicates proposed organization of metropolitan area on an orderly, integrated and neighbourhood unit basis" (Canada Advisory Committee on Reconstruction, 1944: 175). Community planning was the watchword, and in 1946 a group of professionals formed a national association, the Community Planning Association of Canada. Humphrey Carver, who soon became vice-president of the Community Planning Association of Canada (CPAC), had British planning experience (Gordon, 2018). The following year, another group published *A National Housing Policy for Canada*, declaring that "the character of the houses in which Canadian families are raised, and the amenities of the neighbourhoods in which they grow up, will exercise a profound influence upon the coming generation of citizens" (Canadian Welfare Council, 1947: 4). Neighbourhood planning was on the agenda.

The obvious target was the suburbs. This was a period of exceptionally rapid suburban growth, and new subdivisions offered great scope for planning. This usually meant some version of Perry's neighbourhood unit: school-centred, bounded by arterials (Hodge and Gordon, 2008: 83–5). This was what almost everyone was advocating, from guru Lewis Mumford in the United States to expat planner Humphrey Carver. Mumford (1961: 570) articulated the ideal: "the principle of neighbourhood organization was to bring within walking distance all the facilities needed daily by the home and the school," something akin to the modern planner's 15-minute subdivision (see also Kuper, 1951; Mumford, 1954). In a study of Toronto's housing scene, Carver endorsed a City report which favoured neighbourhoods of 1,500–2,000 households. These could support a "well-balanced community" with a school, stores, recreation areas and possibly churches, theatres, and even some light industry (Carver, 1948: 40). In his new capacity as chair of CMHC's research committee, Carver soon promoted such ideas nationwide. In this he was helped by another British expat, Harold Spence-Sales, whose ideas about subdivision planning were published by CPAC (Spence-Sales, 1950).

These ideas mattered. The DHA had been revised by the National Housing Act of 1938, and from 1946 CMHC was tasked with implementing its provisions. At first, the agency offered loans jointly with private lenders. Then, following the Housing Act of 1954, it provided mortgage insurance, but only for properties in subdivisions that its new planning department approved (Harris, 2004: 125). Following the agency's suggested site designs, these followed neighbourhood unit principles (CMHC, 1954). Suburban development was rapidly becoming more regulated. By the early 1950s, all major provinces had passed legislation that required cities to develop master plans, and a new generation of developers and planners was sold on neighbourhood units. Calgary, for example, adopted the neighbourhood plan concept in 1953 and thereafter "almost all subdivisions built [from] the 1960s conform[ed] in some way to Clarence Perry's neighbourhood principles" (Foran, 2010: 73; Townshend, Miller, and Cook, 2020: 195). In Ontario, the Planning Act of 1946 favoured the same model and required subdivisions to be approved by the newly established Ministry of Planning and Development. This was uncontroversial: developers were already on board. In Toronto, Lawrence Manor became the first complete local example of the neighbourhood plan in action, and Humber Valley Village was arguably the best (White, 2016: 99–103). Soon Don Mills, famously but inaccurately described as Canada's first planned suburb, embraced four neighbourhood units, organized as quadrants, bisected by arterials, with a core area that included low-rise apartment buildings and a commercial centre (103–13; c.f. Sewell, 1993: 79–98).

Much the same was happening everywhere, even in Quebec, which resisted federal initiatives. In Montreal, the tradition of owning a single-family home was weak (Choko and Harris, 1990). But in the 1940s secular and religious organizations began to change this as "le rêve du bungalow en banlieue ... faire 'l'objet d'une forte campaigne d'opinion" (Choko, Collin, and Germain, 1987: 247). An early report from the city's Service de l'urbanisme (Town Planning Department) advocated low-density, self-contained subdivisions, focused on a civic and commercial centre (Ville de Montréal, 1944). In different ways, developers carried this through, for example at the Cité-jardin du Tricentaire (1940–) and then Saint-Leonard-de-Port-Maurice (1955–) (Choko, 1989; Collin, 1986; Germain and Rose, 2000: 64). These projects, and indeed Perry's original idea, drew inspiration from the Garden City movement of the early twentieth century (c.f. Carver, 1975: 117). With variations on a theme, postwar suburbs had a recognizable shape.

Neighbourhood planning was also applied to inner districts. Many were the result of "filtering down" across the "housing market," both terms that Carver (1948) used liberally in his study of Toronto in 1948. That was also when "housing market" first appeared in the *Globe and Mail* ("Too liberal mortgages condemned by Reid," 1948). Accepting that filtering was ubiquitous, Toronto's Planning Board grouped the city's 78 planning neighbourhoods into five categories: sound, vulnerable, declining, blighted, and slum, with many in the latter three categories (City of Toronto Planning Board, 1945; Lemon, 1985: 104–5). By the early 1950s, the new thinking had become familiar to *Globe and Mail* readers. There was now no need to explain what a "blighted area" was (e.g., "Blighted area," 1951).

At first, the appropriate response seemed obvious: demolish and redevelop (c.f. Harris, 2020a). The Hamilton Downtown Association argued that clearance and renewal would prevent a flight to the suburbs, and the *Spectator* "acted as [its] public relations wing" (Robick, 2011: 306). In 1943, Toronto's Master Plan wrote off much of the area south of College Street and called for "widespread demolition" (Toronto City Planning Board, 1943: n.p.). This did not happen because redevelopment on that scale was too expensive (Lewis and Hess, 2016). Instead, it was confined to specific sites, a major example being the creation of Regent Park in much of what had been Cabbagetown (Johnson and Johnson, 2017; Rose, 1958). Other cities made new neighbourhoods out of old. Regent Park was held up as a model by Paul Dozois, a Montreal city councillor and chair of an advisory committee on slum clearance. Noting that "residential sections" experience a "varying degree of deterioration," his committee's report argued for wholesale demolition of a "defective residential zone" (City of Montreal, 1954: 4; trans. Maltais). It became a public housing project, Les Habitations Jeanne-Mance (Germain and Rose, 2000: 64). The "Bulldozois report," as planner Hans Blumenfeld referred to it, proved influential, epitomizing the view that older districts must be remade (Mercure-Jolette, 2015). From the outset, it had its opponents. In Vancouver, although the prominent policy adviser Leonard Marsh argued that the Strathcona "district," being a blighted "slum," should be redeveloped as two "neighbourhood units," local opposition prevented this (Marsh, 1950: 2, 35; Loo, 2019: 158–96).

It took time for the new language of neighbourhoods to settle. There was ambiguity about the appropriate scale of the new planning units and, as *Globe and Mail* coverage indicates, the name was still not in common use (table 6.1; bar chart 6.1). As late as 1965, Toronto's Planning Board still talked about "residential areas" and even in 1975 Vancouver's survey of "local areas" did not use "neighbourhood" (City of Toronto Planning Board, 1965; City of Vancouver, 1975). But these were becoming anomalies. By the end of the 1950s, Toronto's Planning Board had divided the city into 25 "planning districts" with an average population of more than 20,000. Most had been developed long before Perry had been heard of. But, although the board used old language, what it had in mind was modern: these "districts" were replete with parks, schools, churches, shops, and community services, altogether "more than a collection of dwelling places" (City of Toronto Planning Board, 1959). The board understood that residents had local attachments, implying that redevelopment would carry social as well as economic costs. For a time, cities like Hamilton barrelled on regardless, taking as much federal money for "urban renewal" as they could get (Patterson, 1993). But others, including Toronto, were rethinking this by the mid-1960s, outlining an "improvement programme for residential areas." Toronto reckoned that, although houses in older districts had been deteriorating, they were worth keeping because "they are mostly composed of houses of a type that meets the needs of the largest part of the population, the families with children" (City of Toronto Planning Board, 1965: 1). Noting the expense of clearance and reconstruction, it concluded that "with reasonable renovation and maintenance they are good for many years to come" (ibid.). The recognition that older neighbourhoods were here to stay was pragmatic, and perhaps grudging. But then a citizen movement made it an article of faith.

The Modern Period

Across the developed world, the 1960s were transformative, most obviously in the fields of culture and politics. The political consequences were apparent locally, changing popular expectations about the role of municipal governments in general and planners in particular. Bound up with this was a new understanding, one which came to be shared by everyone involved, about the role of neighbourhoods in the daily lives of urban residents and in the governance of the city.

Many people, not only the young, were taking to the streets. They protested against the war in Vietnam, racism, poverty, landlords, and, last but not least, city hall. For decades, ratepayers had lobbied against unwanted development, but now the scale and context had changed. Through the urban renewal and public housing programs, the federal government had been sponsoring large-scale redevelopment since the 1950s, and were meeting with local resistance (Loo, 2019; c.f. Cox, 1984). The expansion of funding for public housing in 1964 increased the frequency of opposition. Then, meeting the growing demand for rental housing, privately sponsored redevelopment created an apartment boom which triggered further agitation. Older ratepayer associations had revived after 1945. References to them in the *Globe and Mail* jumped threefold between the 1940s and 1950s, remaining high into the 1960s (table 6.3). New ones, such as Hamilton's Durand, formed (Elman, 2001). Some of the new grassroots organizations included tenants, as in Trefann Court (Fraser, 1972; Vickers, 2020). By the 1970s, many associations were signalling their greater inclusivity. The number of references to "residents' associations" doubled in a decade, while those to "ratepayer associations" dropped by almost exactly the same number (table 6.3). Occasionally, tenants organized – or, as in Kingston, Ontario – were organized (Harris, 1988: 107–8). When proposed highways threatened whole districts in Vancouver, Toronto, and Ottawa, people rebelled (e.g., Taylor, 1986: 196–8). In Edmonton in 1964, only three people had objected to the planned MacKinnon Ravine freeway; seven years later, 22 groups or individuals spoke against the now-expanded proposals (Lightbody, 1983: 269). In some ways, these examples of citizen activism were all parochial, about protecting specific areas (c.f. Fisher, 1982). But their unprecedented number crossed a threshold. Many people came to see themselves as part of a national – and indeed international – movement with wider political goals (Hasson and Ley, 1994: 46–7; Shragge, 2013).

The rhetoric implied that the movement was inclusive. It inspired Sherry Arnstein's (1969) seminal article, eventually cited thousands of times, which spelled out a "ladder of citizen participation" on which any organization or activity could be located. Early reports, for example Graham Fraser's (1972) account of resistance in Trefann Court and Donald Keating's (1975) idealistic take on activism in Riverdale, implied that some groups had already climbed far up the ladder. There were bona fide victories: stopping the Spadina Expressway in 1971 was the most prominent, and for two reasons. It happened in Toronto, the media capital of English Canada, which guaranteed that this battle would reach a national audience. It also engaged the energies of an internationally prominent activist and writer, Jane Jacobs. She had made a name for herself in part as an organizer in New York City, but even more as the author of *The Death and Life of American Cities*, a critical examination of city hall politics and the ideology of modernist planning (Page and Mennell, 2011). She had moved to Toronto in 1968 as a statement against the Vietnam War but, having settled in the Annex, she immediately found herself at the centre of surging neighbourhood opposition to a freeway. She herself did not play much of an active role in the anti-Spadina Expressway movement, but her name and support counted for far more than anything she could have done (White, 2016: 411n140).

But there were snakes on the urban game board, too. As ever, homeowners like Jane Jacobs and other Spadina opponents were more active than tenants, and more influential (Einstein, Glick, and Palmer, 2019). Even when tenants won a battle they could lose the war. Urban renewal was halted in Trefann Court, but soon gentrification accomplished what city hall could not (Vickers, 2020). They were now styling themselves "residents" rather than "ratepayers," but homeowners were still calling the shots (Filion, 1999). What had changed was what they expected, and then accomplished.

Residents were pressing municipalities and planners to do things differently. "Participation" was the watchword. This was supposed to mean something closer to "citizen control" than to polite "consultation" or, certainly, "manipulation," the bottom rung on Arnstein's ladder (Arnstein, 1969). The latter had been the norm. Into the 1960s, Vancouver in effect had a council-manager government whose elected officials had limited power, the ideal of turn-of-the-century urban reformers. Citizen input was sought, but only as a "communications strategy" (Loo, 2019: 173). Indeed, even this minimal gesture was often absent. Consultant Frank Lewinburg (1984: 19) reported that in Toronto until the late 1960s "planners went about their business without consulting residents and presented their final proposals at a public meeting." They saw themselves as experts, acting for the city and, where necessary, overriding local, parochial interests; meanwhile, residents now saw themselves as being concerned, not only with those interests but with the state of civic democracy (Grant, 1994). Their pressure brought changes that affected all neighbourhoods. In Edmonton, grassroots action caused the city to rewrite its transportation bylaw (Lightbody, 1983). In Hamilton, neighbourhood groups won the right to help review and develop policy (Elman, 2001: 23). In Winnipeg, the creation of an amalgamated Unicity came with Resident Advisory Groups (RAGs) which – supposedly – would help shape metro-wide plans (Kiernan and Walker, 1983). As Warren Magnusson (1983a: 34) put it, the political line of division lay "between the defenders of the neighbourhoods ... and the supporters of traditional urban development." It was an old struggle, one which the urban reform movement had tried to settle one way and which was now being rejoined.

The balance of power shifted to (some of) the neighbourhoods. In Ottawa, a reform council was elected in 1974. In Hamilton, the planning commissioner observed – ruefully? – that "it certainly appears there is no way of stopping the citizen involvement in planning" (Elman, 2001: 16). Under local pressures, Halifax developed its first detailed plan for the South End and Peninsula Centre (Ley, 1996: 242). But nowhere did populist reformers have more success than in Toronto (Magnusson, 1981: 81; 1983b). By 1971, the Planning Board was holding neighbourhood meetings to solicit input (Webster, 1970). Two years later, a new Inner Neighbourhoods Division opened 14 site offices. Many of the new local planners had little professional training; they were paid to listen to residents. As Richard White (2016: 320), the historian of Toronto's postwar planning, says, its "guiding principle was that neighbourhoods should decide for themselves what was best." By 1980 the assumption was that participation was always good, that planning should be done locally to address local problems, and that "the existing urban fabric should not be destroyed unless absolutely necessary" (366). Lewinburg (1984: 15) reckoned that activists had "revolutionalised city planning." As the *Globe and Mail*'s coverage shows, and in a way that was quite new, they had put *all* city neighbourhoods on the mental map of urban residents.

Versions of these changes happened in other cities, although later in Montreal. There, almost continuously from 1954 to 1986, municipal policy and planning was centralized under the guidance – more like rule – of Mayor Drapeau. There was a grassroots movement, rooted in working-class neighbourhoods, which laid the groundwork for the province's Quiet Revolution, but at first the effects on civic politics were limited (Fahrni, 2005). By the 1960s there were neighbourhood plans, but they imposed standardized zoning regulations everywhere (Kaplan, 1982: 449). Neighbourhoods were seen as an obstacle to grand projects, including the building of the 1976 Olympic stadium (Legault, 2002), which of course became a fiscal fiasco. Organized opposition grew in the 1970s. Initially it was localized, as when residents in a neighbourhood just north of downtown Montreal banded together to prevent redevelopment in what became "Canada's largest citizen-developer confrontation" (Helman, 1987). A city-wide movement, the Montreal Citizens' Movement (MCM), arose but it was not yet able to win power (Higgins, 1977: 232; Sancton, 1983: 73–5). Initially, this did not affect neighbourhood planning as in many English-Canadian cities. This was testimony to Drapeau's power, and also reflected the fact that Montreal had not been aggressive in promoting urban renewal. Individual projects, notably Milton-Parc and in Little Burgundy, had triggered resistance, but at first the effects remained local (Germain and Rose, 2000: 84–90). It was not until after 1978, thanks to a "localist turn" in civic government led by Yvon Lamarre, Chair of the Executive Council, that local residents and their organizations were given a meaningful role in municipal governance (Léveillée, 1988). Then, when an MCM slate was elected in 1986, some powers were decentralized to the *quartiers.*

Increasingly from the 1970s, the neighbourhood movement and its effects on the planning process were influenced by gentrification. This development expressed the new cultural values of a segment of the middle class who, like Jacobs, appreciated the aesthetic and walkability of homes in inner-city neighbourhoods, and who incidentally took advantage of their often-depreciated market value. As David Ley puts it, some saw the suburbs as "too standardized, too homogeneous, too bland, too conformist, too hierarchical, too conservative, too patriarchal, too straight" (Ley, 1996: 206). Lately, gentrification has affected a few denser, inner suburbs, such as Hochelaga-Maisonneuve in Montreal (Maltais, 2023), but it has always been heavily concentrated in the inner city. The revival and social transformation of inner-city neighbourhoods soon demonstrated that filtering and neighbourhood decline were not inevitable. This, combined with local resistance to the bulldozer, caused the federal government to rethink its urban policies. In the 1960s some cities, notably Toronto and Montreal, had played with the idea of improving rather than flattening residential areas, but only in a half-hearted way; they were going against expert opinion

(City of Toronto Planning Board, 1965; Germain and Rose, 2000: 184). With gentrification, however, and the gathering opposition to urban renewal, the federal government put these programs on hold. Instead, in 1974 it launched a Neighbourhood Improvement Program (NIP), hoping to stabilize some older residential areas. The task of identifying NIP areas was left to municipalities, in consultation with local residents. Moreover, once an area had been chosen, residents got to decide how the federal money would be spent. The program played out differently in each city. Some municipalities, notably Hamilton, showed little interest. In Waterloo, where gentrification had not yet begun, it probably had little effect (Bunting, 1987). In Kingston, its impact was limited because it came after the main surge of interest in grassroots politics (Harris, 1988: 135–7). In Ottawa, however, it contributed to displacement, and in inner-city Halifax it "set some neighbourhoods ... on the path of improvement and gentrification" (Weston, 1982: 16–17; Grant and Ramos, 2020: 175). In Drapeau's Montreal, which was growing slowly, the impact was limited by the centralized administration and because some of the NIP areas consisted mainly of tenants, although renovations and preservation were soon promoted independently by Phyllis Lambert and Heritage Montreal. In Toronto, coinciding with a reform council, the NIP program was actively implemented (Filion, 1988). The first NIP area, Niagara, was also first to get a planning office. In several cases the City's selection of NIP areas contributed to the process of gentrification that was bringing middle-class homeowners into areas such as Cabbagetown. They were the leading edge of the neighbourhood movement.

Vancouver turned out to be the city where the effects have been most visible (c.f. Hasson and Ley, 1994). In 1972, The Electors' Action Movement (TEAM) came to power at city hall. The planning department was reorganized, adding a section on "local area planning." It produced a report endorsing citizen participation and local committees. The city was divided into 21 areas, with each supposedly getting a local planner (Horsman and Raynor, 1978). TEAM hired a new planning director, Ray Spaxman, who for 15 years promoted neighbourhoods and "neighbourliness"

(Ley, Hiebert and Pratt, 1992: 262). The local area focus was re-emphasized by Larry Beasley, co-director 1994–2006, who articulated a planning philosophy of "Vancouverism." According to Beasley (2019: 107), this included the notion that "the city is composed of a constellation of well-established neighbourhoods that residents feel a close affinity with and work within to build networks of mutual support." He saw this arrangement as partly "natural," and partly the result of policy: "consumer preferences and public policy have closely mirrored one another" (108). The combination has worked. A recent study concluded that of all Canadian cities Vancouver has been most successful in creating a walkable environment (Bozikovic, Castaldo, and Webb, 2020).

Behind the rhetoric there was a different reality. Winnipeg's RAGs never got off the ground and were later consolidated (Higgins, 1986: 207–12). Toronto's reform council, along with Mayor John Sewell, was defeated after one term. Even in Vancouver, in the end only 4 of the 21 local areas received a local planner (Horsman and Raynor, 1978). In the 1980s, as development pressures grew, neighbourhood defences were becoming "overrun" (Ley, Hiebert, and Pratt, 1992: 262; Ley, 1996: 245). By the 1990s, Hasson and Ley (1994: 47) discerned the emergence in Vancouver of a "neocorporate" model of governance involving "cooperation between the local state and neighbourhood organizations." Many saw this as a return to the top-down approach of the early postwar decades, in Vancouver as elsewhere. Some critics, such as Tina Loo (2019), believe that structured participation has enabled local governments to retain and even enhance their powers. Residents may think that they have a say, but they are still being manipulated, now more subtly.

There is some truth to this, but the case can be overstated. Lately, a planning mantra has been intensification. Montreal, through Operation 10000 logements, was one of the first to promote this (Ville de Montréal, 1979). There are several arguments in its favour: higher densities mean more efficient use of infrastructure, notably transit; they reduce the environmental footprint of cities; they promote walkability, which in turn benefits public health. Not surprisingly, it is the younger, lower-density suburbs that are the least walkable (Soderstrom, 2008; Turcotte, 2009). That is one of the reasons why the American writer-activists Alan Mallach and Todd Swanstrom (2023) are so passionate in making the case for "good," densely settled, neighbourhoods. Many researchers and agencies have undertaken to document and encourage such ideas, as Toronto Public Health (2012) did in promoting "the walkable city." Although the pressures for redevelopment are strongest in Toronto and Vancouver, planners in smaller centres such as Halifax share the same goals (Grant and Gregory, 2016). As we have seen, property owners have long resisted redevelopment at higher densities but, as municipalities began to throw their weight behind the process, resistance became common enough to merit a new term, "NIMBY" – not-in-my-backyard. Emerging in the 1980s, it has appeared with increasing frequency in the *Globe and Mail*. Now a cliché – to the extent that its opponents do not need to explain "YIMBY" – it speaks of the frustrations of planners and the power of neighbourhood residents. Or perhaps I should say, the power of residents in particular neighbourhoods. In 2023, the City of Vancouver proposed a policy that would allow buildings of four to six units in all areas that had been zoned for single-family dwellings. All areas, that is, except Shaughnessy (Gold, 2023). As I write, in 2024, some provincial governments are trying to encourage – some would say coerce – all major municipalities into general rezonings. Only time will tell whether this happens, and whether it makes much difference to residential densities and housing affordability.

Municipalities now acknowledge the significance of neighbourhoods in another way. In the 1950s and 1960s, they assumed that the best way to deal with rundown areas was to clear them; by the 1970s they were more inclined to encourage upgrading. Since then, in Canada as elsewhere, their emphasis has shifted from the built to the social environment (Harris, 2020a: 17–21). To some extent this reflects a change in who the planners are. For decades, although women were more affected by their home environments, and had been active in ratepayer and residents' groups, few had had much role in shaping the built environment or, when the profession emerged after 1945, become planners (Flanagan, 2018; Hendler and Markovich, 2017). As more entered that profession from the 1970s, planners were better equipped to handle social concerns. But the shift in emphasis has been driven primarily by external forces: the main concern is now "concentrated poverty." In Toronto, two clusters of shootings in 2003 and 2005 highlighted the issue (Horak and Moore, 2015: 196, 198). Established in cooperation with the United Way, a Strong Neighbourhoods Task Force identified nine disadvantaged neighbourhoods with "few local services and supports" (City of Toronto, 2005). Another four were soon added (Horak and Moore, 2015: 199). They were all concentrated in the inner suburbs, areas that had received most of the public housing projects because the land had been cheap, and scattered development guaranteed less opposition (McMahon, 1990: 57). Within two decades, a number were plagued with crime and other social problems, and concentrations of poverty were apparent (Social Planning Council of Metropolitan Toronto, 1979; Cowen and Parlette, 2011; Stapleton, Murphy and Xing, 2012; Polanyi et al., 2014). These areas were settled by low-income, first-generation immigrants but had few newcomer services (Lo, 2011). Toronto was not the only metro area to see these developments. In 2014, a study of eight major metros showed that inner suburbs were the only zone (out of five) that showed a consistent decline in relative incomes (Pavlic and Qian, 2014). A later collection of case studies added detail to, while updating, this picture (Grant, Walks, and Ramos, 2020).

Despite this trend, some inner cities remained a concern. In Winnipeg, the Canadian Centre for Policy Alternatives (2005) argued the need to help three areas with many First Nations residents. In Hamilton, a Neighbourhood Action Strategy designated 10 neighbourhoods in decline, overwhelmingly concentrated in the lower city (Pothier et al., 2019). But regardless of the specific geography, an area-focused strategy was pursued everywhere. In Vancouver, CityPlan became a "framework for deciding city programs, priorities and actions" (Murray, 2011). It proposed "neighbourhood centres" to serve and create community in problem areas, while the Vancouver Agreement coordinated local government, businesses, and community organizations in "areas marked by high levels of disadvantage," and in particular the Downtown Eastside (ibid.: 32).

Montreal took a different approach. In terms of population and wealth, the city had been losing ground to Toronto for decades, and many inner-city neighbourhoods were declining. Continuing uncertainty about Quebec separatism did not help, culminating in provincial referenda in 1980 and 1995. Neighbourhood improvement came to be seen as part of a more general effort to promote the city's economic revival. The program involved development corporations and strengthened neighbourhoods through the "dynamism of commercial streets" (Morin, 1998; Ville de Montréal, 1992: 20, trans. Maltais). Social and cultural considerations were included. In the 1990s, the province rethought immigrant settlement, giving centres a local base, and the city targeted neighbourhoods (Germain, 2013: 5; Dansereau and Germain, 2002: 17). A new cultural plan declared that "citizen participation in cultural life starts in the neighbourhood," while public transportation was aimed at "restoring the appropriate quality of life to Montreal's residential neigbourhoods" (Ville de Montreal, 2005: 23; Ville de Montreal, 2008:42, trans. Maltais). Simultaneously, following a convoluted process of amalgamation (2000) and partial de-amalgamation (2006) with other municipalities on Montreal Island, powers were passed to the boroughs, each with an advisory committee to organize citizen input.

Everywhere, the results of such programs have been criticized, and from various angles. Often, the initiatives created an incoherent "patchwork of place-based frameworks for action on poverty," undermining their effectiveness (Graham, Procyk, and Laflèche, 2020: 244). Typically, in Toronto planners used census data and planning zones to identify target areas so that the results were "not recognized as 'neighbourhoods' by residents" (Horak and Moore, 2015: 199). Despite sustained effort, Hamilton planners have had little success in getting working-class residents and minorities to become engaged (Pothier et al., 2019). The language of slums is no longer used; instead, areas are "disadvantaged," "challenged," or "left behind" (c.f. All-Party Parliamentary Group, 2023). But the very act of designating them that way creates, or more probably reinforces, stigma. It also fosters "an illusion of bounded regions in which social 'problems' are concentrated and emerge from local causation," potentially directing attention away from systemic issues (Leslie and Hunt, 2013: 1,176). In fact, most low-income and disadvantaged households do not live in the designated areas, while those areas typically contain many middle-income families (c.f. Stanger-Ross and Stanger-Ross, 2012; Séguin, Apparacio, and Riva, 2012). The logic of targeting these areas depends on the presence of neighbourhood effects, some of which become apparent only when concentrated disadvantage crosses a threshold (Andersson and Musterd, 2005; Turok, 2004; Harris, 2020a). The case is not open and shut, neither in principle nor in its implementation.

Reflections and Prospects

Cities, and planners, had come a long way from the early twentieth century. Municipalities were now regulating and shaping the built environment, in cities and suburbs. With assistance from upper levels of government, they had absorbed and expanded much of the work of charities and volunteers. They were now responsible for mitigating social problems in specific areas. Everywhere, they were expected to listen to the concerns of local residents and, where necessary, to act. Those concerns included unwanted development which, as ever, entailed balancing local and city-wide concerns. But they now embraced a range of services and supports, tangible and intangible. Municipalities had become responsible for neighbourhoods, not just "residential districts," and neighbourhoods mattered.

The changes affected all cities but were most apparent in the major centres. Vancouver's neighbourhood philosophy, framed within the "Vancouverism" brand, has already been mentioned (Beasley, 2019). In Montreal, by 1992 both the City's planners and the mayor spoke about the city as "a federation of villages," a cosy vision designed as "a strategy to counter the outflow to the suburbs" (Germain and Rose, 2000: 186, 188). It appealed to gentrifiers. Toronto had no need to make the appeal; rather, it eyed an international audience. Richard White (2016: 366) reckons that by 1980 neighbourhoods had become its "defining feature." Stephen

McLaughlin (1984: 4), the city's commissioner of planning and development, declared that "neighbourhoods" – not its financial and cultural institutions – "are Toronto's great strengths." In the same year, Marjorie Harris (1984) published *Toronto: The City of Neighbourhoods*. Soon the draft official plan reckoned that the city's "long-standing reputation" relied on its neighbourhoods, insisting that "this will continue to be at the heart of the Plan's vision for the future" (City of Toronto, 1992). Significantly, the 1990s saw a major uptick in references to neighbourhood in the *Globe and Mail*, a trend which was sustained well into the 2000s (table 6.1; bar chart 6.1). Maintaining and strengthening Toronto's neighbourhoods had become a central municipal goal.

As she helped to launch the city's Strong Neighbourhoods program in the new millennium, Frances Lankin, president and CEO of the United Way, declared that "healthy neighbourhoods are the hallmark of Toronto's civic success" (City of Toronto, 2005: 5). To be sure, this was mostly a rhetorical flourish. Real power lay, as it always had, in the hands of developers and, at another level, Bay Street financiers. But neighbourhoods were on the civic agenda as they had never been before, and residents were catching on. In the 1980s, only one in every three hundred references to local associations indicated that "neighbourhood" was in the organization's name (table 6.3). By the 2000s the proportion had risen to a seventh and a decade later it was a quarter. And everyone was talking about neighbourhoods (table 6.1; bar chart 6.1). The obvious question is, why?

It is not surprising that, over the past century, municipalities have become more involved in the management of all parts of the city, neighbourhoods included. Cities have grown enormously, multiplying the challenges of governance, including land-use regulation and transportation planning. More generally, the Canadian state now plays a much larger role in the lives of all Canadians. But those trends do not explain why popular interest in neighbourhoods has peaked twice, first a century ago and then now. It does not explain the form that civic action has taken, notably with its expanded notion of what residential areas consist of, or the pedestal on which neighbourhoods are now placed. And here lies a paradox. For decades, indeed since at least the 1920s, observers have deplored the demise of local community in the modern city, and yet those communities are celebrated as never before. In his introduction to Marjorie Harris's book, Harold Town (1984) observed that "through the adhesive of cities, neighbourhoods become blocks in the foundation of a nation." A little overdone, perhaps. But clearly, to make sense of what has been happening locally we need to look beyond the city, to wider trends in Canadian society.

chapter seven

The Changing Context

... the right home in the right neighborhood has come to stand for so many things that it almost counts as the answer to the meaning of life. It is an exercise in self-branding, a means of educating the children ... and, let's not forget, an investment strategy and retirement plan.

Matthew Stewart, *The 9.9 Percent*

Death and taxes: perhaps *debt* and taxes would be more to the point.

Avi Friedman and David Krawitz, *Peeking through the Keyhole*

Canadian society has changed enormously since the 1880s. We are still a liberal democracy based on a private, capitalist economy. We are still a mixture of peoples: First Nations, Métis, immigrant settlers, mostly white and English- or French-speaking, but always more than just that. We are still self-consciously a nation in the making, open to change. And important things have changed. The role of women – in politics, at work, and at home has changed. There has been a steady influx of newcomers. Canada is demographically more diverse. Indigenous peoples have moved in greater numbers to the cities, while their rights and experiences are slowly being recognized. Collectively, we are more prosperous, better educated, and far more urban. All of this, and more, is reflected and reproduced through the changing character and role of urban neighbourhoods.

Two changes to that role demand attention. The first are cyclical waves, peaking a century ago and then again today. Both peaks have been defined partly by a concern for poor immigrants, which offers a broad clue to their cause. The second is the growing prominence of neighbourhoods in urban life, which has carried the latest wave to unprecedented heights. This is both more difficult to explain and paradoxical. For decades, observers have bemoaned the demise of neighbourhood community, so why are we talking about it more than ever?

Immigrants and Inequality

The cycles make sense. Since becoming a country, Canada has been a nation of immigrants, but they came in waves, and not all looking the same. Since the 1880s, two waves stand out. The first was stunning, coming after the depressed 1890s when there was a net outflow of migrants, mostly to the United States. Around 1900, there was a sudden shift and by 1912 annual arrivals reached the unprecedented, just recently exceeded peak of 400,000 (Keyfitz, 1961; Boyd and Vickers, 2000). Faltering during World War I, the surge returned, ensuring that immigrants made up more than a fifth of the Canadian population until the Depression. Half came from the British Isles, almost a fifth from the United States, and the balance mostly from Europe. There were job postings that stated "No English need apply," but it was mostly the Southern and Eastern Europeans that concerned the WASP majority. Hence the increase in settlement houses and neighbourhood workers.

Immigration slowed during the 1920s (Boyd and Vickers, 2000). In 1919, regulations barred European "enemy aliens"; in 1923 the Chinese Immigration Act presented another barrier. Then, the Depression threw everything into reverse, as outmigration returned. Only a high birth rate kept the population growing. The proportion of immigrants fell throughout the 1930s and early 1940s, reaching a low of 15 per cent by 1951. Most who had arrived a generation or more earlier had become Canadian citizens, prospering in the 1920s, struggling in the 1930s; in the 1940s, they, or their children, joined the war effort. Many dispersed from the first enclaves of settlement. Immigrant neighbourhoods had ceased to be prominent, and hence a concern.

And then things changed again after 1945. New waves of immigration arose: Italians, Greeks, and more Eastern Europeans. Then, legislation in 1962 and 1967 removed restrictions by national origin, creating instead a points-based system (Boyd and Vickers, 2000). Annual arrivals grew to a quarter of a million or more, stabilizing at that level from the 1990s. By 2011, first-generation immigrants were again a fifth of all Canadians. National origins became ever more diverse. By 2011, arrivals from Britain, Europe, and the United States made up only one-third of the total while 45 per cent were now from Asia. The balance came almost equally from sub-Saharan Africa, North Africa, Latin America, and the Caribbean.

There were some new developments, but familiar problems recurred. A significant minority of postwar immigrants came with skills and savings. The new ethnic enclaves contained a wider range of incomes than those of a century ago, and so many found their way into, or indeed first settled in, "ethnoburbs" (Qadeer, Agrawal, and Lovell, 2010). With marketable skills, many settled in middle-class outer suburbs (Townshend and Murdie, 2020). But, as always, many were poor. Increasingly, they ended up on Vancouver's inner "edge of the suburbs" (Teixeira, 2014), or what in Toronto is now called the inner suburbs. This is typical. In Hamilton in 2016, for example, the census tract with the highest proportion of first-generation immigrants embraced a cluster of 1970s high-rises in suburban Stoney Creek (Harris, 2020b). There, immigrants experience the same sorts of challenges that their predecessors did, several generations earlier: crowding, ill-health, and insecure, poorly paid work. In some respects, they are better off, with public health care for example, but in other respects worse. Inner suburbs commonly lack immigrant services, including language training, and, for those without a car, getting to work, or even to a store, is a challenge. They must make do as best they can. Paul Hess (2019: 301) has shown that those suburbanites without cars, "who dwell between the subdivision and the arterial," have to "forge routes across fence lines, behind strip malls, use private driveway systems as major walking routes, and cross major arterial roadways where there are no traffic signals or crosswalks." Inconvenience is one thing; danger another. Such problems were noted by the Social Planning Council of Metropolitan Toronto as early as 1979, and have grown worse (Lo, 2011). Today, the residents of Toronto's inner suburbs have lower rates of participation in the labour force, and those with jobs spend longer getting to them (Allen and Farber, 2020). The reappearance of immigrant enclaves helps to explain the revived concern for certain types of neighbourhoods.

Racism, always present, has found new targets, while cultural concerns have grown with the appearance of new types of enclaves, some defined by public housing projects. Once mostly the Chinese and Jews were targeted. Now it is Black people, Muslims, and visible minorities as a whole. But the concerns about enclaves during both of the big waves of immigration have involved more than racism, or cultural disquiet. An even more important type of wave has driven this, one with the same pair of peaks.

Income Inequality

On 24 June 2015, Craig Heron spoke at the Hamilton launch of his *Lunch-Bucket Lives*, the most detailed and wide-ranging account of the experience of workers in any Canadian city in the early twentieth century.[1] For those who wondered about the relevance of such historical works Heron spelled out the parallels between the 1910s and the 2010s. Both are periods of precarious work when large corporations shun unions, of families struggling to make ends meet, of consumerist advertising that prods people to buy happiness in the form of products they can't afford, and of tensions between white, English-speaking workers and thousands of newcomers. There are, he suggested, many parallels between past and present.

The most telling parallel, which underlies many of the others, is income inequality. This peaked in the early twentieth century, fell until the 1970s, and has lately risen again (Alvaredo et al., 2013: 6; see also Green, Riddell, and St. Hilaire, 2016; Piketty, 2014). Across

---

1 Speech at the Workers' Arts and Heritage Centre, Hamilton, 24 June. See Heron (2015). I am grateful to Craig Heron for sharing the written version. I have paraphrased his comments.

North America, the late nineteenth and early twentieth centuries saw rampant inequality. This fuelled the urban reform movement and the rise of volunteer activism. Historical data for the United States reach back into the late 1800s, showing how income inequality rose steadily to a peak in the 1920s, when the top one per cent received a quarter of all personal incomes. Canadian data are only available since the 1920s, but they show that the Canadian one-per cent's proportion had reached 18 per cent, dipped slightly, and rose to that level again during the Depression (Heisz, 2015). It took a war to change things. Inequality fell precipitously – or, to put it more positively, social equality boomed – during the 1940s and through the early postwar decades. By the 1970s, in Canada the one per cent's share had fallen to 10 per cent. Since then it has risen, reaching 15 per cent by 2010 (Osberg, 2018; c.f. "Topsy Turvy," 2023). Canada has almost always been a more equal society than the United States, but its cycle of inequality has tracked that of its southern neighbour.

These cycles have been reflected in urban space: the rise, fall, and resurgence of social segregation, and its flipside, the variable polarization of neighbourhoods. As yet, we can't say anything definitive about trends in segregation in the 1800s and early 1900s, especially because they are complicated by big changes in urban size. All we know is that contemporaries were struck that areas of low-income settlement were emerging. The past half century is a different story. Census data show that, especially since the 1990s, growing inequality has found geographical expression; indeed, neighbourhood inequality has grown more rapidly than that between individuals or households (Bourne and Hulchanski, 2020: 28; Chen, Myles, and Picot, 2012; Walks, 2010, 2020). Neighbourhoods have polarized. David Hulchanski dramatized this by identifying the "three cities" of Toronto: poor, middle class, and upper-income neighbourhoods. In the 1970s, most tract-scale neighbourhoods fell into the middle category, but since then they have been eroded from both sides (Hulchanski, 2010). In different degrees, the same has happened for most major Canadian cities (Grant, Walks, and Ramos, 2020). There are close parallels in the United States (Bischoff and Reardon, 2013). In both countries, as social disparities have grown, so have the contrasts between neighbourhoods.

The cycles of immigration and inequality have tracked each other quite closely. A century ago, immigrants tried to make do and get ahead in a very unequal society, and the same is true today. In 1980, across seven major Canadian urban areas, the income of recent immigrants (arrivals within the previous five years) was 77 per cent of that of other Canadians (Walks, 2020: 179). By 2015 the figure had fallen to 50 per cent. The trend for all visible minorities – 92 per cent falling to 66 per cent – was the same. This shaped how newcomers and minorities lived. Until the 1970s, first-generation immigrants in Toronto, Montreal, and Vancouver were more likely to own their own home than were native-born Canadians (Haan, 2005). This reflected the long-standing, almost stereotypical, desire of immigrants to build security and put down roots, often by living with extended family. By 2001, however, even with such strategies, they were lagging. They have been drawn into another type of polarization, that between owners and tenants. In 1975, the average income of tenant households was two-thirds that of homeowners (Walks, 2020: 186). By 2015 it was barely half. Given that some recent immigrants have brought substantial capital and settled comfortably, a very large number have obviously faced major challenges. It was the combination of great inequality and high rates of immigration which created the "foreign slums" of the early twentieth century and which defines the "disadvantaged neighbourhoods" of the early twenty-first. Plus ça change.

The Decline of Neighbourhood Community

But we have not just reinvented the past; major trends, including secularization, have been reshaping our neighbourhoods, too. For over a century, across the Western world, writers have bemoaned the loss of community in urban areas (Bender, 1978; Putnam, 2000: 24–7). The impact has been most obvious at the local scale. A hundred years ago, the influential urban sociologist Ernest Burgess (1925: 154) commented: "the social forces of city life seem, from our studies, to be destroying the city neighbourhood." Others agreed. Sidney Dillick (1953: 68) reckoned that by the 1920s "many sociologists" had come to believe that "the local community as a unit of social organization was booked for discard," a view also expressed in Canada. At an early postwar conference in Toronto, Hugo Wolter (1948: 28–30) stated simply, "The geographic unity of neighbourhoods no longer exists." That is why, as noted in chapter 2, when sociologists studied working-class neighbourhoods in the 1950s and 1960s, they adopted a declensionist narrative (Garrioch and Peel, 2006; Topalov, 2003), while historians emphasized how different things had been in the past (e.g., Von Hoffman, 1994).

The reasons seemed obvious. Historically, churches had played a vital role in local communities but, for more than a century, observers have been talking about – and in many cases deploring – the secularizing influence of city living. Until at least the Depression, the point can be overstated. Using the examples of the Pointe-Saint-Charles Congregational church in Montreal and the Barton Street Methodist church in Hamilton, Michael Gauvreau (2006) has argued that in the early twentieth century Protestant churches responded effectively to immigrant settlement. Central funding enabled self-governing congregations to play a significant role in those working-class communities. As noted earlier, Catholic parishes in urban Quebec were even more prominent locally, and were indispensable to neighbourhood life into the early postwar years. Indeed, especially in Ontario, in the 1950s and 1960s the arrival of Southern European immigrants rejuvenated, and created new, Catholic congregations, while later immigrants from other parts of the world have founded synagogues, mosques, and temples that play a similar local role.

But these postwar developments have been happening with a different backdrop. Christian churches, the main institutions in the early 1960s and still nominally important, have mostly been fading. The most striking decline has occurred in Quebec since the late 1960s. The Catholic Church conceded its powers over education to the government, and from 1965 the civil registration of marriages became possible, and then popular. As a result, "the idea of the parish as a community centre lost its purpose" (Christie and

Gauvreau, 2010: 192). Nationwide attendance in most Christian denominations, always lower in Canada than in the United States, has dropped steadily. And overall, since 1960 the proportion of those with no religious affiliation has risen dramatically from 4 per cent to 35 per cent ("Religion in Canada," 2022). In specific immigrant communities, religious institutions embody and sustain a neighbourhood spirit, but elsewhere their role has become marginal.

As Suzanne Keller (1968: 58) noted, by the 1960s rising prosperity, growing everyday mobility, and access to a widening array of media meant that people needed the local neighbourhood less, whether physically, socially, or psychologically. A generation later, Claude Fischer (1991: 85) pointed out that, as more women went out to work, there were fewer people around to maintain local networks. Since then, most people have become even more mobile, finding, building, and maintaining community over great distances. For a century, the erosion of neighbourhood life has seemed inevitable (Cockayne, 2012).

This is no myth. Barrett Lee and his colleagues compared the activities of neighbourhood organizations in Seattle in 1929 and 1979. They found that political concerns, notably over land use, remained important but that social functions had waned: by 1979, organizations met less frequently and organized fewer social events (Lee et al., 1984). The decline seems to have been steady. In a study of one neighbourhood in Rochester, New York, Donald Foley (1952) found that neighbouring and "sense of community" in the early 1950s was weakening. Then, a national survey of household behaviour found that socialization within the neighbourhood declined in both absolute and relative terms between 1974 and 1996 (Guest and Wierzbicki, 1999). A vital function of social activities is that of finding a mate. For this, lately, the internet has become more important, not only in relation to neighbourhood and school, but also family, workplace, and friends (Rosenfeld and Thomas, 2012). Drawing on this sort of research, and presenting his own, Robert Putnam's *Bowling Alone* (2000) has famously documented the decline of community in the United States. In *The Lonely American* Jacqueline Olds and Richard Schwartz (2009: 21) quote a friend's pithy observation: "Being neighborly used to mean visiting people. Now being nice to your neighbors means not bothering them." This had long been middle-class practice, or at least one version of it, and it became an approach which working-class households could adopt in the postwar decades of rising prosperity. Using the terms coined by Peter Mann in 1954, neighbourliness is now more latent than manifest.

Similar forces have been at work in Canada, with broadly similar results. Barry Wellman, a sociologist at the University of Toronto who is widely recognized as a student of neighbouring, has shown that the internet can build local connections (Hampton and Wellman, 2003). Today, Facebook and NextDoor underline the point. But the internet also fosters long-distance linkages, and the most important social ties are not local (Wellman, 1979). This is a theme that many have picked up on, including Brian Bethune (2014) in an article for *Maclean's*, entitled "The End of Neighbours." One indicator has been the decline of block parent programs. These started in London, Ontario, in 1968 but lately numbers have declined and many local groups have disbanded. After all, in many blocks there are few if any parents present during the day. A spokesman for Street Smart Kids has said, "This is no longer a world where neighbours can rely on neighbours to keep a watch over every kid on the street," adding, "I'm nostalgic for that way of life, but I realize that it's gone" (McGinn, 2013b). In a similar vein, the executive director of the Missing Children Society of Canada commented that "people are looking to be a connected community in a very different way" (ibid.).

People, developers, and planners have fought this trend in various ways. One has been through physical design. Almost a century ago, Clarence Perry hoped that neighbourhood units would encourage the "reappearance of the local community, differing from the village prototype in the absence of occupational basis" (Perry, 1926: 141). Many planners and urbanists believed the same. Writing at a time when developers were putting Perry's idea into practice around every North American city, Lewis Mumford (1961: 571) suggested that "Clarence Perry had in effect restored, with modern ideas and modern facilities … one of the oldest components of the city, the quarter." But he then conceded that the ideal had not been realized.

Perry's model design went international. So has the idea of investing in an imaginary reality, hoping that it might come true. In marketing subdivisions, developers and real estate agents have long been in that business; the same has become true for inner-city projects (Martindale, 1977). In Vancouver, for example, Caroline Mills (1993) found that advertisements for redevelopment in Fairview Slopes highlighted the "urbane" lifestyle, which is what Leslie Kern (2010) found in the promotion of Toronto's downtown condos. A variant, inspired by Manhattan's SoHo district, has been the mythology of loft living. In Montreal it has been spread through newspapers, magazines, and films as well as advertising (Podmore, 1998). How closely this ideal corresponds with reality has always been a question, even in SoHo. But the image has been powerful enough that, from the 1990s, it encouraged a business association to try rebranding a warehouse district in Regina. It has met with challenges (Mathews, 2019).

But increasingly, in loft districts as elsewhere, residents have thrown their weight behind the project of image-making. Thirty years after Donald Foley's study, Albert Hunter went back to the same Rochester neighbourhood. As expected, he found that people relied less on local facilities, travelling further to work and shop. But he also found that more residents were now identifying with the area by consciously creating an "ideological community" (Hunter, 1975). It was becoming gentrified, and those who have studied similar districts have found much the same. In four districts of Manchester, England, researchers found what they call "elective belonging." What mattered most to local residents was the image, not the reality: "rather than a concern with the quality and nature of local ties and personal relationships, it is th[e] ability to place oneself in an imaginary landscape which is central to people's sense of belonging" (Savage, Bagnall, and Longhurst, 2005: 90).[2] This is consistent with what Alan Walks found in the

---

2 There are obvious parallels here with Benedict Anderson's (1983) conception of nations as "imagined communities." In both contexts, "imagined" implies that reality might differ, but that the imagining itself can have a powerful impact.

east end of Toronto. Those who moved into city and suburban neighbourhoods were looking for different things. Those choosing the Beaches – the more urban area – wanted "community," but of a particular kind. It meant above all "the aesthetics of prewar neighbourhood forms," because, after all, the residents "rarely talked to the neighbours" (Walks, 2006: 409). For some, the image mattered more than the reality.

It is again easy to overstate the point. Many who buy into a brand want the lifestyle that goes with it, whether that be urbane or neighbourly. Good examples are the Little Burgundy and Pointe-Saint-Charles districts of Montreal, historically poor but now gentrifying. Steven High and colleagues found that middle-class households were attracted to those areas for their affordability but also for a sense of belonging that, as one interviewee put it, is rooted in the "vie de quartier" (High et al., 2020: 507). The same is true of those who buy into gated communities. Jill Grant (2007: 483) reports that many have responded to "the perceived loss of a sense of community in industrial cities." Her interpretation is telling. Yes, it seems that these buyers want "civility," which includes old-style neighbourliness, but she also sees this as the expression of a "desire for an imaginary or imagined community" (ibid.). Increasingly, we yearn for the image of community as much as the thing itself.

One consequence is that, as with property booms, the image can become self-fulfilling, or at least profitable. When shared and promoted, imaginary landscapes attract like-minded others. After spending decades studying Cincinnati, his hometown, Zane Miller (2001: 69) detected a trend that he dated from the 1950s, whereby residents busy themselves in "redesigning their neighbourhoods for the purpose of attracting the right kind of residents." Those gentrifying areas in Rochester, Toronto, and Montreal are prime examples. The modern planning ideal of the walkable, 15-minute neighbourhood has created an additional selling point, one which, with many people confined to home territory, the pandemic underlined (Grabar, 2020; Talen and Koschinsky, 2013; Talen, 2018). There are great variations, even at the city-wide scale. A recent study calculated the proportion of people in various metro areas who are living on "amenity-rich" blocks, that is, close to services such as a grocery store, library, school, or pharmacy. It turns out that the top-ranked places are the largest, with Vancouver – "the paradise city" – leading the pack, followed by Toronto and Montreal (Bozikovic, Castaldo, and Webb, 2020). Calgary, Halifax, and St. John's lag well behind.

Many observers have implied that local self-promotion is confined to the inner city (e.g., Betancur and Smith, 2016). Hunter (1987), however, sees something similar in the suburbs. The difference is that in newer districts the branding has already been started by developers, or by suburban municipalities themselves. Everywhere, people are working to create a sense of place at a time when our social networks – not to mention flows of capital and information – erode the reality. As Doreen Massey (1995b: 48) has observed, we could view this either as a paradox or as a natural, defensive response to globalization.

The Rise of Homeownership

Back in the 1920s, when sociologists were predicting the decline of neighbourhood community, Clarence Perry disagreed. He noted the growth of parent-teacher associations and the "emergence of property-owning and property-buying associations" (Perry, 1930: 558). This was prescient, probably based on observation of just a few middle-class neighbourhoods. In their study of Seattle, Barrett Lee and his colleagues (1984: 1178) found that, in 1929, land use and public services preoccupied only one-fifth of all local organizations. In the next half century, however, the proportion had risen to four-fifths. In part, this reflected the growing role of municipal governments and the emergence of neighbourhood planning. But, given who, typically, gets involved in local politics, it is surely also due to the growing significance of homeownership.

Homeownership has become more important for neighbourhoods in two ways. The first is simply that more households own their own homes. At the turn of the twentieth century, in Canada as in the United States, only about a third of households were owner-occupiers (Harris and Hamnett, 1987). To contemporaries, the consequences were obvious. As Louis Wirth, a prominent Chicago sociologist, commented, "overwhelmingly the city-dweller is not a home-owner, and since a transitory habitat does not generate binding traditions and sentiments, only rarely is he a neighbor" (Wirth, 1938: 17). Today, after some ups and downs, the proportion is about two-thirds. In 1900, the average neighbourhood was one of renters while, today, homeowners rule. Given that owners are much more likely to get involved in neighbourhood activities, this simple fact explains much of the growing prominence of neighbourhood organizations.

The expansion of homeownership has reflected rising prosperity, and its major reversals have been associated with economic depressions. Kingston's homeownership rate, for example, dropped markedly during the 1890s and so, probably, did those in other cities – the evidence is thin (Harris, Levine, and Osborne, 1981). Rates fell almost everywhere across the continent during the 1930s, to be followed by an extraordinary boom after 1945 (Harris and Hamnett, 1987). But the growth and spread of homeownership has also reflected a hugely significant change in attitudes, and the stakes involved.

Today we assume that almost everyone who can afford to buy will do so, but this did not become generally true until after 1945. In the late nineteenth and early twentieth centuries, the desire to own was felt much more strongly by immigrant workers than by anyone else. Rates of homeownership among skilled workers were typically higher than those of professionals and other white-collar workers. This was the situation in Montreal in the third quarter of the nineteenth century, as it was in Kingston in its latter two decades (Katz, Doucet, and Stern, 1982: 134; Lewis, 1990; Harris, Levine, and Osborne, 1981). That was especially true when economic differences are considered. In nineteenth-century Hamilton, controlling for wealth, blue-collar workers were much more likely to own their home than were merchants, "gentlemen," or professionals (Katz, Doucet, and Stern, 1982: 134). This remained true for decades. Across the United States in 1940, the urban ownership rate among professionals (30 per cent) was only four percentage points higher than for unskilled workers (Harris, 1990). More significantly, among those earning $1,500–$1,999, a fairly average income at that time, the order was reversed: 21 per cent and 32 per cent, respectively. This sort of contrast persisted into the early postwar period. In Toronto in 1950, for example, in the modal (most common) income category ($2,000–$2,499), labourers were much more likely to own their own home (62 per cent) than professionals (41 per cent) (Harris, 1996: 132). Two years later, Morley Callaghan published a short story in which a prominent Toronto lawyer briefly hosted a carpenter's assistant, the father of his son's friend. His home was an apartment in a "row of fine apartment houses" (Callaghan, 1959: 32). Today, that lawyer and his family would be living in a fine home in North Toronto or, perhaps, a downtown condo penthouse. In the early 1950s, however, a nice, rented apartment was still plausible.

Homeownership priorities varied by social class. In an era when the law regarding rental tenure was weak or non-existent, and employment was unpredictable, workers craved security (Dean, 1951: 66). From their study of nineteenth-century Hamilton, Michael Katz and colleagues (1982: 155) concluded that workers "sought to buy homes whenever they could." If they could not buy they built their own. Those in the building trade had the skills and connections, while others reckoned to learn on the job, perhaps with a little help from friends and neighbours (Harris, 1996; 2012a: 24–30). Many immigrants were eager to own property, more so than the native born. The leading exception seems always to have been Jews, for whom education has been the top priority: in 1961, they were greatly overrepresented in school attendance while Italians, for example, were underrepresented (Porter, 1965: 89). From his study of St. Louis in the 1880s–1890s, Gary Mormino (1986: 112) concluded that "home ownership became … a capstone to the immigrants' relentless labors. No home, no community; no property, no dignity." In urban areas across the United States in 1900, the ownership rate among immigrant households (35 per cent) was higher than for native-born whites (26 per cent) (Kirk and Kirk, 1981: 476). In that year the contrast was perhaps unusually large, but a difference persisted for decades. Across Canada, ownership rates among immigrants remained higher than among the native born, and this was certainly not because their incomes were higher (Haan, 2005). Tying the North American class and the immigrant experience together, in his study of early twentieth-century Detroit, Olivier Zunz (1982: 161) declares simply that homeownership, however modest, was an "emblem of immigrant working-class culture."

For many decades, this was demonstrably untrue for the middle classes and managerial elites. In Hamilton, Michael Katz and colleagues (1982) found that it was servants that mattered: women aspired to be household managers, not homemakers. Couples cared that the home be impressive, or at least substantial. Peter Baskerville (2001) found that, in early twentieth-century Canada, the number of rooms per person varied greatly by class. So, in late nineteenth-century Kingston, did the value of homes. Only 1 per cent of the homes owned by skilled and semi-skilled workers were worth more than $2,500, but among professionals the proportion was 26 per cent.

The elite could afford to buy an impressive home and to hire the servants to keep it running. The larger middle classes had to set priorities, and homeownership was the lesser marker of status. The best clues come from advice manuals. None were aimed at a Canadian audience, but several were read widely north of the border. Addressing women in *The Modern Household*, Marion Talbot and Sophonisba Breckenridge (1912: 15) were sceptical about ownership: they conceded some advantages but emphasized that it brings "uncertain costs of operating" and a "greater amount of servicing." Three years later, in a series of lectures entitled "The Normal Life," Edward Devine (1924) made no mention of homeownership. LeeAnn Lands sums it up well. From a study of middle-class families in Atlanta, Georgia, she concluded that, before World War I, "buying simply was not that important to most families" and that high levels of homeownership were not a good guide to neighbourhood "quality" (Lands, 2008: 944).

Things began to change in the 1920s. It became difficult to get servants, and middle-class homes grew smaller as wives became homemakers. In the United States, backed by the federal government, the real estate industry mounted a vigorous campaign to promote homeownership (Harris, 2012a: 33–8; Lands, 2008). Side-effects were felt in Canada. Attitudes began to shift, although not dramatically. To the long-standing view that ownership provided security, some now added that it also offered control (Harris, 2012a: 38–40). In *Economic Problems of the Family*, published just before the Depression, Hazel Kyrck (1929: 418) pointed out that owners were better able to shape their environment to suit their tastes, while "home ownership furnishes ways by which men and boys especially can make themselves useful when they would otherwise be idle." Here, in the first stirrings of the DIY movement, an economic argument is given a moral tone: save money by improving yourself as well as your home. But even Kyrck (1929: 420) was agnostic about whether buying a home made economic sense, noting simply that sometimes you had to buy in order to get a "properly equipped dwelling" with play space for children and a "good environment." Many remained unpersuaded. In a study of the apartment boom of the 1920s, American real estate experts George Wehrwein and Coleman Woodbury (1930: 193) found that many men, instead of caring for the furnace and "making repairs," were getting into golf and "motoring," the craze of the decade. Indeed, they claimed that "there is little special prestige attached to the ownership of a single-family home" (194).

The decisive shift happened after 1945, but even then it took time. As late as 1953, an American manual advised that no one should buy a home for its investment possibilities: "Your house will tend to decline in value as time goes on" (Daniels, 1953: 179–80). This thinking was becoming outdated. In Canada as well as the United States, homeownership was encouraged by federal policy from the mid-1930s on. It favoured 20- to 25-year mortgages for 80 per cent, and then 90 per cent, of the value of the home. Increasingly (almost) everyone aspired to own, so that savings and income alone determined who would buy. By the mid-1950s in Forest Hill, for example, "male and female, at a certain stage of physical maturity, must acquire a house" (Seeley, Sim, and Loosley, 1956: 59). Across Canada in 1931, despite wide income differences, the homeownership rate of the middle class (41 per cent) was only three percentage points higher than for workers. By 1979, the rates had risen for both groups, but the gap had widened, to 14 points.

Slowly, financial incentives became more important. Although income tax had been introduced during World War I, it was only in the late 1940s that most Canadians and Americans began to pay it and, because of a decision made decades earlier, no capital gains tax was levied on people's principal residence. This, and other tax benefits, created a significant incentive to own (Clayton, 2010). It took a while for this to sink in. In 1948, a rare study showed motives of middle-class buyers as they emerged into the postwar boom. Irving Rosow interviewed professionals who had recently bought homes in the Detroit area. Feelings of security were a consideration, but "ego satisfaction" and "creativity/self-expression" also figured prominently (Rosow, 1948: 753). Owning was becoming a mark of achievement, while an emerging DIY boom rested on a new model of (limited) male domesticity. *Better Homes and Gardens* published a how-to manual for would-be owners. It listed several reasons to buy, and none to rent. These echoed the motives of Rosow's professionals: security ("peace of mind"), the expression of individuality, development of character, and so forth (Morris, 1948). Both the manual and Rosow's buyers mentioned financial considerations, but not prominently and only because buying a home – and taking on debt – encouraged savings. Neither used the word "investment." Neither, apparently, did any of the residents of Forest Hill interviewed by John Seeley and colleagues. For them, the purpose of owning was "to transform the house into a home" (Seeley, Sim, and Loosley, 1956: 59).

In Canada there was a local twist to these considerations. By North American standards, ownership rates in Montreal had been abnormally low since the late nineteenth century (Choko and Harris, 1990). Renting was the norm for everyone – working, middle-class, and even the elite. During the 1940s and early 1950s, local campaigns promoted homeownership and, partly for this reason, the idea caught on. The character of Montreal's housing stock prevented a dramatic change: condominium tenure did not exist yet, and so the ownership rate in duplexes and triplexes had a low limit. That is one reason why neighbourhood activism remained relatively subdued in Montreal into the 1970s. Even so, ownership rates within the city almost doubled from 11.5 per cent in 1941 to 20.7 per cent in 1961. Growing by 60 per cent in those decades, the suburbs offered more scope for change, enabling the ownership rate to jump from 28.4 per cent to 51.2 per cent (Choko and Harris, 1990). Everyone participated. A study of *les nouveaux espaces résidentiels* found that by the early 1970s the physical and social character of these new suburbs was converging on the Canadian norm (Turcotte, 1980). Admittedly, some working-class families were buying plexes, a building type with which they were accustomed. But their ownership rate was high, while the majority of the new owners were white-collar and middle-class families living in single-family homes. Montrealers were joining the North American club.

The shift in ownership aspirations of the middle class was vitally important for neighbourhoods in many ways, including politically. It raised the overall ownership rate, and hence the number of people with a strong incentive to advocate for, or to defend, their neighbourhood. More particularly, it increased the likelihood that residents would speak up. It has usually been middle-class property owners who have been most inclined to act: better-educated and with higher status, they expect to be listened to and many have organizational skills. There were several reasons for the surge of neighbourhood activism in the late 1960s, notably that redevelopments were provocative. But the presence of a large, new generation of middle-class owners was also important. The evidence in the *Globe and Mail* backs this up. For decades, from the 1880s to the 1950s, very few articles with "neighbourhood" in the title made reference to homeowners in the text – never more than 4 per cent (table 7.1; bar chart 7.1). This jumped to 21 per cent in the 1960s and, after dipping in the 1970s, rose again, indeed rising further in the 2010s.

That is not the end of the story. People who bought homes in the 1950s and 1960s were not thinking like investors. They hoped and expected that the value of their new possession – by far the most valuable thing they owned – would hold steady, and indeed rise with inflation. But inflation was modest and house prices moved in step. There was no understanding that buying a home was not just a safe investment but a shrewd one. Slowly, however, this changed. Historically, hardly any *Globe and Mail* articles with "neighbourhood" in the title had mentioned "real estate," but these began to appear in the 1960s, rising in the next three decades, and surging from the early 2000s (table 7.1; bar chart 7.1). They reflected a concern that price trends until the spring of 2022 underlined: that house price inflation exceeded the general rate and that those hoping to buy must do so now or be left behind. This spawned a new acronym, FOMO (fear of missing out). In February 2022, the *Globe and Mail*'s real estate reporter observed that paying off a mortgage was no longer a priority for many: buying was simply an investment, and as long as interest rates were lower than house price inflation there was no incentive to eliminate debt. Middle-class homeowners had always fretted that poor neighbourhood services or inappropriate nearby developments might affect the value of their home, but over the past half century that value has become an obsession.

There is a vital link between this obsession and the growth of inequality since the 1970s. A recent study has shown that, in the United States, inequality in house prices has paralleled that in incomes, declining to a postwar valley in the 1960s and rising thereafter (Albouy and Zabek, 2016). The obvious explanation is that widening price disparities have reflected what has been happening to household incomes. But the causality is two-way. Except among the elite, homes are a family's most valuable asset. Rising prices have benefited most the families with higher incomes: when value doubles, the $250,000 home earns $250,000 in tax-free capital gains, but the $500,000 home generates double that. This widens the wealth gap between households. Furthermore, the same study showed that the rate of price appreciation has been greatest at the upper end of the market. Although no comparable study has been undertaken in Canada, all the indications are that the same forces and trends have been at work here. As Lars Osberg (2018: 140–1) has observed, "wealth acquisition for Canada's middle class is now primarily capital gains in the housing market – a lottery whose prizes depend on where and when you buy." Such thinking is perhaps clearest in the condominium market, which has been fuelled not only by those keen to get into the real estate market but also those who see a unit or two as a smart investment (Harris, 2011;

c.f. Webster, 2003). Housing, then, has contributed significantly to the growth of social inequality.

As Osberg hints, neighbourhoods have been an indispensable part of this story, and not just because valuable homes are concentrated in expensive areas. The American study found that the size and quality of homes only explained part of the growing inequality in house prices. Other considerations, pertaining to the geographical setting, including the quality of schools, have accentuated the trends. It is no wonder that NIMBYism has grown, along with inequality, since the 1980s, as those with assets struggle to keep and enhance them (c.f. Stewart, 2021).

It may be true that, socially, neighbours matter less now than a century ago. But financially, as homeownership rates and the financial stakes have risen, neighbourhoods have come to matter more. This is another reason why more residents have learned to care about their neighbourhood's image. The right name, with the right associations, is psychologically reassuring. It also serves as a brand, attracting attention and raising demand in ways that guarantee and enhance one's investment. Most urban residents would probably not make the case so bluntly, but few buyers are oblivious to such considerations. When a municipality declares that it is a "city of neighbourhoods," it evokes the image of community. But real estate agents and planners know that this is underpinned by careful and unrelenting calculation.

The Growing Importance of Schools

For homeowners who have children, and increasingly for those who don't, schools have become a major element in those calculations. Like the growth of homeownership, this is a circumstance many decades in the making, helping to ensure that urban neighbourhoods matter more than ever.

Some parents – and some socio-economic classes – have always cared more about their children's education. In England, notoriously, the elite have preferred private schools. Exclusive academies impart high-quality education and the right "brand" to the privileged few, so that places like Eton and Harrow channel young men through Oxbridge to the City, the Inns of Court, or Parliament. Private schools have never played quite that role in Canada – currently, along with independent schools, they educate 7 per cent of the nation's children – but they do carry cachet. In late nineteenth-century Quebec, although the Catholic Church opposed mandatory education, upper-middle-class families sent their children to private schools run by priests or nuns (Gagnon, 1996). Around 1900, Vancouver's West End elite supported the Granville and Miss Gordon's, where children were guaranteed a "sound English education"(Robertson, 1977: 47). They got what they paid for. When, in *Two Solitudes*, Paul's mother could no longer afford his private education, he really felt the difference (MacLennan, 1945: 248).

Class in education has also been apparent in the public system where the overwhelming majority of pupils have attended local schools. That is why schools have figured prominently in reports about neighbourhoods. In the early twentieth century, more than a fifth of all *Globe and Mail* articles with "neighbourhood" in the title also mentioned schools (table 7.1; bar chart 7.1). Several spoke of how schools such as Clinton Canadianized immigrants (c.f. Vipond, 2017: 2). This was an important function of schools everywhere, notably in Winnipeg's North End, where, it seems, most children were keen to assimilate (Usiskin, 1980: 86). In that period, it is the *Toronto Star Weekly* that provides the most systematic information. Between 11 November 1916 and 15 June 1918, its *Weekly* published a series of reports by W.A. Craick, a writer and associate editor of the *Financial Post*. Craick visited every public school in Toronto. Clinton was mixed, socially as well as ethnically, and fared better than many (Vipond, 2017: 46–7). Others lay towards the extremes. At one end, in an "aristocratic district," Craick reported that the parents in Brown School took "a great interest in the progress children are making," while "boys and girls generally stay until they have gone through every grade" (Craick, 1916a). The contrast was with Earlscourt, which he had visited three weeks earlier. There, in an area settled by British immigrants, children were "poorly-clad and ill-fed, and sometimes illiterate" (Craick, 1916b). Things had improved somewhat, but most were still leaving school as soon as they could: "I've known children who were fourteen on Saturday starting to work on Monday" (ibid.). This was with their parents' encouragement: "There are plenty of cases where parents try to make out that their children are over fourteen, when they're not" (ibid.). In a similar vein, introducing a short article on Toronto's Hess Street school, Irving Abella and David Millar (1978: 104) observe that "the drop out rate at Grade 8 seems to have been typical throughout English Canada," and then add that "the children of French-Canadian workers seldom proceeded even that far."

Throughout the late nineteenth and early twentieth centuries, working-class parents were sceptical of formal education. Hamilton illustrates the point. The working-class people whom Jane Synge interviewed, who grew up in the early twentieth century, emphasized that children started work as soon as they could leave school, at age 14. One man, born in 1901, stated that "as soon as you got the rudiments of schooling, it was up to you to start working" (Synge, 1979: 258). There was no sharp transition, nothing like a high school graduation. School attendance was iffy, varying by season. It typically "peaked in late fall, as students apparently dribbled in after the annual seasonal layoff of less-skilled workers" (Heron, 2015: 135). It declined again in May-June. In 1920, on the average day between one-third and two-fifths of pupils were absent. Widespread absenteeism persisted into the 1930s. In technical high schools, attendance was one-quarter lower in spring than in September, varying year to year depending on the economy (144). Especially during the Depression, working-class families often needed all hands on deck.

Such behaviour made sense. All sorts of jobs were available to those with a minimal formal education, and when whatever skills were needed could be learned on the job. Even basic literacy was an option. In an influential study of what he calls "the literacy myth," Harvey Graff has shown that illiterate workers were at little disadvantage in their ability to acquire homes. In Hamilton in 1861 the rates for literate and illiterate household heads were 27.2 per cent and 24.1 per cent respectively (Graff, 1979: 95). Heron (2015: 134) reckons that, half a century later, "few working-class parents in

Hamilton ... saw secondary schooling as a valid avenue for their children's education," a view that still made sense in the 1930s. His analysis of job ads suggests that "extensive schooling would not make much difference for most jobs" (144). Hamilton was, and for some time remained, unusual. Canada's most industrial city, in the early twentieth century it had the lowest rate of school attendance among 14 cities across English Canada (135). But, almost everywhere, boys and girls with limited education could get work, and their parents knew it.

The results were apparent, even in cities not noted for manufacturing. In 1902, when it was a booming town of 5,500, a local newspaper observed that "Calgary furnishes excellent business opportunities for boys of the High School age, and most youths are lured away from the paths of education for positions in offices and stores" (Stamp, 1977: 292). The most telling research, however, pertains to Vancouver. In her superb study of working-class schooling on the West Coast in the 1920s, Jean Barman mined the BC Department of Education's reports. These say a lot about parental attitudes as well as children's behaviour. During the 1920s, there was a series of votes on funding bylaws to enable the purchase of school equipment. Support was invariably lower, sometimes much lower, than average in the working-class East End and higher in middle-class areas such as Kitsilano (Barman, 1988: 62). Indeed, in some middle-class areas, parents lobbied for special fees to buy resources that might give their children a leg up. Attendance levels followed the same pattern (59).

This geography reflected, and inevitably reinforced, class differences. In 1921, only 13.7 per cent of the children of unskilled labourers were still in school at age 15. This compares unfavourably with the rates for the children of clerical employees (24.8 per cent) and professionals (31.7 per cent) (Barman 1988: 66). It wasn't that the middle classes attached enormous weight to education, just enough to make a difference. There were just a few for whom it was a stepping-stone to higher education and professional careers. After all, in 1930 only 2.8 per cent of Canadians aged 20–24 pursued higher education, more than in Britain (1.9 per cent) but less than in the United States (11.3 per cent) (Axelrod, 1990: 22). They were not a social cross-section. More than half of those students had fathers who were professionals, even though such occupations accounted for only 12 per cent of the labour force (23). Who your parents were and where you grew up determined which schools you attended and where your education took you.

That is still true, of course, but educational pressures have increased as the stakes have changed. It should be clear that we are talking about formal qualifications, signalled by certificates. In *The Intruders*, Hugh Garner (1976: 117) has Syd Tedland exclaim "we're all self-taught, Rawley. Schools give us schooling, but living gives us our education." That has always been true but, in addition to, and sometimes regardless of, the knowledge that is imparted in schools and colleges, certificates have come to matter a lot. As a result, school attendance has risen steadily. In English Canada in 1911, the proportion of 15- to 19-year-olds in school was a measly 18.7 per cent (Gidney and Millar, 2012: 46). The proportion rose steadily thereafter. In Ontario in 1921, the leaving age was raised to 16, and changes were made in other provinces. Not surprisingly, the rising proportion picked up pace in the 1920s, reaching 33.7 per cent by 1931 (ibid.; c.f. Stamp, 1978). Significantly, that was when American research showed that middle-class parents were becoming eager for their children "to associate with children from homes which hold standards similar to their own," and when the idea of the neighbourhood unit, centred on a primary school, was gaining traction (quoted in Fischler, 1998: 707). No wonder, then, that in Canada as well as the United States, primary schools commonly "stood out as the most impressive buildings in their neighbourhoods" (Sutherland, 1986: 178).

The increase in school attendance at age 15 and above slowed during the 1930s and early 1940s. Local job markets had a big impact. Hamilton still lagged: in 1942 the proportion there who made it to Grade 11 was 33.7 per cent, lower than Toronto (39.2), Vancouver (47.3) and, mostly notably, Ottawa (67.5) (Gidney and Millar, 2012: 61). But then, along with the suburban boom, postwar school expansion came "helter skelter" (Gidney, 1999: 27). Between 1946 and 1961 the Canadian population grew by 50 per cent but school enrolment jumped, by 116 per cent at the primary level and

142 per cent at the secondary. There was a baby boom, but the big change at the secondary level came because many more children were staying on to Grades 10, 11, and then 12. Across Ontario in those 15 years, the proportion of 15- to 19-year-olds in school leaped from 35 per cent to 74 per cent (Curtis, Livingstone, and Smaller, 1992). Completing high school was becoming the norm.

In this rapidly changing context, local schools attracted a lot of attention. From the 1940s to the 1960s, almost two-fifths of all *Globe and Mail* articles about neighbourhoods referred to schools (table 7.1; bar chart 7.1). Class sizes at the primary level, which in Vancouver had been in the 40s during the 1920s and 1930s, had dropped to 34 by the mid-1950s (Sutherland, 1986: 207). Increasingly, curricula and teaching methods were standardized (208). In an era when social and neighbourhood inequality was declining, the average quality of education rose. But differences did not disappear, and because most parents were coming to believe that a high school diploma mattered, middle-class parents began to pay more attention to something they had taken for granted – their children's peers. This could not be standardized, and middle-class parents found new ways to ensure that schools as well as neighbourhoods stayed different. Their strategies varied. In suburban Etobicoke after 1945, councillors and ratepayers used school taxes to exclude tenants and working-class homeowners (Ellis, 2019). Excluding lower-income households helped keep taxes low while ensuring appropriate playmates and friends. Their strategy was foiled, in part deliberately, by the creation of Metropolitan Toronto in 1954. But their efforts signalled the emergence of a new era.

Since the origins of the public school system, the neighbourhood has always shaped the character of the school. For decades, everyone took this for granted. No longer. Growing competition – initially for a high school diploma, then for good grades, then for excellent grades from a good school, and then excellent grades from an excellent school – has made parents increasingly choosy. This dynamic picked up pace as income inequality grew from the 1970s onwards (Putnam, 2015). Parents choose a neighbourhood for its school, and then defend their turf. By the turn of this century, a study of Hamilton showed what everyone knew: the best schools, measured in terms of children's performance on standard-ized tests, were in the higher-income (and higher priced) areas (Harris and Mercier, 2008).

In this regard, Canadians are like their American and British cousins. Reflecting on how America changed in the past century, Claude Fischer and Michael Hout (2006: 6) observe that "educa-tional attainment increasingly shaped and distinguished Ameri-cans' fates and fortunes." As a result, "parents agonize over which school and community will be best for their children" (Altonji and Mansfield, 2011: 340). They consider the influence of neighbours and peers as much as they do the schools themselves (Bayer, 2007). Those with a college education are more demanding; those with-out often prefer not to live in areas dominated by college grads (ibid., 626). Municipalities know all of this. Recently, for example, Philadelphia used school investments to attract middle-class pro-fessionals back to the inner city: it recognized that "the decision parents make about where to send their children to school is a criti-cal moment in the social reproduction process" (Cucchiara, 2013: 97). The same is true in Britain. Speaking about how the gentrify-ing middle class transformed inner London, Tim Butler and Garry Robson (2003: 4) comment that "it is with respect to th[e] crunch issue of education that much of the talk about 'strategies' arises."

Schools matter more than anything. In a study of Bristol, Gary Bridge (2006) found that, if it came down to a decision between picking an area with a traditional gentrification aesthetic and one with a good school the latter would win. The good school is likely to be in an area where the middle class is, or is becoming, dominant. A comparative study of Bristol, Montreal, and Paris underlined "the disinclination of middle-class parents to commit themselves to neighbourhoods whose public schools serve mainly poor … children" (Rose et al., 2013: 446). No wonder, then, that of all the urban places where people meet, schools are the least likely to see a mixing of classes (Massenkoff and Wilmers, 2023). (The most inclusive places are libraries and chain restaurants.) Dam-aris Rose's (2013) focused study of Montreal concluded that neigh-bourhoods with large ethnic minorities are also viewed sceptically. The ideal of social mix, promoted by planners in recent years, has its sceptics (August, 2007). Ambitious parents are not fools. As discussed in chapter 5, there is abundant evidence that neighbours, peers, teachers, and school resources all have a substantial, even determining, effect on a child's health and prospects in life. And the stakes have never been higher.

Families with young children pay careful attention to school catchment areas, which have become important markers of neighbourhood boundaries. Their significance is clear whether they are enforced or not. Jesse Butler and colleagues (2019) tell the story of an Ottawa high school whose catchment included a disadvantaged neighbourhood. Because it had a bad reputation and a lax transfer policy, middle-class parents within its catchment moved their children elsewhere. This reduced enrolment forced the school to offer fewer course options and reinforced its decline. Catchments are sometimes crossed for other reasons. Ambitious parents may use their insistence on French immersion for their child as a way to transcend school boundary restrictions if they do not live in the right area. Ed Keenan managed this for his son: living in the less expensive part of Toronto's Junction, he was able to enrol his son in a school on the more expensive side, "where they have a French immersion program and the students score well above the provincial average on standardized tests" (Keenan, 2013: n.p.). That is risky, however, because children are generally expected to attend the local school.

When a school closes, all hell can break loose. On a large scale, when parish and school boundaries were changed wholesale in the Sainte-Foy suburb of Quebec City in the 1970s, parents' associations and citizen groups mobilized to define neighbourhood boundaries and regain control over their children's social and learning environment (Hulbert, 1981). No wonder: a map shows numerous revisions to the previous catchments (figure 20). On a still larger scale, in 2015 Ontario moved to fast-track the procedure for closing "under-utilized" schools. The policy director for the Association of Municipalities spoke up, emphasizing "the impact closing a school could have on residential real estate values" (Howlett, 2015). The example of KCVI in Kingston, mentioned earlier, shows that the concern was real. It boasted the highest test scores in the city. The threat of closure aroused great opposition because it meant that children from middle-class Sydenham ward would have to travel farther to attend a new high school in the North End, along with the children of workers and public housing tenants. Residents feared that this would reduce the "livability" of the Sydenham neighbourhood, encouraging middle-class parents to up stakes for the suburbs, and undermine property values (Collins, Allman, and Irwin, 2019). Parents knew very well what was at stake.

Increasingly, then, they have been willing to pay more for homes in the catchments of good schools. A decade ago, a survey of 50 studies in eight countries found that schools with higher test scores were consistently associated with higher house prices (Nguyen-Huang and Yinger, 2011). This relationship is one of mutual cause and effect. Most obviously, parents are attracted to areas that have good schools. From her Australian research, Emma Rowe (2017: 132–3) observes that middle-class gentrifiers reject private schools as elitist, instead seeking the "brand community" of a neighbourhood with a good public school. She provides a nice example of the effects on real estate. After a desirable new school, Thompson High, was opened, "there has been a scramble to buy into the zone" [i.e., catchment], and real estate agents market homes as being "inside the coveted THS zone" (Rowe, 2017: 139). The neighbourhood then recorded the highest home price increase in all of Australia. Michael Insler and Kurtis Swope (2016) argue that it was this sort of competition which explains why the increase in house prices in the United States between 2001 and 2009 was especially strong in the areas with good schools. Indeed, they suggest that the search for such areas contributed to the housing bubble.

As always with neighbourhood effects, however, there is a reciprocal influence. Patrick Bayer (2007) found that the short-run effect of school quality on housing demand and neighbourhood preferences was real, but limited. It was reinforced by the preference of educated middle-class parents for neighbours and children's peers who are similar to themselves. And then there is a compounding, multiplier effect. Anything that causes, or creates the appearance of, an increase in school quality will precipitate moves that change the character of the neighbourhood, with an effect on house prices. Intriguingly, it seems that even those households without school-age children recognize this dynamic and support expenditures that improve the local school (Fischel, 2014). The net effect, as Matthew Lassiter (2012: 199) puts it, is that public education has become "another consumer product in the capitalist marketplace," one which, because it is rooted in place, becomes integrated with the housing market.

Over the past century or so, the significance of schools and homeownership has grown. Becoming more tightly intertwined, they have made neighbourhoods progressively more important. Home turf has come to be defined as the place where homeowners make crucial, calculated investments. The most important concerns the local effects, real or imagined, on the daily lives and future prospects of school-age children. Everyone involved – parents, school officials, developers, real estate agents, planners, and politicians – is aware of this, or if they are not they pay the price. Neighbourhoods have become less important as places where people find community. But, in the modern era of social inequality, they matter more than ever.

PART 3

Reflections

chapter eight

Looking Forward, and Beyond Canada

Democracy doesn't require perfect equality, but it does require that people from different walks of life encounter one another in the course of their everyday lives.

Michael Sandel

For individuals, numerical ages, and their associated milestones, have special significance. We become old enough to get a driver's licence, to buy a drink – legally, that is – or to qualify for our Canada pension. With societies, too, quantitative change can have qualitative effects. Today's paper brought news that the Supreme Court of Canada had made a significant legal decision on a close 5–4 split; if one judge had changed their mind it would have gone the other way, with major social effects. Securing an extra seat in the House of Commons can give a government a majority. Small incremental changes can have major effects. The Canadian-born writer Malcolm Gladwell (2000) popularized this insight in *The Tipping Point*. Since the late nineteenth century, the trajectory of Canadian neighbourhoods, too, has been seriously affected by the crossing of certain thresholds. The first came when the average neighbourhood came to be dominated by owners, not tenants, the second when most adolescents began to graduate from high school. The first created a new dynamic for neighbourhoods, implicitly putting municipalities, and their planning departments, on guard. The second reflected new expectations among employers and parents, with ratcheting consequences. Together, they suggest that the 1950s laid the groundwork for what we now take for granted.

Building on these, a third threshold was crossed late in the 1960s, when neighbourhood issues came to dominate civic politics. Various elements contributed to this change, some having nothing to do with local issues. In addition to rising homeownership, these included economic prosperity, a broadening cohort of university students, and youth rebellion, along with the side-effects of the civil rights movement and popular resistance to America's war in Vietnam. But once the smoke had cleared, citizen participation, variously organized, showed that it was here to stay.

These tipping points did not arrive with simple majorities. Homeowners have always been more active than tenants in neighbourhood affairs, and it may be that an average ratio of 40 per cent, or even lower, can tip the scales of local activism. Then, too, for municipal affairs, a national average is not very meaningful. The shift to homeownership came later in Montreal than in other urban centres. This may help to explain the later emergence of its neighbourhood movement. Or not. After all, for many decades, and especially in working-class districts, Montreal tenants moved frequently but very locally. Each stayed in an area for many years, maintained local connections, and perhaps cared almost as much about the neighbourhood as any property owner. And, more generally, the homeowner threshold was reached earlier in the suburbs than in the city, which helps to explain the varied priorities of residents and their associations. The tipping point varied according to place and time.

Because those thresholds have been crossed, the public conversation about neighbourhoods is now more prominent, and diversified, than in the past. In part, it is again focused on issues of poverty, inequalities of income and service provision, and the difficulties faced by immigrants. But it now also engages planners on the issues of traffic, infrastructure, public health, services, development, and redevelopment, while vexing school board officials faced with the allocation of resources and the possible closure of schools. The 1960s put neighbourhoods on the local political agenda and they show no signs of falling off it. Indeed, everything suggests that their importance is secure. It is difficult to imagine that investing in a home or getting a "good education" is going to matter less in the future, or that either will be less tied to local territory. Following the pandemic, more people are working from home, at least part of the time, and so their local setting – what Carlos Moreno (2024) calls the 15-minute city – has become more important. That is especially true if the use of online shopping continues. A decade ago, Ray Godfrey recalled his childhood in an inner-city neighbourhood. He recalled all the "delivery figures – breadmen, coalmen, icemen, milkman, and of course our own 'postie'," and commented "in retrospect we lived in a kind of neighbourhood one seldom sees in Toronto any more" (Godfrey, 2013). True, up to a point. But FedEx, Amazon.ca and DoorDash were reinventing some of it even before the pandemic hit. Prediction is a mugs game, but it seems unlikely that that trend will be reversed.

Although I have often phrased this narrative as fact, there are many subjects about which we remain quite ignorant. For example, as more women have gone out to work and men are more active in the home, is this affecting the dynamics of neighbourhood networks? Is neighbouring different in an apartment building than in a condominium, as we might expect? In terms of their desire and ability to form enclaves, how different are Indigenous people from immigrants, and with what consequences? Exactly how different, not only in scale but also in significance, are the neighbourhoods in towns as opposed to larger cities? Would parents be more amenable to living in socially mixed neighbourhoods if the quality of local schools did not vary so much? Would homeowners moderate their NIMBY tendencies if the investment stakes associated with ownership were lower? And does greater walkability actually lead to more people walking, and does that in turn really help build meaningful local social connections?

We remain most ignorant about the past, making it more difficult to figure out what makes the present different. We know little, for example, about how people made decisions about where to live before the 1960s: is it really true, as I have argued, that on average they cared less about where they lived than most people do today? In particular, what were the experiences and views of social and economic elites, the elusive group that has always been most neglected by urban and historical researchers? Then, too, historically, when tenants moved around a lot, but perhaps only within a limited area, did they develop an attachment to place which was equivalent to that of the modern homeowner? In that vein, can Montreal's high tenancy rate really explain the initial weakness of its neighbourhood movement? And were those immigrant, working-class neighbourhoods of the pre–World War II era really as tight-knit and supportive as some oral histories suggest, and as later observers have supposed? There are conflicting views on these, and much else, but as yet no solid basis for judgment.

There is also much about the recent past, and indeed the present, that remains unclear. What about all the towns and small cities that I have not mentioned because – as far as I am aware – writers and researchers have paid little attention to their neighbourhoods? What of Barrie, Yellowknife, and Fredericton, not to mention Charlottetown, Abbotsford, and Baie-Comeau, the sorts of places where, collectively, millions of urban Canadians live? In these smaller centres segregation occurs at a smaller scale and may take different forms, so that neighbourhoods matter in different ways. And we need to know more about the Indigenous experience in urban areas, ideally from insider perspectives. Urban reserves are especially intriguing: contradicting the prevailing stereotype of northern isolation, those that are residential in character function as a unique type of Canadian neighbourhood. How they function within their wider urban context might tell us something important about the possibilities for reconciliation.

Faced with many uncertainties, in telling the Canadian story I have tapped research on the United States, which on most topics is more abundant. This is reasonable because, for the most part, our national experiences have been similar. We have all become more mobile and able to maintain community over long distances. Many more women now go out to work. On average, adults spend less time in their neighbourhood and, when there, are more likely to hang out indoors or in their backyard. The level of urban home-ownership in the two countries has always been similar, moving on a parallel track. House prices have boomed and become more unequal in ways that both reflect and enhance social inequality. Trends in formal education have also followed much the same path. There are differences. More American than Canadian parents send their children to private school – closer to 10 than to 5 per cent – and rather more Americans have obtained some form of higher education certificate. But these are variations on a shared theme.

Indeed, a few Americans writers have noted the value of taking the long-run view and have sketched some elements of the argument that I have been making for Canada. Forty years ago, Patricia Mooney-Melvin (1985: 366) noted that we have few studies that take as their subject "the changing consequences of locality," and little has changed. Almost a century ago, Clarence Perry (1930: 558) anticipated one of the themes emphasized here when he noted the growth of parent-teacher associations and the "emergence of property-owning and property-buying associations." In 1978, reviewing the period since Perry wrote, Morris Janowitz (1978) argued that locally provided public services, notably schools, had enhanced the value of local residential settings. Soon, Claude Fischer (1991) amplified this, emphasizing the importance of home-ownership, not least for reducing mobility (c.f. Fischer, 2002). The millions of American subscribers to Nextdoor would agree. This social media site for neighbours was launched in 2008. Within ten years it was established in two thousand neighbourhoods across the country. By then, in online conversations, residents talked about two things above all: property values and children (Stross, 2012). As Nirav Tolia, its chief executive, observed "the neighbourhood is where you buy a home, where your kids go to school, where you spend the majority of your physical life." He was speaking about his subscribers, probably for the most part middle-class, and probably not typical even of them. Not everyone has children or can afford to buy a home. But for most people at some point in their lives, and aspirationally for many more, he had a point.

Another parallel between our countries has to do with patterns of neighbourhood change. To be sure, on average Canadian cities have been more densely settled, with healthier transit systems, and more inner-city neighbourhoods that have appealed to middle-class families (Goldberg and Mercer, 1986; Tomalty and Mallach, 2015). Despite our climate, we cycle more than Americans do (Pucher and Buehler, 2006). But our cities, too, experienced postwar inner-city decline and saw efforts at slum clearance and urban redevelopment. Since the 1970s, gentrification has had a greater impact in Canada, but even Detroit has tasted inner-city revival and (admittedly very localized) gentrification. Meanwhile, federal policy in both countries began to transform housing and neighbourhoods in the same decade and on the same principle: favour homeownership and the suburbs, and let the market rule. Land-use control, and then city planning, emerged at the same time, and have taken similar forms, endorsing versions of the "neighbourhood unit." There is one difference at the metropolitan scale. American suburbs have had the constitutional authority to resist amalgamation, and within limits exclude the poor, while – as many Torontonians, Montrealers, and Hamiltonians know to their chagrin – Canadian municipalities are creatures of the provinces. That is why Toronto has so many public housing projects in its inner suburbs, and why those erstwhile suburbs are now part of the city (c.f. Saunders, 2013). These differences are not trivial, but none change the overall narrative (c.f. Harris, 2024). The same is true of the forces that have shaped the cyclical emergence of concentrated poverty. Across the continent, income inequality peaked in the early decades of the twentieth century, and has reasserted itself since the 1970s. Then, as now, those peaks coincided with high levels of immigration.

Here, however, we come to some significant differences. Hispanic immigrants are a larger presence in some American cities than they, or any ethnic group, are in any Canadian city. There is no Canadian parallel to the historical experience of African Americans, including, in the twentieth century, that of the Great Migration and its effects on the formation and growth of Black ghettoes in northern and western cities. Yes, some Canadians once owned slaves, and communities of African Canadians have existed in some Canadian cities, notably Halifax and Montreal, for more than a century (e.g. Rutland, 2018). But the scale, meaning, and impact of these settlements are wholly different from those of Harlem or Chicago's South Side, or for that matter the ghetto in Peoria, Illinois. They are one reason why there is no Detroit in Canada (Hackworth, 2016; c.f. Filion, 1987). Related to this, social inequality has always been greater in the United States and, despite "land of opportunity" rhetoric, rates of social mobility have been lower there (Heisz, 2015; Bourne and Hulchanski, 2020). In part, more inequality south of the border reflects the greater social supports offered by the Canadian state, notably in the form of health care.

One consequence is that the nature, degree, and effects of both privilege and of concentrated poverty are more marked in US cities (Bradford, 2013; Oreopoulos, 2008). That is why Ingrid Ellen and Justin Steil (2019: 12) have stated what in the United States is something of a truism: "unequal neighbourhoods that have both short- and long-term impacts on individual access to opportunity is perhaps the most glaring reason to be concerned about segregation." Coupled with the gun culture and high rate of violent crime – aspects of American life that most amaze and repel Canadians – these socio-geographical contrasts translate into greater stress, poorer health, and a pervasive concern, often tipping into fear, of public space (Wilkinson, and Pickett, 2019: 3). Conversations on Nextdoor show that, even in middle-class American neighbourhoods, the dominant concern is crime, an issue that is routinely racialized (Molla, 2019). Canadian cities are simply safer, especially for women (Kluch and Gordon, 2018; Goldberg and Mercer, 1986). When Phyllis Rudin announced on social media that she planned to walk every street in Montreal many Americans, fearing for her life, went online to dissuade her. Afterwards, however, she reported that "my project wasn't dangerous at all ... Never did I walk through any look-over-your-shoulder neighbourhoods" (Rudin, 2022). She might have had a different story to tell if she had travelled by night, but the contrast holds. All of this accounts for the greater prevalence of gated communities in the United States, and the determination of many suburbs to exclude lower-income and racialized minorities. Greater inequality, segregation, neighbourhood disparities in health and crime – these are some of the reasons to believe that neighbourhoods matter even more in the United States than they do in Canada. In *A Nation of*

Neighborhoods, Benjamin Looker (2015: 3) concluded that, in post-war America, neighbourhoods "came to occupy a prominent place in the political imaginary of many of the twentieth-century's most consequential social and intellectual movements." For all their significance in Canada, it is impossible to make that same claim here.

Our experience has therefore been more modest, centred on home and school, everyday life and local politics. When Keith Hampton (2016) "lost" a bunch of letters on random city streets, more Canadians than Americans picked them up and mailed them. In the fall of 2019, Nextdoor expanded its service into Canada. Contrasting its purpose with that of other media, Christopher Doyle, its Canadian lead, argued that its main purpose was to promote trust: "we are more interested in rewarding offline interactions and kindness"; instead of a "liking" button, Nextdoor has "thank" (Grainger, 2020). It turned out that Canadians used this button "more than anyone else" (ibid.). It seems we are indeed – yes, that dreaded word – nicer.

But not invariably. In his book on Toronto's Annex neighbourhood, Jack Batten (2004: 14–16, 23) tells of a "prickly" next-door neighbour who fired his BB gun at cats and provoked another neighbour to file a lawsuit. More recently, while conceding that there was "plenty of banal and even heart-warming stuff" on Nextdoor, Marsha Lederman (2023) noted that there was also a risk of being "mobbed by un-self-aware, self-righteous, self-important snitches," sometimes making this site a "battleground posing as a playground." We like to think we are nice. On 14 November 2015, there was a firebomb attack on a mosque in Peterborough, Ontario. The *Globe and Mail* was careful to report that Larry Gillman immediately offered the Beth Israel Synagogue a temporary space where local Muslims could worship. "This is what neighbours do," he explained, "this is what Canadians do" (MacGregor, 2017). J.S. Woodsworth would have approved. Alas, that cannot be the last word. After all, some other neighbour planted that bomb. Every day, journalists and politicians receive abuse and threats on social media. For every child that gets into a "good" school another does not. For every family that owns a home and joins the neighbourhood association, another frets about the rent, and maybe renoviction. Canadians, and their neighbourhoods, are unequal, and have been becoming more so. Given that those neighbourhoods matter more than ever, we should care about that – and do something. But what?

One thing is to strengthen public education. Among other things, that means hiring more teachers so that all schools can offer the full range of skills- and knowledge-training, including physical education, music, and, yes, shop. That way, the better-off among us wouldn't expend so much effort, and money, looking for the "best" schools in the most expensive neighbourhoods. Meanwhile, children of the less well-off would have better life prospects. Another thing we need to do is to downplay the incentives for homeownership while strengthening the alternatives, including more secure rentals, along with non-profits, and better-run public housing. That should make housing more affordable, reduce resistance to the rezonings that enable densification, and take away some of the competition for "better" neighbourhoods. Such measures would reduce the financial significance of neighbourhoods while increasing their social role, as residents find schools more inclusive and denser neighbourhoods more friendly.

I know, this sounds like pie-in-the-sky, nostalgia for a lost past. Not so. We cannot turn the clock back, nor should we try. Education will continue to matter more than ever; the task is to improve it by levelling the playing field, and in some cases that means actually providing inner-city schools with playing fields. Physically and in terms of ownership, the types of housing Canadians need now are different from those we occupied in the past. Fewer units for families; more, and denser dwellings, for singles and couples without children. In the future, cafés and corner stores may flourish, especially in Ontario now that they can sell alcohol, and if people are working from home they may have more opportunities to connect with their neighbours. None of this will happen overnight. It took decades for us to arrive at our present place. Even with new federal and provincial policies, the built environment will change direction even more slowly than the proverbial oil tanker. And anyway, I am not making a prediction. Change is not inevitable. That's up to us, our children, and our children's children.

Appendix: Google Ngram Viewer and the *Globe and Mail*

Digital online records provide a cornucopia of information. Many make it easy to track changes in the coverage, associations, and, by inference, the significance, of particular subjects. For present purposes, two record sets were of particular value: Google's Ngram Viewer and the digital archives of the *Globe and Mail* since 1844.

The Ngram Viewer contains extensive records of a variety of material, above all books, published in English and seven other languages from 1500 to 2019. It has been used by researchers interested in long-run historical trends, a notable example being Robert Putnam and Shaylyn Garrett's (2020) study of income inequality. Based on place of publication, it distinguishes between American and British English. Keyword searches can be limited by periods, and results are reported graphically. Simple counts are influenced by the amount of material published, which has increased enormously over time. Usefully, graphed results show *relative* not absolute word frequency. It is freely accessible to anyone who has internet access. For present purposes, its value is that of providing a context for Canadian material in the *Globe and Mail.*

ProQuest's *Globe and Mail* database covers the *Globe* from 1844 and then the *Globe and Mail* after a merger in 1936. Because of its historical depth, it is invaluable, making it possible to track changes in the usage, meaning, and associations between several words or phrases of relevance to the present study. A few Canadian urbanists have exploited its strengths, for example in a recent study of gentrification (Tolfo and Doucet, 2021). For present purposes, relevant terms included "neighbourhood," "residential area," and "residential district," as well as "home owner"/"homeowner," "planner," "planning," "schools," "suburbs," and "resident," "ratepayer," and "neighbourhood associations." Although the British spelling of "neighbourhood" is now the norm in Canada, the American version was quite common until at least the 1980s. Conveniently, the search tool automatically includes both in searches for either. As with Ngram, searches can be limited by period, with resulting counts presented in a bar chart. Searches can be limited in various ways, and for present purposes only "articles" were examined. Links to the original articles, in which search words are highlighted, can be listed by "relevance" or date of publication.

A useful feature of this database is that it supports searches of the full text of articles or just the titles. This was crucial for "neighbourhood" because the word has been used in various contexts, many having nothing to do with residential areas. For example, "in the neighbourhood" has both literal and metaphorical senses. By confining searches of "neighbourhood" only to the titles of articles it was possible to eliminate a high proportion of, although not all, material of no relevance to the present study. Conversely, phrases such as "residential area" are invariably relevant but rarely appear in titles. Here, full-text searches were used. The useful "advanced search" feature of the search tool made it possible to combine search terms and in different ways. Thus, for example, it was possible to identify the changing incidence of articles with "neighbourhood" in the title that contained "planning" or "homeowners" in the full text.

Unlike the Ngram Viewer, ProQuest's search tool does not adjust for the relative frequency with which words appear. Simple counts could be misleading because over the decades the newspaper varied in length. One way to assess this is to use "and" as a keyword, on the assumption that it has always been used with the same relative frequency. "And" appeared 36,581 times in the *Globe* in the 1860s, rising steadily through 126,166 in the 1890s before levelling off in the 1920s, when the count was 423,637. After minor fluctuations, it peaked at 493,930 in 1980s before falling in the 2010s. Although the search tool suggests that the database covers the entire decade of the 2010s, at the time the research was undertaken it only covered up to and including 2017. Accordingly, all counts for the 2010s have been multiplied by 10/8 in order to render them comparable to previous decades. Indexed, using the 1950s as a base, values ranged from a high of 123 in the 1980s to lows of 24 in the 1880s and 72 in the 2010s. This should be a reliable point of comparison: a search of "the" yielded similar results.

Using the counts of "and," the results for all other keywords were standardized. This made it possible to provide a meaningful interpretation of historical trends. Two approaches were taken. The first was to construct an adjusted index for each keyword. Again using the 1950s as a base, interim index values were calculated, and then adjusted by the "and" index. The procedure for "neighbourhood" illustrates the method (Table A1). In the 1950s, 26 articles in the *Globe and Mail* contained "neighbourhood" in the title, receiving an interim index value of 1.00. There were 37 such articles in the 1890s, earning an interim value of 1.42, and 379 in the 2000s, earning an interim value of 14.58. But a comparison of these three values – in historical sequence 1.42, 1.00, and 14.59 – would be misleading because, as noted above, the newspaper was much shorter in the 1890s than in later decades: index values for "and" in those decades were 0.32, 1.00, and 0.99. And so, to enable meaningful comparison, the interim values for neighbourhood were adjusted by those for "and." For the three decades in question, this yielded a sequence of 4.44, 1.00, and 14.74. This procedure made no significant difference to a comparison of the 1950s with the 2000s but a substantial difference to our understanding of all decades prior to the 1920s, and also the 2010s.

A second approach adjusted raw counts. Index values give a clear impression of historical trends but disguise the absolute numbers on which they are based. Those numbers themselves are also of interest. For that reason, in the second approach, absolute numbers (respectively 37, 26, 379) were adjusted by dividing by the appropriate index values for "and," in this case 0.32, 1.00, 0.99. The results are counts, in this case 115, 26, and 382, that might be termed hypothetical. Their purpose is to indicate that, if the newspaper in the 1890s had been as substantial as it was in the 1950s then there would have been 115 articles with "neighbourhood" in the title, not 37. Obviously, such numbers need to be interpreted with care, especially before the 1920s, after which the discrepancy between reported and adjusted counts in generally quite small (table 6.1).

Numbers and indexes, however they are reported or adjusted, need to be interpreted. By searching only titles it was possible to eliminate many irrelevant usages of "neighbourhood" but not all, and it did not prevent inclusion of articles on other cities or countries. Such "noise" was always present and, except where mentioned in the text, had little impact on the interpretation of trends. More importantly, counts say nothing about the meaning of words and phrases, or how meanings change. For that, it was necessary to read the articles in question, or at least a representative sample of them. I adopted an informal procedure. In each case, I selected decades that were especially notable for each word or phrase – when they appeared, were common, or went into decline – and then read those that ProQuest flagged as most "relevant" until a clear pattern of meaning or association emerged. The alternative, of reading all articles, was unfeasible. Life is too short.

Databases of other local newspapers, including the *Toronto Star* and Montreal's *La Presse*, are available. Unfortunately, search options on these sources are more limited. Notably, it is not possible to search titles and full text separately, or to combine search terms for these newspapers. In addition, results are not summarized by time period. For those reasons, these sources were not used in any systematic manner.

Table A1. The Indexing of "Neighbourhood"

	Raw counts		Counts indexed		Adjusted index for "neighbourhood"
	"And"	"Neighbourhood"	"And"	"Neighbourhood"	
1880s	96,085	28	24	108	450
1890s	126,166	37	32	142	444
1900s	200,262	28	50	108	216
1910s	289,188	80	72	308	427
1920s	423,637	128	105	492	469
1930s	440,802	68	110	261	237
1940s	456,626	31	114	119	105
1950s	400,491	26	100	100	100
1960s	441,859	24	110	92	84
1970s	481,997	73	120	281	234
1980s	443,930	74	123	285	232
1990s	387,973	154	97	592	610
2000s	398,053	379	99	1459	1,474
2010s	231,519	311	–	–	–
2010s adj[1]	289,399	389	72	1495	2,076

Note: Row counts for "and" are based on the full text of articles. Those of "neighbourhood" are based on titles only. The search engine automatically includes British and American spelling of "neighbourhood."
1 Counts adjusted to obtain comparable estimate for whole decade.
Source: ProQuest *Globe and Mail* database.

Tables

Table 4.1. Population of the Largest Canadian Urban Centres 1881–2020[1]
Ranked by size of urban area in 1951

		1881	1901	1951	2020	Top ten rank in 2020
1	Montreal	140,700	267,700	1,395,400	4,364,200	2
2	Toronto	113,100	238,100	1,117,500	6,552,200	1
3	Vancouver	–	29,400	530,700	2,737,700	3
4	Winnipeg	12,200	48,000	354,100	850,100	7
5	Ottawa	27,400	59,900	281,900	1,461,500	6
6	Quebec City	62,400	68,800	274,900	832,300	8
7	Hamilton	36,000	52,600	259,700	804,700	9
8	Edmonton	–	2,600	173,100	1,465,900	5
9	Windsor	6,600	12,200	157,700	356,900	–
10	Calgary	–	4,400	139,100	1,543,300	4
11	Halifax	36,100	40,800	133,900	448,500	10
12	London	19,700	38,000	121,500	511,000	
13	Victoria	5,900	20,900	104,300	408,900	
14	Saint John	41,400	40,700	78,300	131,800	
15	Regina	–	2,300	71,300	236,500	
16	Thunder Bay[2]	1,300	6,800	71,200	126,900	
17	Sudbury	–	2,000	70,900	172,500	
	Selected others					
	St. John's	31,600	29,600	67,800	214,000	
	St. Catharines	9,000	9,900	67,100	437,100	
	Saskatoon	–	113	53,300	336,600	
	Kingston	14,100	18,000	49,300	176,200	
	Galt	5,200	7,900	19,200	–	
	Charlottetown	11,500	12,100	15,900	42,400	
	Paris	3,200	3,300	5,300	12,300	
	Hanover	?	1,400	3,500	7,700	

1 Larger than 70,000 in 1951, and selected others
2 To 1951, Port Arthur and Fort William
Source: Census of Canada

Table 6.1. The Language of Neighbourhood in the *Globe and Mail*, 1880–2019 Counts

	Raw counts				Adjusted counts[1]		
	RD	RA	N		RD	RA	N
1880s	2	–	28		8	–	108
1890s	13	–	37		41	–	115
1900s	78	4	28		156	8	56
1910s	72	10	80		100	14	111
1920s	93	34	128		89	32	122
1930s	64	40	68		58	36	62
1940s	67	62	31		59	54	27
1950s	92	246	26		92	246	26
1960s	66	331	24		60	301	22
1970s	42	271	73		35	226	61
1980s	55	267	74		45	217	60
1990s	21	158	154		22	163	159
2000s	22	165	379		22	166	382
2010s	16	65	311		22	90	432
2010s adj[2]	20	81	389		28	112	540

RD: Residential district (full text search)
RA: Residential area (full text search)
N: Neighbourhood (title search only)
1 Counts adjusted by the relative frequency of "and" in full text
2 Counts adjusted to obtain comparable estimate for whole decade (Appendix)
Source: ProQuest *Globe and Mail* database

Bar chart 6.1. The Language of Neighbourhood in the *Globe and Mail*,
1880–2019

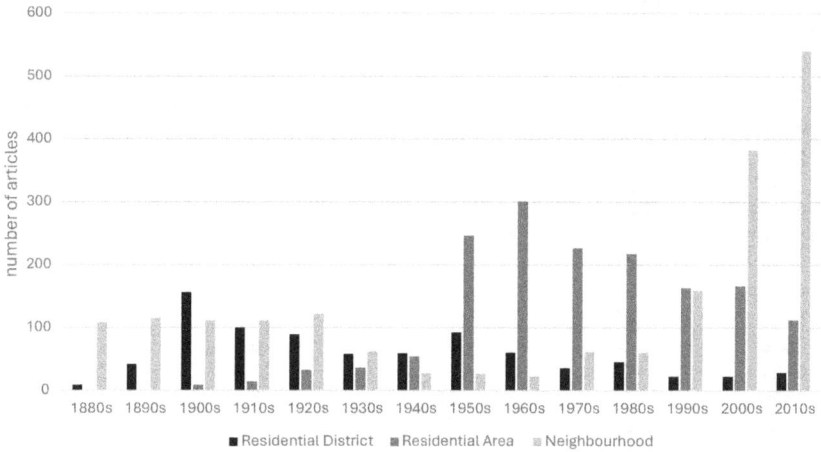

Number of articles where these terms are mentioned in the full text ("residential district," "residential area") or in the title only ("neighbourhood")
Source: ProQuest *Globe and Mail* database

Table 6.2. The Associations of "Neighbourhood"

The proportion of articles with "neighbourhood" in the title that contain the following words in the full text

	No. of articles	"planner" or "planning"	"worker"	"poor" or "immigrant"
1880s	28	0	0	25
1890s	37	0	8	16
1900s	28	0	4	14
1910s	80	3	33	13
1920s	128	2	56	13
1930s	68	7	74	15
1940s	31	10	36	13
1950s	26	12	15	8
1960s	24	25	8	8
1970s	73	38	11	12
1980s	74	23	5	12
1990s	154	12	14	16
2000s	379	14	15	11
2010s	311	21	10	10
2010s adj[1]	389	–	–	–

1 Counts adjusted to obtain comparable estimate for whole decade (Appendix)
Source: ProQuest *Globe and Mail* database

Table 6.3. The Language of Residential Organizations in the *Globe and Mail*,
1880–2019: Counts

	Raw counts			Adjusted counts[1]		
	RA	ResA	NA	RA	ResA	NA
1880s	0	0	0	0	0	0
1890s	151	3	0	472	9	0
1900s	66	1	0	132	2	0
1910s	268	1	0	372	1	0
1920s	667	0	0	635	0	0
1930s	402	1	0	365	1	0
1940s	102	7	1	90	6	1
1950s	360	19	0	360	19	0
1960s	339	110	0	308	100	0
1970s	221	206	0	184	172	0
1980s	117	230	1	95	187	1
1990s	24	54	4	25	56	4
2000s	31	70	15	31	71	15
2010s	9	46	18	16	79	31
2010s adj[2]	11	58	23	20	99	39

RA: Ratepayers' associations
ResA: Residents' associations
NA: Neighbourhood associations
1 Counts adjusted by the relative frequency of "and" in full text
2 Counts adjusted to obtain comparable estimate for whole decade (Appendix)
Source: ProQuest *Globe and Mail* database

Table 7.1. The Further Associations of "Neighbourhood"

The proportion of articles with "neighbourhood" in the title that contain the following words in the full text

	No. of articles	Suburb	Homeowner	Real estate	School
1880s	28	0	7	11	18
1890s	37	0	3	0	14
1900s	28	4	0	0	21
1910s	80	3	4	1	25
1920s	128	0	0	1	23
1930s	68	3	0	0	19
1940s	31	0	3	0	39
1950s	26	0	4	0	31
1960s	24	13	21	4	42
1970s	73	7	10	8	29
1980s	74	10	19	7	27
1990s	154	15	21	11	27
2000s	379	10	23	25	31
2010s	311	10	31	20	31
2010s adj[1]	389	–	–	–	–

1 Count adjusted to obtain comparable estimate for whole decade (Appendix)
Source: ProQuest *Globe and Mail* database

Bar chart 7.1. The Further Associations of "Neighbourhood"

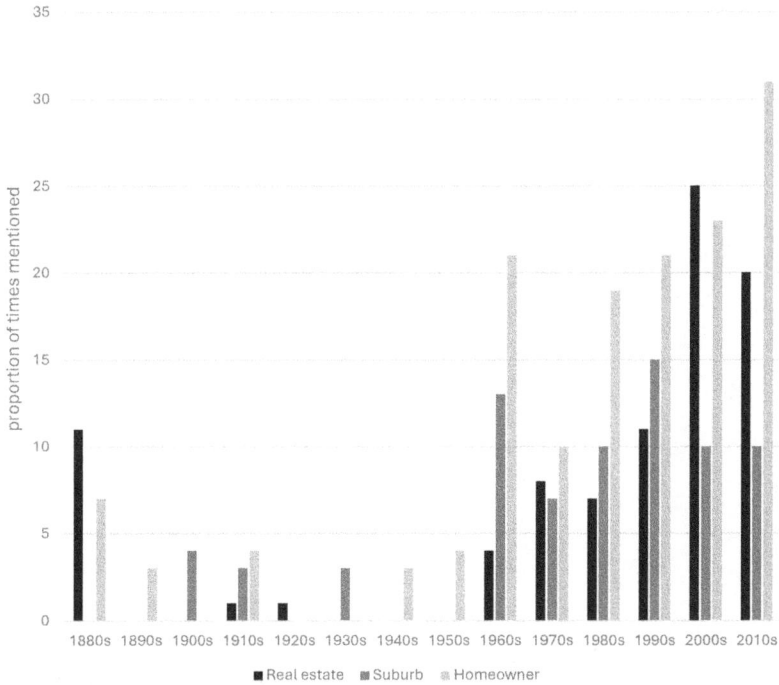

The proportion of articles with "neighbourhood" in the title that contain the above words in the full text
Source: ProQuest *Globe and Mail* database

Figures

Figure 1. When gentrification came to Cabbagetown: Average price of residential property in Don Vale, 1945–1982.

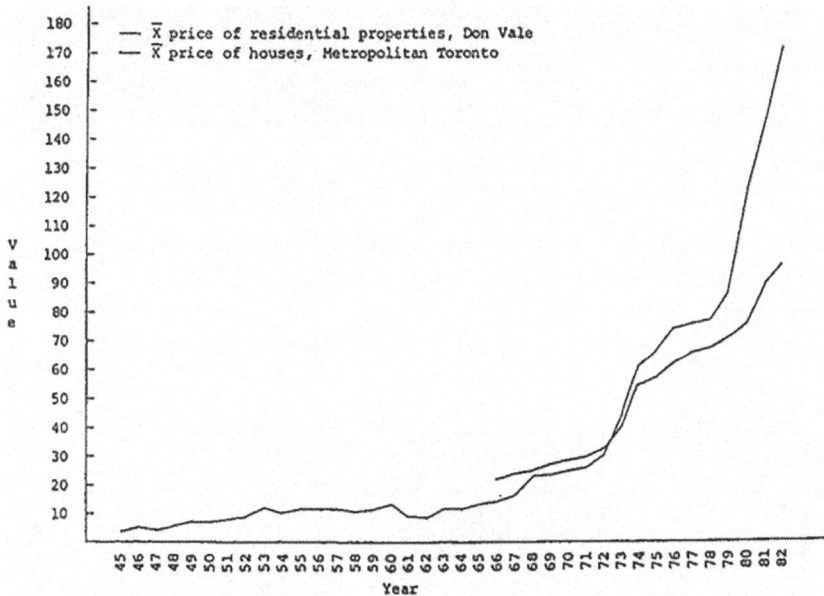

Source: Deeds, Land Registry Office. Toronto Real Estate Board, *House Price Trends*.

Source: Joanne Sabourin, "The Process of Gentrification: Lessons from an Inner-City Neighbourhood," in Frances Frisken, ed., *The Changing Canadian Metropolis: A Public Policy Perspective,* vol.1 (Berkeley and Toronto: Institute of Government Studies Press, UC Berkeley and Canadian Urban Institute, 1974), 273.

Figure 2. Although Montreal renters moved frequently until well into the twentieth century, their neighbours often included family.

Ann Beattie
the grandmother lives
at 351 rue Charron,

Her son John lives in the
white house on Liverpool,
visible from Ann's window

Her son William & family
live one block away.

Source: Sherry Olson.

Figure 3. Located on Toronto's fringe, Jean Bradley's childhood stomping grounds in the 1920s and 1930s necessarily included a wide territory.

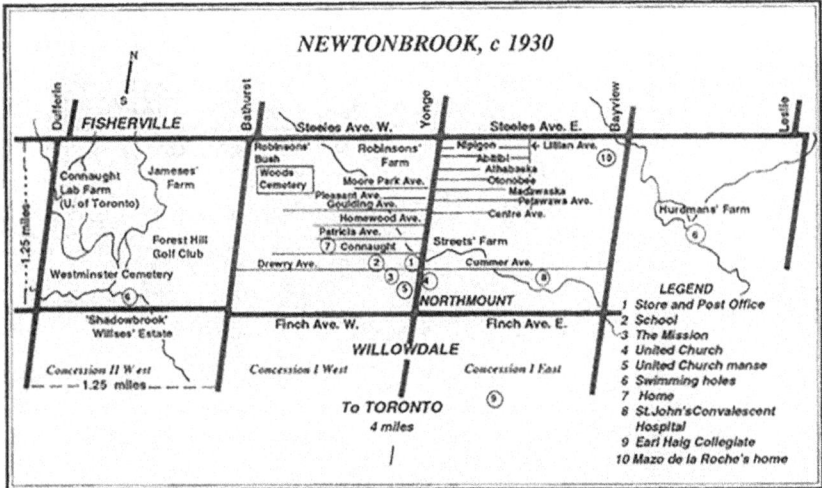

NEWTONBROOK, c 1930

LEGEND
1 Store and Post Office
2 School
3 The Mission
4 United Church
5 United Church manse
6 Swimming holes
7 Home
8 St. John's Convalescent Hospital
9 Earl Haig Collegiate
10 Mazo de la Roche's home

Source: Jean W. Bradley, *A Home across the Water: A Memoir* (North York, ON: Braeward, 1996).

Figure 4. In an early postwar subdivision of Grimsby, Ontario, neighbourhood children moved in packs from one unfenced backyard to another.

Source: B. Atlee.

Figure 5. In Vancouver's elite West End in 1908, women organized "at homes" for neighbours on adjacent blocks.

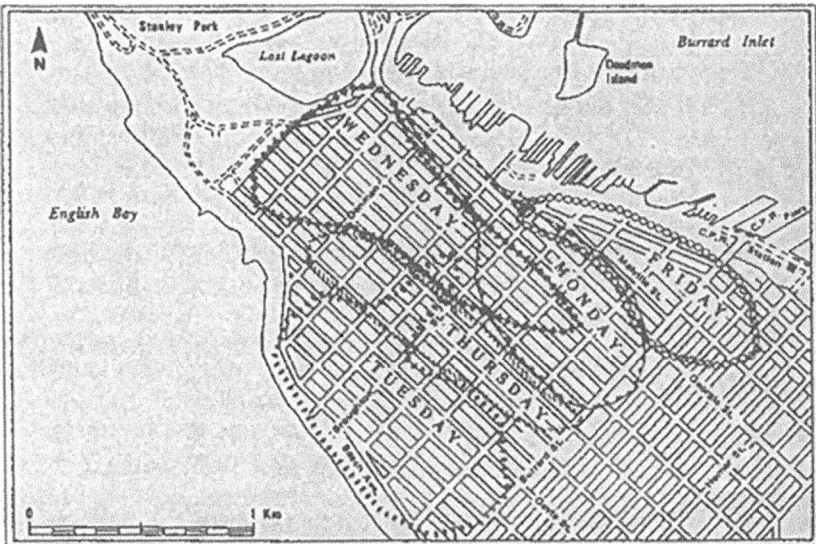

Source: Deryck Holdsworth, "House and Home in Vancouver: Images of West Coast Urbanism," in Gilbert A. Stelter and Alan F.J. Artibise, eds., *The Canadian City: Essays in Urban History* (Toronto: McClelland and Stewart, 1977), 199.

Figure 6. Winnipeg's skid row, 1976. Although the term is not used much now, many cities still have a cluster of blocks with homeless shelters, rooming houses, and social service agencies.

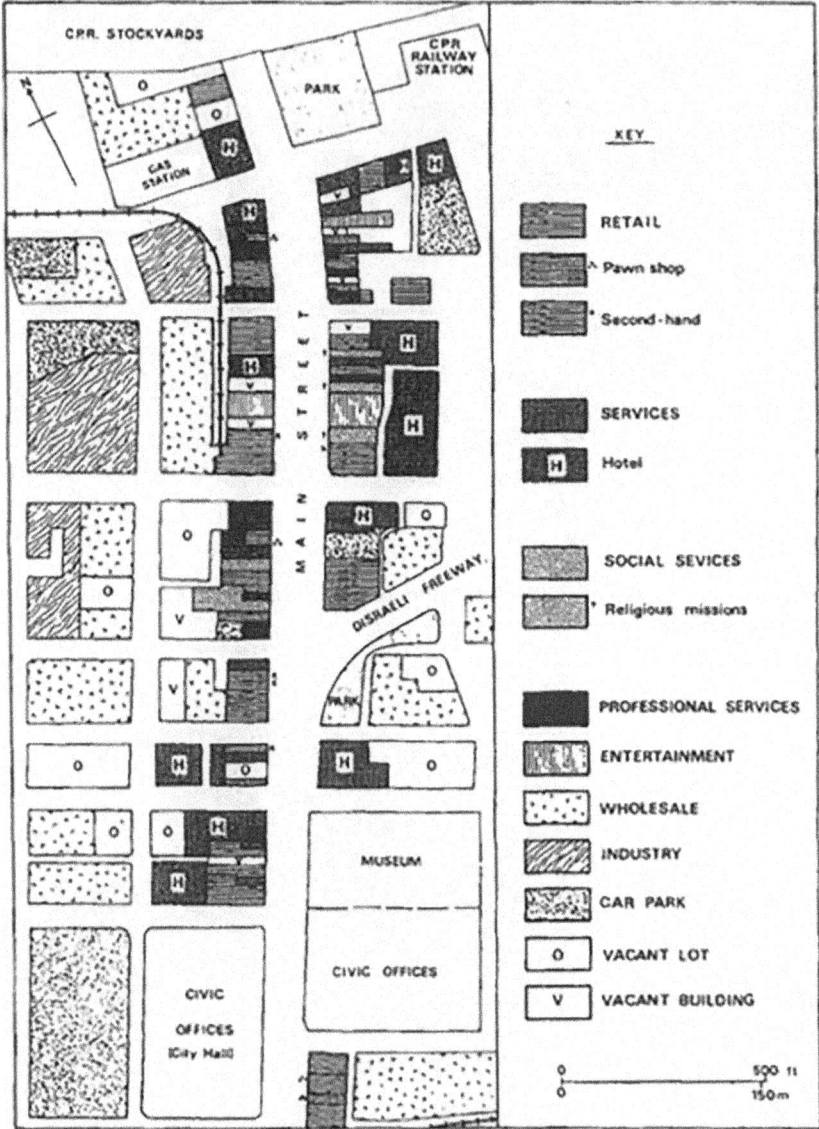

Source: Gwyn Rowley, "'Plus ça change …': A Canadian Skid Row," *Canadian Geographer* 22, no. 3 (1978): 214.

Figure 7. In 1910, facing discrimination, Chinese Canadians settled in a tight cluster in downtown Vancouver.

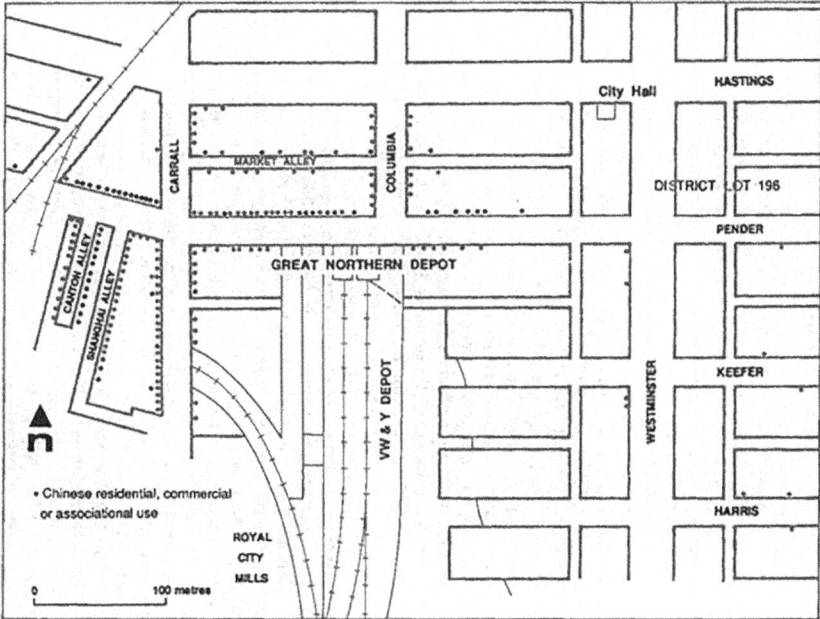

Source: Kay Anderson, *Vancouver's Chinatown: Racial Discourse in Canada, 1875–1980* (Montreal and Kingston: McGill-Queen's University Press, 1991).

Figure 8. Shunned by white colonial settler society, and with limited resources, Métis formed a small community on Winnipeg's fringe in the early twentieth century.

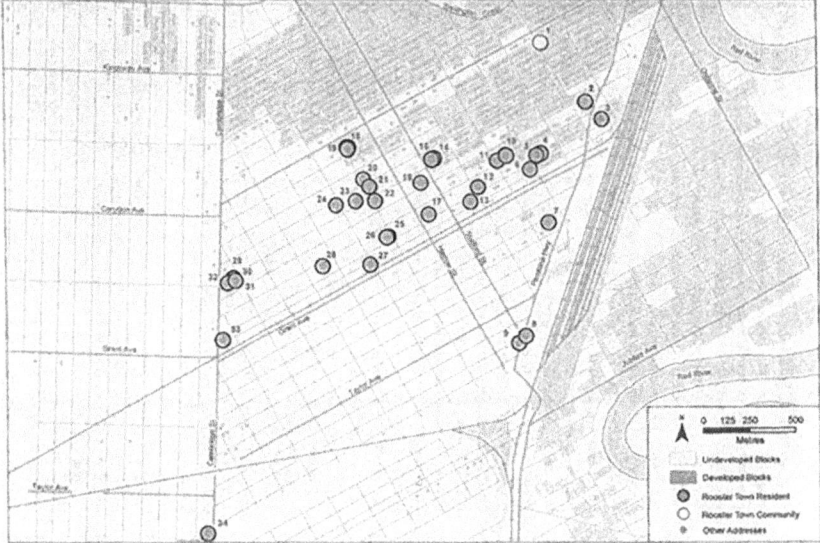

Source: Evelyn Peters, Matthew Stock, and Adrian Werner, *Rooster Town: The History of an Urban Métis Community, 1901–1961* (Winnipeg: University of Manitoba Press, 2018), fig. 31, p. 59. Map produced by Adrian Werner.

Figure 9. Montreal, 1942. In the dense urban environments that once characterized many eastern and central Canadian cities, children played on sidewalks and streets.

Source: *Montreal Gazette* Collection, National Archives of Canada. NAC 108313.

Figure 10. Gangs were usually male-dominated and defended a well-defined urban turf. In the 1950s, Burnsiders were an exception.

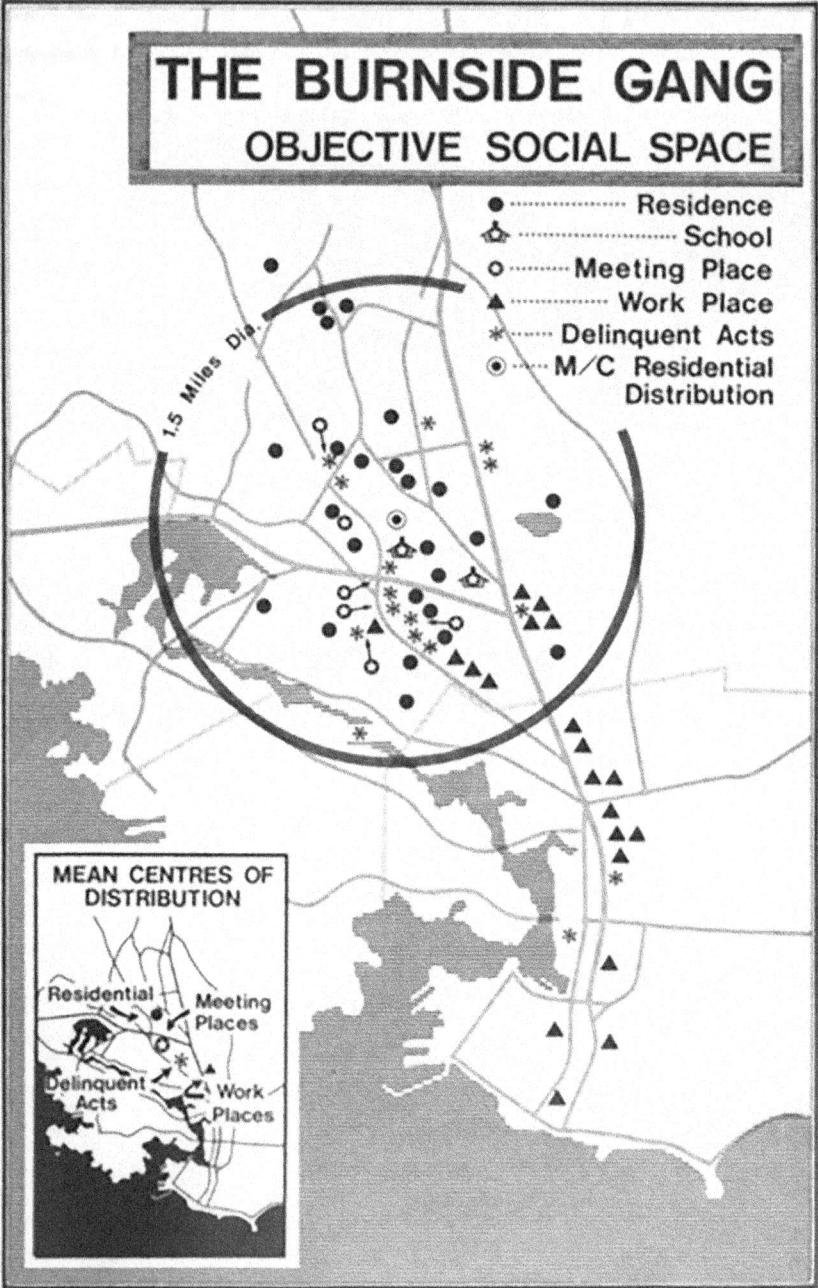

Source: Douglas Porteous, "The Burnside Teenage Gang: Territoriality, Social Space, and Community Planning, in Charles N. Forward, ed., *Residential and Neighbourhood Studies in Victoria*, Western Geographical Studies, vol. 5 (Victoria, BC: Department of Geography, University of Victoria, 1973).

Figure 11. Sydney, Nova Scotia, c. 1913. Even at the block scale, in immigrant districts there was a mixture of ethnicities.

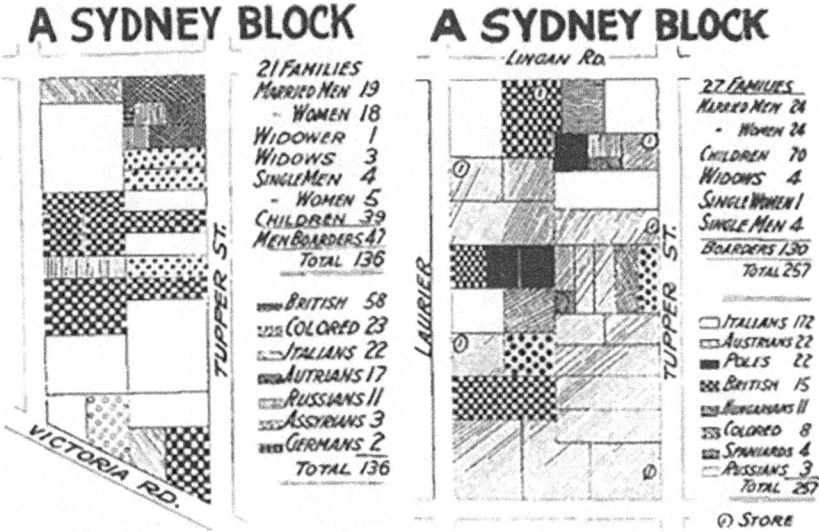

Source: Bryce M. Stewart, "The Housing of Our immigrant Workers," in *Papers and Proceedings: Canadian Political Science Association* 1 (1913): 98–111. Reproduced in Paul Rutherford, ed., *Saving the Canadian City: The First Phase 1880–1920* (Toronto: University of Toronto Press, 1974), 137–54.

Figure 12. Calvert, Newfoundland, 1980s. Even within the smallest communities, geographical distinctions are made.

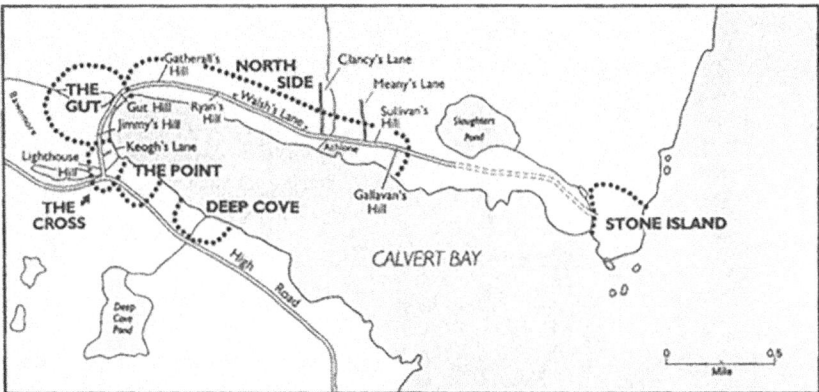

Source: Gerald Pocius, *A Place to Belong: Community Order and Everyday Space in Calvert, Newfoundland* (Montreal and Kingston: McGill-Queen's University Press), 71.

Figure 13. In Montreal, where ethnic segregation was the norm, growth in the late nineteenth century enabled class segregation to emerge within each group.

1881

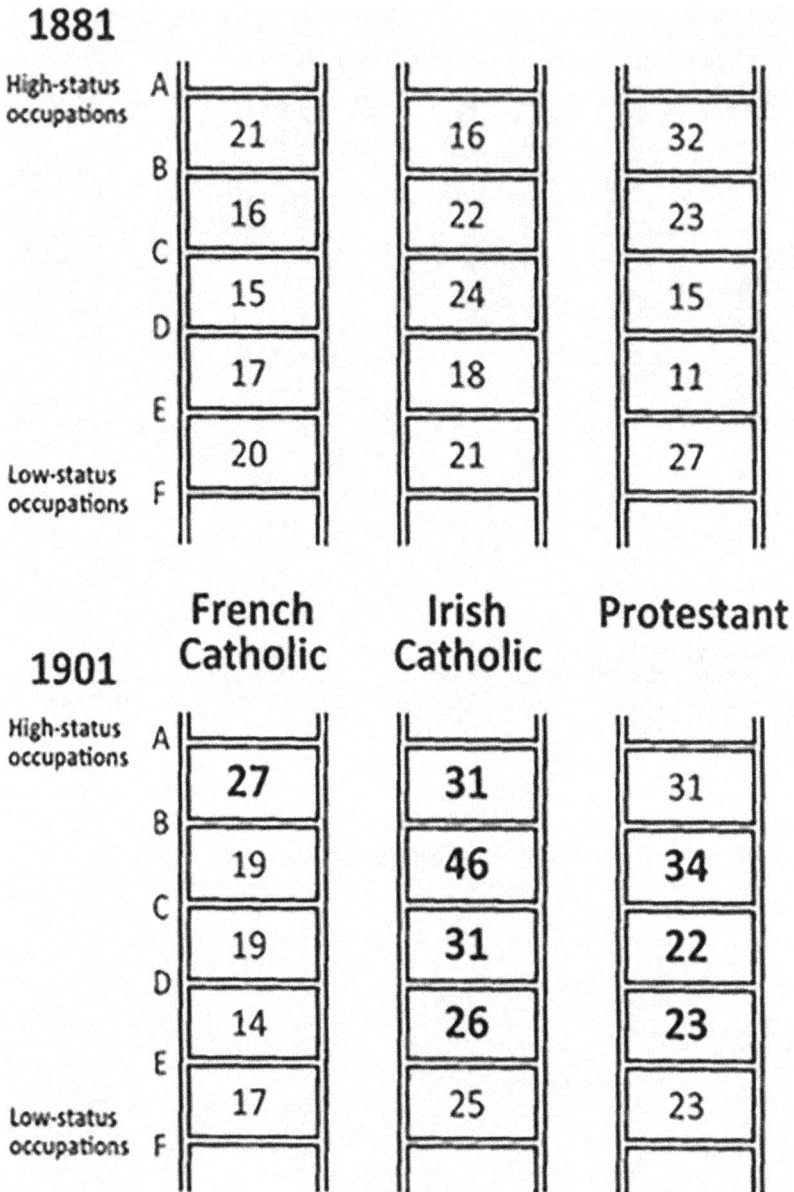

	1881	French Catholic	Irish Catholic	Protestant
High-status occupations	A	21	16	32
	B	16	22	23
	C	15	24	15
	D	17	18	11
	E	20	21	27
Low-status occupations	F			

French Catholic **Irish Catholic** **Protestant**

1901

	1901	French Catholic	Irish Catholic	Protestant
High-status occupations	A	27	31	31
	B	19	46	34
	C	19	31	22
	D	14	26	23
	E	17	25	23
Low-status occupations	F			

Figure 5 Social distances between people of same ethnicity but different occupational status appear farther apart in 1901 than in 1881

Sources: SI measures, census of 1881 and 1901 sample.

Note: Occupations are ranked A (high) to F (low). Bold numbers in 1901 highlight a substantial increase in the spread between rungs of occupation, 1881–1901.

Source: Jason Gilliland, Sherry Olson, and Danielle Gauvreau, "Did Segregation Increase as the City Expanded? The Case of Montreal, 1881–1901," *Social Science History* 35, no 4 (2011).

Figure 14. Africville, Halifax, in 1958, shortly before its demolition. Several other Black neighbourhoods existed before migration from the Caribbean picked up in the 1960s.

Source: Donald Clairmont and Dennis W. Magill, *Africville: The Life and Death of a Canadian Black Community*, 3rd ed. (Toronto: Canadian Scholars Press, 1999).

Figure 15. Carol Talbot's mental map of the tight-knit Black community in Windsor in the 1940s.

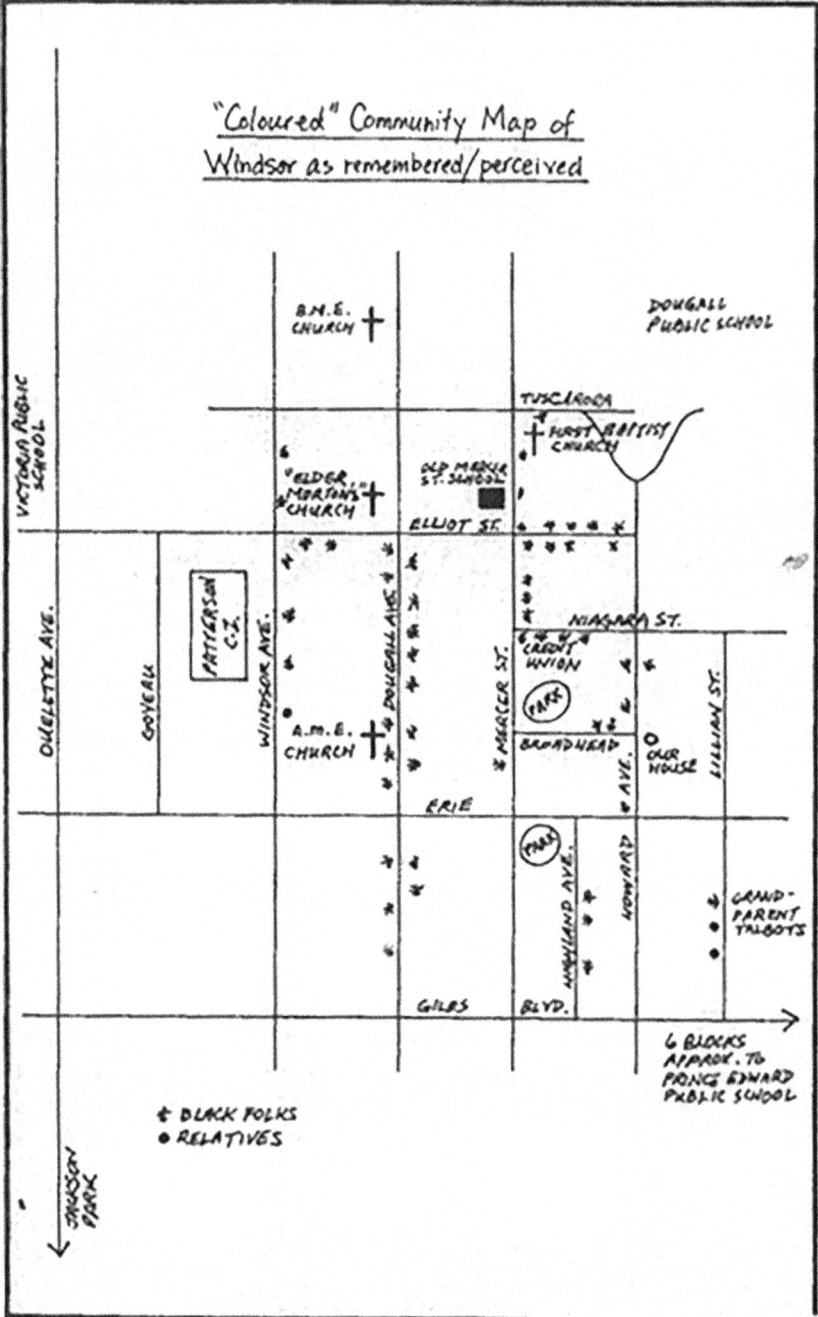

Source: Carol Talbot, *Growing Up Black in Canada* (Toronto: Williams-Wallace, 1984), 18.

Figure 16. By the early twenty-first century, Calgary's common interest developments were distributed widely across its suburbs.

LEGEND

CBD area

Nose Hill park

Industrial area

Distinctive community district (city-designated)

Community districts with Common Interest Development (CID)

Golf course community (Pre-1992)

Lake community (Pre-1992)

Golf course community (Post 1992)

Lake community (Post 1992)

"Environment" community (Post 1992)

"New Urbanism" community (Post 1992)

Wired or "E-community" (Post 1992)

Other or combined themes (Post 1992)

Community district with at least one retirement village in 1994

Community district with at least one retirement village built between 1994 and 2004

ARB Arbour Lake
BRI Bridlewood
CGR Cougar Ridge
CHA Lake Chaparral
COR Coral Springs
COU Country Hills
COV Coventry Hills
CPF Copperfield
CRA Cranston
CRM Crestmont
DIS Discovery Ridge
DOU Douglasdale Estates
EVE Evergreen
EVN Evanston
HAM Hamptons
HAR Harvest Hills
LKB Lake Bonavista
MCK McKenzie Lake
MCT McKenzie Towne
MID Midnapore
MRT Martindale
NEB New Brighton
PAN Panorama Hills
ROC Rocky Ridge
ROY Royal Oak
SCA Scarboro
SDC Lake Sundance
SHS Shawnee Slopes
SYV Symons Valley
TUS Tuscany
UMR Upper Mount Royal
VAL Valley Ridge
VAR Varsity
WIL Willow Ridge
WSP West Springs

0   4   8 km

Source: Ivan Townshend, "From Public Neighbourhood to Multi-tier Private Neighbourhoods: The Evolving Ecology of Neighbourhood Privatization in Calgary," *GeoJournal* 66 (2006): 100.

Figure 17. A map that shows the high degree of segregation of Montreal's Jewish community. Rabbi Glazer was rarely required to travel far from home.

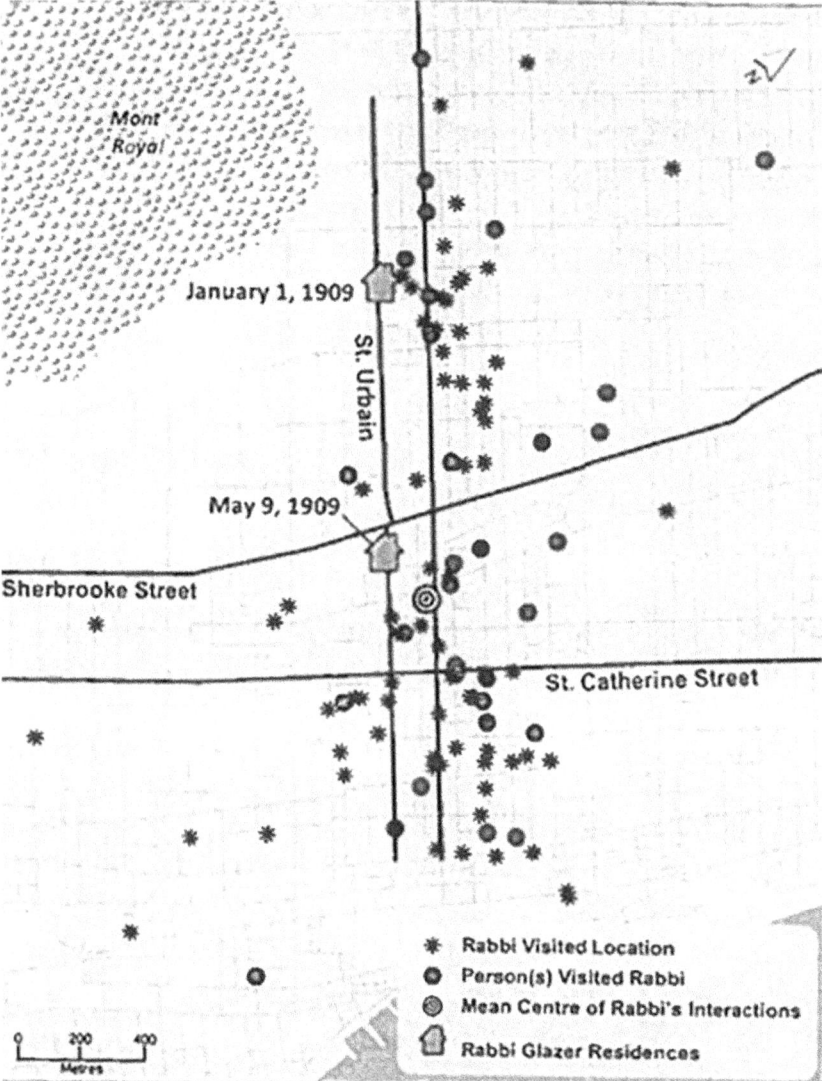

Source: Mary Poutanen and Jason Gilliland, "Mapping Work in Early Twentieth-Century Montreal: A Rabbi, a Neighbourhood and a Community, *Urban History Review* 45, no. 2 (2017): 13.

Figure 18. Montreal, 1880. Reflecting the city's social geography, infant
mortality rates ranged widely in the late 1800s and early 1900s.

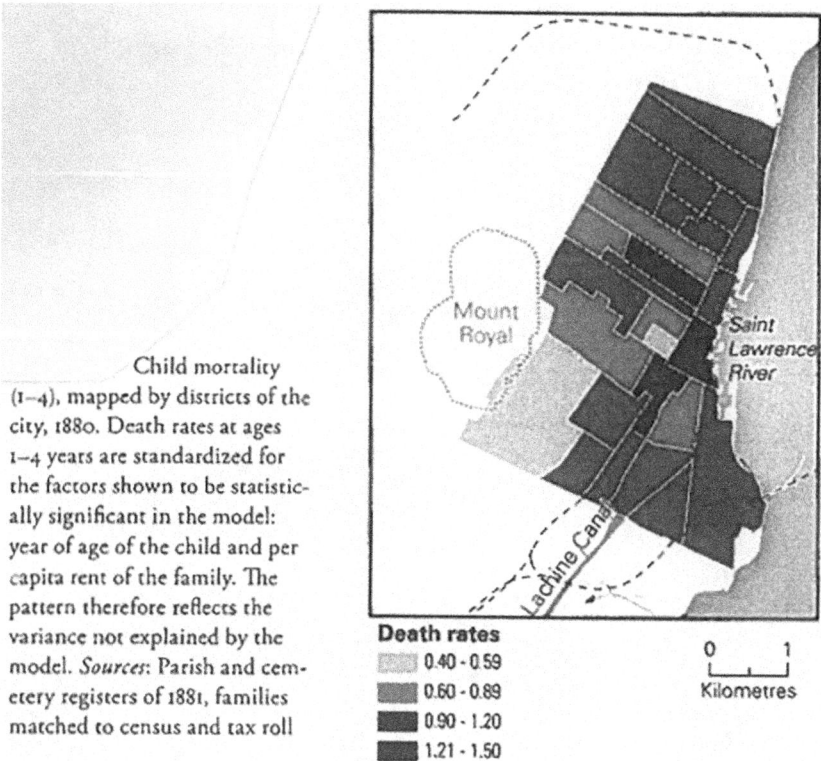

Child mortality
(1–4), mapped by districts of the
city, 1880. Death rates at ages
1–4 years are standardized for
the factors shown to be statistically significant in the model:
year of age of the child and per
capita rent of the family. The
pattern therefore reflects the
variance not explained by the
model. *Sources*: Parish and cem-
etery registers of 1881, families
matched to census and tax roll

Death rates

- 0.40 - 0.59
- 0.60 - 0.89
- 0.90 - 1.20
- 1.21 - 1.50

0   1
Kilometres

Source: Sherry Olson and Patricia Thornton, *Peopling the North American City:
Montreal 1840–1900* (Montreal and Kingston: McGill-Queen's University Press,
2011), 119

Figure 19. Ad hoc zoning. Over half a century, the petitions of local residents created an intricate map of restrictions that excluded non-residential land use from large parts of Toronto.

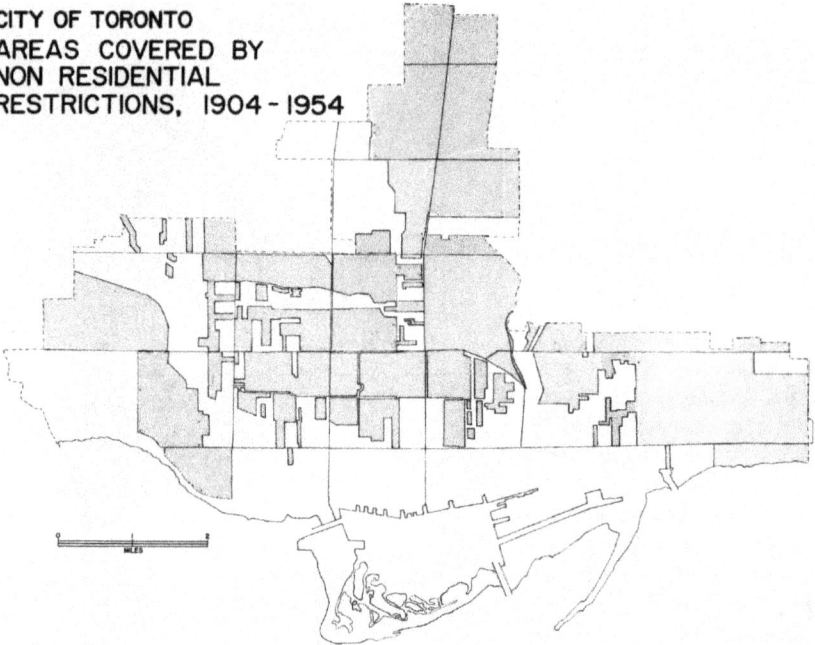

CITY OF TORONTO
AREAS COVERED BY
NON RESIDENTIAL
RESTRICTIONS, 1904 - 1954

Source: Peter Moore, "Zoning and Planning: The Toronto Experience, 1904–1970," in Alan F.J. Artibise and Gilbert Stelter, eds., *The Usable Urban Past* (Toronto: Macmillan, 1979), 321.

Figure 20. Sainte-Foy, Quebec City, 1970s. Rearrangements of school catchment areas aroused widespread concern and opposition.

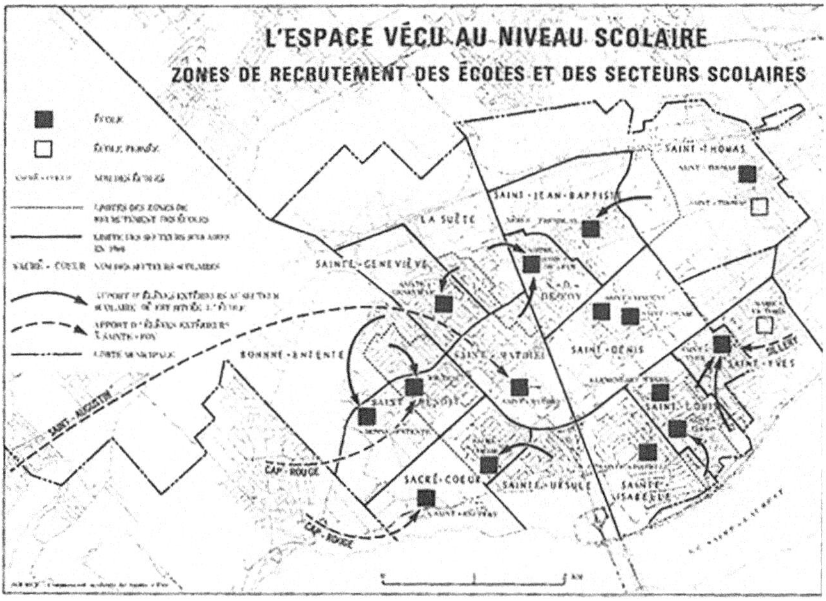

Source: François Hulbert, "Pouvoir municipal et développement urbain: Le cas de Sainte-Foy en banlieue de Québec," *Cahiers de géographie de Québec* 25, no. 66 (1981): 394.

References

Abella, Irving, and David Millar, eds. 1978. *The Canadian Worker in the Twentieth Century.* Toronto: Oxford University Press.

Adams, Thomas. (1923) 1974. Modern city planning: Its meaning and methods. In P. Rutherford, ed., *Saving the Canadian City: The First Phase 1880–1920.* Toronto: University of Toronto Press, 247–73. https://doi.org/10.3138/9781487583446-026

Africville Genealogy Society. 2010. *The Spirit of Africville.* 2nd ed. Halifax: Formac.

Albouy, David, and Mike Zabek. 2016. Housing inequality. National Bureau of Economic Research Working Paper 21916. Cambridge, MA: NBER. https://doi.org/10.3386/w21916

All-Party Parliamentary Group. 2023. *A Neighbourhood Strategy for National Renewal.* London: APPG. https://reform.uk/wp-content/uploads/2023/10/A-Neighbourhood-Strategy-for-National-Renewal.pdf

Allen, Jeff, and Steven Farber. 2020. Suburbanization of transport poverty. *Annals of the Association of American Geographers* 111, no. 6. https://doi.org/10.31235/osf.io/hkpfj

Altonji, Joseph G., and Richard K. Mansfield. 2011. The role of family, school and community characteristics in inequality in education and labour-market outcomes. In Greg J. Duncan and Richard J. Murnane, eds., *Whither Opportunity? Rising Inequality, Schools, and Children's Life Chances,* 339–58. New York: Russell Sage.

Alvaredo, Facundo, Anthony B. Atkinson, Thomas Piketty, and Emmanuel Saez. 2013. The top 1 percent in international and historical perspective. *Journal of Economic Perspectives* 27, no. 3: 3–20. https://doi.org/10.1257/jep.27.3.3

Ames, Herbert W. (1897) 1972. *The City below the Hill.* Toronto: University of Toronto Press. https://doi.org/10.3138/9781442656291

270 References

Anderson, Alan. 2017. *Home in the City: Urban Aboriginal Housing and Living Conditions*. Toronto: University of Toronto Press.
Anderson, Benedict. 1991. *Imagined Communities: Reflections on the Origins and Spread of Nationalism*. London: Verso.
Anderson, James. 1979. The municipal government reform movement in Western Canada. In Alan Artibise and Gilbert A. Stelter, eds., *The Usable Urban Past*, 73–111. Toronto: Macmillan. https://doi.org/10.1515 /9780773580640-007
Anderson, Kay. 1991. *Vancouver's Chinatown: Racial Discourse in Canada, 1875–1980*. Montreal and Kingston: McGill-Queen's University Press. https://doi.org/10.1515/9780773562974
Andersson, Roger, and Sako Musterd. 2005. Area-based policies: A critical appraisal. *Tijdschrift voor Economische en Sociale Geografie* 96: 377–89. https://doi.org/10.1111/j.1467-9663.2005.00470.x
Andrzcjewski, Anna V. 2009. Building privacy and community: Surveillance in a postwar American suburban development in Madison, Wisconsin. *Landscape Journal* 28, no. 1: 40–56. https://doi.org/10.3368/lj.28.1.40
Armstrong, Pat, and Hugh Armstrong. 2010. *The Double Ghetto*. Toronto: McClelland and Stewart.
Armstrong, W.S.B. 1938. Toronto's pressing traffic problem. *Globe and Mail*, 29 March.
Arnason, David, and Mhari Mackintosh. 2005. *The Imagined City: A Literary History of Winnipeg*. Winnipeg: Turnstone Press.
Arnstein, Sherry. 1969. A ladder of citizen participation. *Journal of the American Planning Association* 35, no. 4: 216–24. https://doi.org /10.1080/01944366908977225
Artibise, Alan F.J. 1975. *Winnipeg: A Social History of Urban Growth, 1874– 1914*. Montreal and Kingston: McGill-Queen's University Press. https:// doi.org/10.1515/9780773580633
Ask many improvements. 1911. *The Globe* 20 March.
Ask to operate in home zones. 1954. *Globe and Mail*, 3 June.
August, Martine. 2007. Social mix and Canadian public housing redevelopment: Experiences in Toronto. *Canadian Journal of Urban Research* 17, no. 1: 82–100.
August, Martine. 2014. Challenging the rhetoric of stigmatization: The benefits of concentrated poverty in Toronto's Regent Park. *Environment and Planning A*. 46, no. 6: 1317–33. https://doi.org/10.1068/a45635
Aulakh, Raveena. 2006. Spate of killings doesn't faze violence-weary Jane-Finch. *Globe and Mail*, 29 December.

Axelrod, Paul. 1990. *Making a Middle Class: Student Life in English Canada during the Thirties*. Montreal and Kingston: McGill-Queen's University Press. https://doi.org/10.1515/9780773562424

Bacher, John. 1993. *Keeping to the Marketplace: The Evolution of Canadian Housing Policy*. Montreal and Kingston: McGill-Queen's University Press. https://doi.org/10.1515/9780773563827

Bailey, Bill, ed. 1980. *Stories of York*. Toronto: York Historical Society.

Baillargeon, Denyse. 1999. *Making Do: Women, Family and Home. Montreal during the Great Depression*. Waterloo, ON: University of Waterloo Press.

Baillie, David. 2015. *What We Salvage*. Toronto: ChiZine.

Bain, Alison, and Julie A. Podmore. 2021. Relocating queer: Comparing suburban LGBTQ2S activism on Vancouver's periphery. *Urban Studies* 58, no. 7: 1500–19. https://doi.org/10.1177/0042098020931282

Balakrishnan, J.R. 1976. Ethnic residential segregation in the metropolitan areas of Canada. *Canadian Journal of Sociology* 1, no. 4: 481–98. https://doi.org/10.2307/3339757

Balakrishnan, J.R. 1982. Changing patterns of ethnic residential segregation in the metropolitan areas of Canada. *Canadian Review of Sociology and Anthropology* 19, no. 1: 92–111. https://doi.org/10.1111/j.1755-618X.1982.tb01323.x

Bania, Melania. 2009. Gang violence among youth and young adults: (Dis)affiliation and the potential for prevention. *IPC Review* 3: 89–116.

Barlow, Matthew. 2016. *Griffintown: Identity and Memory in an Irish Diaspora Neighbourhood*. Vancouver: UBC Press. https://doi.org/10.59962/9780774834353

Barman, Jean. 1986. Neighbourhood and community in interwar Vancouver: Residential differentiation and civic voting behaviour. *BC Studies* 69–70: 97–141. https://doi.org/10.59962/9780774857079-006

Barman, Jean. 1988. "Knowledge is essential for universal progress but fatal to class privilege": Working people and the schools in Vancouver during the 1920s. *Labour/Le Travail* 22: 9–66. https://doi.org/10.2307/25143027

Baskerville, Peter. 2001. Home ownership and spacious homes: Equity under stress in early twentieth-century Canada. *Journal of Family History* 26, no. 2: 272–88. https://doi.org/10.1177/036319900102600206

Bator, Paul A. 1979. "The struggle to raise the lower classes": Public health reform and the problem of poverty in Toronto, 1910 to 1921. *Journal of Canadian Studies* 14, no. 1: 43–8. https://doi.org/10.3138/jcs.14.1.43

Batten, Jack. 2004. *The Annex: The Story of a Toronto Neighbourhood*. Erin, ON: Boston Mills Press.

Baum, Kathryn, and Matthew McClearn. 2021. Extreme deadly heat in Canada. *Globe and Mail*, 25 Sept. https://www.theglobeandmail.com/canada/article-extreme-deadly-heat-in-canada-is-going-to-come-back-and-worse-will-we/

Bayer, Patrick, Fernando Ferreira, and Robert McMillan. 2007. A unified framework for measuring preferences for schools and neighborhoods. *Journal of Political Economy* 115, no. 4: 588–632. https://doi.org/10.1086/522381

Beasley, Larry. 2019. *Vancouverism*. Vancouver: UBC Press. https://doi.org/10.59962/9780774890328

Beauchemin, Yves. (1981) 1986. *The Alley Cat*. Toronto: McClelland and Stewart.

Beauchemin, Yves. (1989) 1993. *Juliette*. Toronto: McClelland and Stewart.

Belec, John. 1997. The Dominion Housing Act. *Urban History Review* 25, no. 2: 53–62.

Bell, Margaret. 1913. Toronto's melting pot. *Canadian Magazine* 41: 234–42.

Bellavance, Claude, and France Normand. 2014. Trois-Rivières and its people: A portrait of a smaller city in transition at the beginning of the twentieth century. In Gordon Darroch, ed., *Canada's Century. Hidden Histories*, 171–96. Montreal and Kingston: McGill-Queen's University Press. https://doi.org/10.1515/9780773589391-013

Bender, Thomas. 1978. *Community and Social Change in America*. Baltimore: Johns Hopkins University Press.

Berardi, Luca. 2021a. Neighborhood wisdom: An ethnographic study of localized street knowledge. *Qualitative Sociology* 44: 103–24. https://doi.org/10.1007/s11133-020-09454-z

Berardi, Luca. 2021b. Personal communication. 18 February.

Bérubé, Harold. 2015. *Des Sociétés Distinctes: Gouverner les Banlieus Bourgeoises de Montréal, 1880–1939*. Montreal and Kingston: McGill-Queen's University Press. https://doi.org/10.1515/9780773596146

Bérubé, Harold. 2016. La ville, quartier par quartier. *Labour/Le Travail* 78: 265–279. https://doi.org/10.1353/llt.2016.0059

Bérubé, Harold. 2019. Les "villages" de Montréal, ou la metropole comme comunauté de communautés: Reflexions sur l'utilisation de la notion de quartier en histoire urbaine. In Joanne Burgess and Paul-André Linteau, eds., *Histoire et Patrimoine: Pistes de Recherche et de Mise en Valeur*, 57–76. Montreal: Presses de l'université Laval. https://doi.org/10.1515/9782763743301-004

Betancur, John, and Janet Smith. 2016. *Claiming Neighborhood: New Ways of Understanding Urban Change*. Chicago: University of Illinois Press. https://doi.org/10.5406/illinois/9780252040504.001.0001

Bethune, Brian. 2014. The end of neighbours: How our increasingly closed-off lives are poisoning our politics and endangering our health. *Maclean's*, 8 August.

Bherer, Laurence, and Jean-Pierre Collin. 2013. Enjeux urbains et mobilisation politique: De la subsidiarité à la gouvernance institutionnalisée. In Narcisee Perron and Dany Fougères, eds., *Histoire de la Region Montréalaise, des Origines à Nos Jours, Tome 2: 1921–2010*, 1169–208. Montreal: Éditions PUL/IQRC.

Bischoff, Kendra, and Sean F. Reardon. 2013. Residential segregation by income, 1970–2009. In John R. Logan, ed., *The Lost Decade? Social Change in the U.S. after 2000*. New York: Russell Sage.

Bissonette, Laura, Kathi Wilson, Scott Bell, and Tayyab Shah. 2012. Neighborhoods and potential access to health care: The role of spatial and aspatial factors. *Health and Place* 18: 841–53. https://doi.org/10.1016/j.healthplace.2012.03.007

Blighted area: Shabbiness is spreading northward. 1951*Globe and Mail*, 11 August.

Blokland, Talja. 2001. Bricks, mortar, and memories: Neighbourhoods and networks in collective acts of remembering. *International Journal of Urban and Regional Research* 25, no. 2: 268–83. https://doi.org/10.1111/1468-2427.00311

Blokland, Talja. 2003. *Urban Bonds: Social Relationships in an Inner City Neighbourhood*. Cambridge: Polity Press.

Bloomfield, Elizabeth. 1982. Reshaping the urban landscape? Town planning efforts in Kitchener-Waterloo, 1912–1925. In Gilbert A. Stelter and Alan F.J. Artibise, eds., *Shaping the Urban Landscape: Aspects of the Canadian City-Building Process*, 256–303. Ottawa: Carleton University Press. https://doi.org/10.1515/9780773584860-011

Boivin, Remi, and Silas N. de Melo. 2019. The concentration of crime at place in Montreal and Toronto. *Canadian Journal of Criminology and Criminal Justice* 61, no. 2: 46–65. https://doi.org/10.3138/cjccj.2018-0007

Boone, Christopher. 2011. The political ecology of floods in the late nineteenth century. In Stéphane Castonguay and Michèle Dagenais, eds., *Metropolitan Natures. Environmental Histories of Montreal*, 133–45. Pittsburgh: University of Pittsburgh Press. https://doi.org/10.2307/j.ctv111jfg2.13

Boschman, Robert. 2021. *White Coal City: A Memoir of Place and Family*. Regina: University of Regina Press. https://doi.org/10.1515/9780889777989

Bossard, James H.S. 1932. Residential propinquity as a factor in marriage selection. *American Journal of Sociology* 38, no. 2: 219–24. https://doi.org /10.1086/216031

Bottomley, John. 1977. Experience, Ideology and the Landscape: The Business Community, Urban Reform and the Establishment of Town Planning in Vancouver, B.C., 1900–1940. PhD diss., University of British Columbia.

Bouchier, Nancy, and Ken Cruikshank. 2020. Look on the Brightside, 1910-present. In Paul Weinberg, ed., *Reclaiming Hamilton. Essays from the New Ambitious City*, 45–82. Hamilton, ON: James Street North Books82.

Bourne, Larry, and J. David Hulchanski. 2020. Inequality and neighbourhood change: Context, concept and process. In Jill Grant, Alan Walks, and Howard Ramos, eds., *Changing Neighbourhoods. Social and Spatial Polarization in Canadian Cities*, 5–30. Vancouver: UBC Press. https://doi.org/10.59962/9780774862042-005

Bouthillette, Anne-Marie. 1997. Queer and gendered housing. A tale of two neighborhoods in Vancouver. In Gordon Ingram, Anne-Marie Bouthillette, and Yolanda Retter, eds., *Queers in Space: Communities/ Public Places/ Sites of Resistance*, 213–32. Seattle, WA: Bay Press.

Boyd, Monica, and Michael Vickers. 2000. 100 years of immigration to Canada. *Canadian Social Trends* (Autumn). Ottawa: Statistics Canada.

Bozikovic, Alex. 2017. The Storeys Margaret Atwood condemns. *Globe and Mail*, 30 August. https://www.theglobeandmail.com/opinion /the-storeys-margaret-atwood-condemns/article36123420/

Bozikovic, Alex, Joe Castaldo, and Danielle Webb. 2020. The 15-minute city aims to build more livable neighbourhoods. *Globe and Mail*, 23 November. https://www.theglobeandmail.com/canada/article-when-it-comes-to -liveable-neighbourhoods-theres-a-wide-divide-in/

Bradbury, Bettina. 1993. *Working Families: Age, Gender, and Daily Survival in Industrializing Montreal*. Toronto: McClelland and Stewart.

Bradford, Neil. 2013. Neighbourhood revitalization in Canada: Towards place-based policy solutions. In David Manley, Maarten van Ham, Nick Bailey, Ludi Simpson, and Duncan Maclennan, eds., *Neighbourhood Effects or Neighbourhood Based Problems? A Policy Context*, 156–76. Dordrecht: Springer. https://doi.org/10.1007/978-94-007-6695-2_8

Bradley, Jean W. 1996. *A Home across the Water: A Memoir*. North York, ON: Braeward.

Breton, Raymond, Wsevolod Isajiw, and Warren Kalbach. 1990. *Ethnic Identity and Equality. Varieties of Experience in a Canadian City*. Toronto: University of Toronto Press.

Bridge, Gary. 2006. It's not just a question of taste: Gentrification, the neighbourhood, and cultural capital. *Environment and Planning A*. 38, no. 10: 1965–18. https://doi.org/10.1068/a3853

Buddle, Kathleen. 2011. Urban aboriginal gangs and street sociality in the Canadian West: Places, performances and predicaments of transition. In Heather A. Howard and Craig Proulx, eds., *Aboriginal People in Canadian Cities: Transformations and Continuities*, 171–202. Waterloo, ON: Wilfrid Laurier University Press. https://doi.org/10.51644/9781554583140-010

Buffel, Tine, and Chris Phillipson. 2018. Urban ageing: New agendas for geographical gerontology. In Mark Skinner, Gavin J. Andrews, and Malcolm Cutchin, eds., *Geographical Gerontology*, 123–35. London: Routledge. https://doi.org/10.4324/9781315281216-10

Bula, Frances. 2024. How to build a friendly building. *Globe and Mail* 26 April.

Bunting, Trudi. 1987. Invisible upgrading in inner cities: Homeowners' reinvestment behaviour in central Kitchener. *Canadian Geographer* 31, no. 3: 209–22. https://doi.org/10.1111/j.1541-0064.1987.tb01235.x

Bureau of Municipal Research. 1918. *What Is 'The Ward' Going to Do with Toronto?* Toronto: The Bureau.

Burgess, Ernest W. 1925. Can neighborhood work have a scientific basis? In Robert E. Park and Ernest W. Burgess, eds., *The City*, 142–55. Chicago: University of Chicago Press.

Burley, David. 2013. Rooster Town: Winnipeg's lost Métis suburb, 1900–1960. *Urban History Review* 42, no. 1: 3–25. https://doi.org/10.3138/uhr .42.01.01

Burley, David, and Mike Maunder. 2008. *Living on Furby: Narratives of Home, Winnipeg, Manitoba, 1880–2005*. Winnipeg: Winnipeg Inner-City Research Alliance.

Business lively in the neighbourhood of the Limestone city. 1881. *The Globe*, 6 October.

Butler, Jesse K., Ruth Kane, and Fiona R. Cooligan. 2019. The closure of Rideau High School: A case study in the political economy of urban education in Ontario. *Canadian Journal of Educational Administration and Policy* 191: 83–105.

Butler, Tim, and Garry Robson. 2003. *London Calling: The Middle Classes and the Re-making of Inner London*. Oxford: Berg.

Cahuas, Madelaine C., Mannat Malik, and Sarah Wakefield. 2016. When is helping hurting? Understanding and challenging the (re)production of dominance in narratives of health, place and difference in Hamilton, Ontario. In Melissa Giesbrecht and Valorie Crooks, eds., *Place, Health* and Diversity. Learning from the Canadian Experience, 141–62. London: Routledge.

Callaghan, Morley. 1959. A cap for Steve. In Morley Callaghan, *Morley Callaghan's Stories*, 26–37. Toronto: Macmillan.

Canada Advisory Committee on Reconstruction. 1944. *Housing and Community Planning. Subcommittee Report No.4*. Ottawa: King's Printer.

Canadian Centre for Policy Alternatives. 2005. *The Promise of Investment in Community-Led Renewal: State of the Inner City Report – 2005*. Winnipeg: The Centre.

Canadian Welfare Council. 1947. *A National Housing Policy for Canada*. Ottawa: Canadian Welfare Council.

Capps, Kriston. 2015. How many neighborhoods is too many for one map? *CityLab* 3 (September). https://www.bloomberg.com/news/articles/2015-09-03/how-many-neighborhoods-is-too-many-for-one-map

Careless, James M.S. 1985. The Emergence of Cabbagetown in Victorian Toronto. In Robert Harney, ed., *Gathering Place. Peoples and Neighbourhoods in Toronto, 1834–1945*, 25–45. Toronto: Multicultural History Society of Ontario.

Carpiano, Richard, and Perry W. Hystad. 2012. "Sense of community belonging" in health surveys: What social capital is measuring. *Health and Place* 17: 606–17. https://doi.org/10.1016/j.healthplace.2010.12.018

Carr, Adriane. 1980. The development of neighbourhood in Kitsilano: Ideas, actors and the landscape. MA thesis, University of British Columbia.

Carter, Thomes S., Chesya Polevychuk, and John Osborne. 2009. The role of settlement housing and neighbourhood in the re-settlement process: A case study of refugee households in Winnipeg. *The Canadian Geographer* 53, no. 3: 305–22. https://doi.org/10.1111/j.1541-0064.2009.00265.x

Carver, Humphrey. 1948. *Houses for Canadians: A Study of Housing Problems in the Toronto Area*. Toronto: University of Toronto Press.

Carver, Humphrey. 1975. *Compassionate Landscape*. Toronto: University of Toronto Press. https://doi.org/10.3138/9781442652583

Chafe, Ronald E. N.d. A Story about the Grandview Co-op Housing Group. Unpublished manuscript, 42 pp.

Chaskin, Robert J. 1998. Neighborhood as a unit of planning and action: A heuristic approach. *Journal of Planning Literature* 13, no. 1: 11–30. https://doi.org/10.1177/088541229801300102

Chen, Wen-Hao, John Myles, and Garnett Picot. 2012. Why have poorer neighbourhoods stagnated economically while the richer have flourished? Neighbourhood income inequality in Canadian cities. *Urban Studies* 49, no. 4: 877–96. https://doi.org/10.1177/0042098011408142

Choko, Marc. 1989. *Une Cité-Jardin a Montréal: La Cité-Jardin de Tricentenaire 1940–1947*. Montreal: Méridien.

Choko, Marc. 1998. Ethnicity and homeownership in Montreal, 1921–1951. *Urban History Review* 26, no. 2: 32–41. https://doi.org/10.7202/1016657ar

Choko, Marc, Jean-Pierre Collin, and Annick Germain. 1987. Le logement et les enjeux de la transformation de l'espace urbain: Montréal, 1940–1960, Deuxième partie. *Urban History Review* 15, no. 3: 243–52. https://doi.org/10.7202/1018018ar

Choko, Marc, and Richard Harris. 1990. The local culture of property: A comparative history of housing tenure in Montreal and Toronto since 1862. *Annals of the Association of American Geographers* 80, no. 1: 73–95. https://doi.org/10.1111/j.1467-8306.1990.tb00004.x

Christie, Nancy, and Michael Gauvreau. 1996. *A Full-Orbed Christianity: The Protestant Churches and Social Welfare in Canada, 1900–1940*. Montreal and Kingston: McGill-Queen's University Press. https://doi.org/10.1515/9780773565944

Christie, Nancy, and Michael Gauvreau. 2010. *Christian Churches and Their Peoples, 1940–1965: A Social History of Religion*. Toronto: University of Toronto Press. https://doi.org/10.3138/9781442660007

City of Montreal, Advisory Committee on Slum Clearance and Low Rental Housing. 1954. *Project to Renovate a Defective Residential Zone and to Build Low-Rental Dwellings*. Author.

City of Toronto. 1992. *Draft Official Plan Part I Consolidation: City Plan Final Recommendations*. Toronto: Planning and Development Department.

City of Toronto Department of Health. 1911. *Report of the Medical Health Officer Dealing with the Recent Investigation of Slum Conditions in Toronto*. Author.

City of Toronto, Strong Neighbourhoods Task Force and United Way of Greater Toronto. 2005. *Strong Neighbourhoods: A Call to Action*. Author.

City of Toronto Planning Board. 1945. *Third Annual Report of the City Planning Board*. 30 December, 1944. Author.

City of Toronto Planning Board. 1959. *The Changing City*. Author.

City of Toronto Planning Board. 1965. *Improvement Programme for Residential Areas*. Author

City of Vancouver. 1986. *New Neighbours: How Vancouver's Single-Family Residents Feel About Higher-Density Housing*. Author.

City of Vancouver, City Planning Department. 1975. *Vancouver Local Areas*. Author.

City of Vancouver, City Planning Department. 1978. *Housing Families at High Densities*. Author.

City of Winnipeg, City Planning Commission. 1913. *City Planning Commission Report*. Author.

City of Winnipeg, Health Department. 1921. *Report on Housing: Survey of Certain Selected Areas Made March and April 1921*. Author.

Clairmont, Donald, and Dennis W. Magill. 1999. *Africville: The Life and Death of a Canadian Black Community*, 3rd ed. Toronto: Canadian Scholars Press.

Clark, Samuel D. 1966. *The Suburban Society*. Toronto: University of Toronto Press.

Clark, Samuel D. 1978. *The New Urban Poor*. Toronto: McGraw-Hill Ryerson.

Clarke, Austin. 1967. *The Meeting Point*. Toronto: Macmillan.

Clarke, Austin. 1986. *Nine Men Who Laughed*. Markham, ON: Penguin.

Clarke, Mary J. (1917) 1974. Report of the Standing Committee on Neighbourhood Work. In Paul Rutherford, ed., *Saving the Canadian City: The First Phase 1880–1920*, 171–93. Toronto: University of Toronto Press. https://doi.org/10.3138/9781487583446-019

Clayton, Frank A. 2010. Government subsidies to homeowners versus renters in Ontario and Canada. http://neighbourhoodchange.ca/documents/2014/09/clayton-2010-subsidies-owners-and-renters.pdf

CMHC (Central Mortgage and Housing Corporation). 1954. *Principles of Small House Grouping*. Ottawa: CMHC.

CMHC (Canada Mortgage and Housing Corporation). 1974. *New from Old: A Pilot Study of Housing Rehabilitation and Neighbourhood Change*. Toronto: CMHC.

Cockayne, Emily. 2012. *Cheek by Jowl: A History of Neighbours*. London: The Bodley Head.

Collin, Jean-Pierre. 1986. *La Cité Cooperative Canadienne-Française: Saint-Leonard-de-Port-Maurice, 1955–1963*. Montreal: INRS-Urbanisation.

Collin, Jean-Pierre. 1998. A housing model for lower and middle-class wage earners in a Montreal suburb: Saint-Léonard, 1955–1967. *Journal of Urban History* 24, no. 4: 468–90. https://doi.org/10.1177/009614429802400402

Collins, Patricia, Lindsay Allman, and Bill Irwin. 2019. Exploring the perceived impacts of a public high school closure for urban livability in a Canadian mid-sized city. *Local Environment* 24, no. 8: 678–95. https://doi.org/10.1080/13549839.2019.1631774

Colvin, J. 2019. *Africville*. Toronto: HarperCollins.

Comacchio, Cynthia. 1999. *The Infinite Bonds of Family: Domesticity in Canada, 1850–1940*. Toronto: University of Toronto Press. https://doi.org/10.3138/9781442681491

Comacchio, Cynthia. 2006. *The Dominion of Youth: Adolescence and the Making of a Modern Canada, 1920–1950*. Waterloo, ON: Wilfrid Laurier University Press.

Copp, Terry. 1974. *The Anatomy of Poverty: The Condition of the Working Class in Montreal, 1897–1929*. Toronto: McClelland and Stewart.

Coulton, Claudia J., Jill Korbin, Tsui Chan, and Marilyn Su. 2001. Mapping residents' perceptions of neighborhood boundaries: A methodological note. *American Journal of Community Psychology* 29: 371–83. https://doi.org/10.1023/A:1010303419034

Cowen, Deborah, and Vanessa Parlette. 2011. *Inner Suburbs at Stake: Investing in Social Infrastructure in Scarborough*. Research Paper No. 220, Cities Centre, University of Toronto. http://neighbourhoodchange.ca/wp-content/uploads/2011/06/Cowen-220-Social-Infrastructure-Inner-Suburbs-June-2011.pdf

Cox, Kevin. 1982. Housing tenure and neighborhood activism. *Urban Affairs Quarterly* 18, no. 1: 107–29. https://doi.org/10.1177/004208168201800109

Cox, Kevin. 1984. Social change, turf politics and concepts of turf politics. In Andrew M. Kirby, Paul Knox, and Steven Pinch, eds., *Public Service Provision and Urban Development*, 283–315. London: Croom Helm. https://doi.org/10.4324/9781315168777-12

Craick, William A. 1916a. Brown school on the hill essentially a public school though situated in aristocratic district. *Globe and Mail*, 23 December.

Craick, William A. 1916b. Earlscourt school had its beginnings just 26 years ago. *Toronto Star Weekly*, 2 December, 17.

Crane, Jonathan. 1991. The epidemic theory of ghettoes and neighborhood effects on dropping out and teenage childbearing. *American Journal of Sociology* 96: 1226–59. https://doi.org/10.1086/229654

Cruikshank, Ken, and Nancy B. Bouchier. 2004. Blighted areas and obnoxious industries: Constructing environmental inequality on an industrial waterfront, Hamilton, Ontario, 1890–1960. *Environmental History* 9, no. 3: 464–96. https://doi.org/10.2307/3985769

Crysdale, Stuart. 1970. Family and kinship in Riverdale. In William E. Mann, ed., *The Underside of Toronto*, 95–108. Toronto: McClelland and Stewart.

Cucchiara, Maia B. 2013. *Marketing Schools, Marketing Cities: Who Wins and Who Loses when Schools Become Urban Amenities*. Chicago: University of Chicago Press. https://doi.org/10.7208/chicago/9780226016962.001.0001

Curtis, Bruce, D.W. Livingstone, and Harry Smaller. 1992. *Stacking the Deck: The Streaming of Working-Class Kids in Ontario Schools*. Toronto: Our Schools/Ourselves Education Foundation.

Cushon, Jennifer A., T.H. Lan Vu, Bonnie L. Jansen, and Nazeem Muhajarine. 2011. Neighborhood poverty impacts children's physical health and well-being over time: Evidence from the early development

instrument. *Early Education and Development* 22, no. 2: 183–205. https://
doi.org/10.1080/10409280902915861

Dalzell, A.G. 1926. Town planning problems in Toronto. *The Municipal
Review of Canada* 22, no. 7: 238.

Dalzell, A.G. 1927. *Housing in Canada, I: Housing in Relation to Land
Development*. Toronto: Social Services Council of Canada.

Daniels, George. 1953. *How to Build or Remodel Your House*. New York:
Greystone Press.

Dansereau, Francine, and Annick Germain. 2002. Fin où renaissance des
quartiers? Les significations des territoires de proximité dans une ville
pluriethnique. *Espaces et sociétés* 108–9: 11–28. https://doi.org/10.3917
/esp.g2002.108n1.0011

Darroch, Gordon, and Wilfrid G. Marston. 1969. Ethnic differentiation:
Ecological aspects of a multidimensional concept. *International Migration
Review* 4: 71–95. https://doi.org/10.1177/019791836900400105

Davey, Ian, and Michael Doucet. 1975. The social geography of a commercial
city. In Michael Katz, The People of Hamilton, Canada West, Appendix 1.
Cambridge, MA: Harvard University Press.

Davy, Denise. 2021. *Her Name Was Margaret: Life and Death on the Streets*.
Hamilton: Wolsak and Wynn.

Dean, J.P. 1951. The ghosts of homeownership. *Journal of Social Issues* 7: 59–
68. https://doi.org/10.1111/j.1540-4560.1951.tb02222.x

Dear, Michael J., and Jennifer R. Wolch. 1987. *Landscapes of Despair: From
De-institutionalisation to Homelessness*. Cambridge: Polity Press.

Del Guidice, Luisa. 1993. The "Archvilla": An Italian Canadian architectural
archetype. In Luisa Del Guidice, ed., *Studies in Italian American Folklore*,
53–105. Logan, UT: Utah State University Press.

Delaney, Jill. 1991. The Garden Suburb of Lindenlea, Ottawa: A model
suburb for the first federal housing policy, 1918–1924, *Urban History
Review* 19, no. 3: 151–65. https://doi.org/10.7202/1017590ar

Delegran, W.A. 1970. Life in the Heights. In William E. Mann, ed., *The
Underside of Toronto*, 75–94. Toronto: McClelland and Stewart 94.

Deluca, Patrick F., Steve Buist, and Neil Johnson. 2012. The Code Red
project: Engaging communities in health system change in Hamilton,
Canada. *Social Indicators Research* 108: 317–27. https://doi.org/10.1007
/s11205-012-0068-y

Deluca, Patrick F., and Pavlos Kanaroglou. 2015. Code Red: Explaining
average age of death in the City of Hamilton. *AIMS Public Health* 2, no. 4:
730–45. https://doi.org/10.3934/publichealth.2015.4.730

Dennis, Richard. 1989. Toronto's first apartment-house boom: An historical geography, 1900–1920. *Research Paper No.177*, Centre for Urban and Community Studies, University of Toronto.

Dennis, Richard. 1994. Interpreting the apartment house: Modernity and metropolitanism in Toronto, 1900–1930. *Journal of Historical Geography* 20, no. 3: 305–22. https://doi.org/10.1006/jhge.1994.1023

Devine, Edward. 1924. *The Normal Life*. New York: Survey Associates.

Dillick, Sidney. 1953. *Community Organization for Neighborhood Development Past and Present*. New York: William Morrow.

Distasio, Jino, and Sarah Zell. 2020. People, policies, and place: Indigenous and immigrant population shifts in Winnipeg's inner city neighbourhoods. In Jill Grant, Alan Walks, and Howard Ramos, eds., *Changing Neighbourhoods. Social and Spatial Polarization in Canadian Cities*, 214–34. Vancouver: UBC Press. https://doi.org/10.59962/9780774862042-014

Distinctive houses mid natural beauty lure homemakers. 1925. The Globe, 18 December.

Dobson, Kathy. 2011. *With a Closed Fist: Growing Up in Canada's Toughest Neighbourhood*. Montreal: Véhicule Press.

Donnelly, Catherine, Paul Nguyen, Simone Parniak, and Vincent DePaul. 2020. Beyond long-term care: The benefits of seniors' communities that evolve on their own. The Conversation, 8 September. https://theconversation.com/beyond-long-term-care-the-benefits-of-seniors-communities-that-evolve-on-their-own-144269

Doucet, Michael. 1976. Working class housing in a small nineteenth century Canadian city: Hamilton, Ontario, 1852–1881. In Greg Kealey and Peter Warrian, eds., *Essays in Working-Class History*, 83–105. Toronto: McClelland and Stewart.

Doucet, Michael, and John Weaver. 1991. *Housing the North American City*. Montreal and Kingston: McGill-Queen's University Press. https://doi.org/10.1515/9780773562820

Downs, Anthony. 1981. *Neighborhoods and Urban Development*. Washington, DC: Brookings Institution.

Driedger, Leo, and Glenn Church. 1974. Residential segregation and institutional completeness: A comparison of ethnic minorities. *Canadian Review of Sociology and Anthropology* 11, no. 1: 30–52. https://doi.org/10.1111/j.1755-618X.1974.tb00002.x

Dunn, James R., and Michael V. Hayes. 2000. Social inequality, population health and housing: A study of two Vancouver neighborhoods. *Social

Science and Medicine 51, no. 4: 563–87. https://doi.org/10.1016/S0277
-9536(99)00496-7

Dunphy, Bill. 2006. Beasley: Portrait of a neighbourhood. *Hamilton Spectator*,
28 January.

Earlscourt ratepayers. 1909. The Globe, 28 September.

Ecker, John, and Tim Aubry. 2017. A mixed methods analysis of housing
and neighborhood impacts on community integration among vulnerably
housed and homeless individuals. *Journal of Community Psychology* 45, no.
4: 528–42. https://doi.org/10.1002/jcop.21864

Einstein, Katherine L., David Glick, and Maxwell Palmer. 2019. *Neighborhood
Defenders: Participatory Politics and America's Housing Crisis*. Cambridge:
Cambridge University Press. https://doi.org/10.1017/9781108769495

Einstein, Katherine L., Maxwell Palmer, and David Glick. 2019. Who
participates in local government? Evidence from meeting minutes.
Perspectives on Politics 17, no. 1: 28–46. https://doi.org/10.1017
/S153759271800213X

Ellen, Ingrid G., and Justin P. Steil. Eds. 2019. *The Dream Revisited:
Contemporary Debates about Housing, Segregation, and Opportunity*. New
York: Columbia University Press. https://doi.org/10.7312/elle18362

Ellen, Ingrid G., and Margery Turner. 1997. Does neighborhood matter?
Assessing recent evidence. *Housing Policy Debate* 8: 833–66. https://doi
.org/10.1080/10511482.1997.9521280

Elliott, Bruce S. 1991. *The City Beyond: A History of Nepean, Birthplace of
Canada's Capital, 1792–1990*. Nepean, ON: City of Nepean.

Elliott, Susan, Donald Cole, Paul Krueger, Nancy Voorberg, and Sarah
Wakefield. 1999. The power of perception: Health risk attributed to air
pollution in an urban industrial neighborhood. *Risk Analysis* 19: 615–28.
https://doi.org/10.1111/j.1539-6924.1999.tb00433.x

Ellis, Jason. 2019. Public school taxes and the remaking of suburban space
and history: Etobicoke, 1945–1954. *Journal of the Canadian Historical
Association* 30, no. 2: 1–36. https://doi.org/10.7202/1074375ar

Elman, Russell. 2001. *Durand: A Neighbourhood Reclaimed*. Hamilton, ON: NA
Group.

Erickson, Paul. 2004. *Historic North End Halifax*. Halifax, NS: Nimbus.

Ethnic food: The next generation. 2017. *Globe and Mail*, 14 January.

Fahrni, Magda. 2005. *Household Politics: Montreal Families and Postwar
Reconstruction*. Toronto: University of Toronto Press. https://doi
.org/10.3138/9781442627451

Fahrni, Magda. 2011. Influenza and the urban environment, 1918–1920. In
Stéphane Castonguay and Michèle Dagenais, eds., *Metropolitan Natures:*

Environmental Histories of Montreal, 68–81. Pittsburgh: University of Pittsburgh Press. https://doi.org/10.2307/j.ctv111jfg2.9

Falardeau, Jean-C. 1949. The parish as an institutional type. *Canadian Journal of Economics and Political Science* 15: 353–67. https://doi.org/10.2307 /138096

Farrell, Susan, Tim Aubry, and Daniel Coulombe. 2004. Neighborhoods and neighbors: Do they contribute to personal well-being? *Journal of Community Psychology* 32, no. 1: 9–25. https://doi.org/10.1002/jcop.10082

Ferretti, Lucia. 1992. *Entre Voisins: La Société Paroissiale en Milieu Urbain, Saint-Pierre-Apôtre de Montréal, 1848–1930*. Montreal: Boréal.

Ferretti, Lucia. 2001. La paroisse urbaine comme communauté sociale, 1870– 1914. In Serge Courville and Normand Séguin, eds., *Atlas Historique du Québec*, vol.7, *La Paroisse*, 219–28. Sainte-Foy, QC: Presses de L'Université Laval.

Ferretti, Lucia. 2021. Personal communication. 9 June.

Filion, Pierre. 1987. Concepts of inner city and recent trends in Canada. *Canadian Geographer* 31, no. 3: 223–32. https://doi.org/10.1111 /j.1541-0064.1987.tb01236.x

Filion, Pierre. 1988. The Neighbourhood Improvement Plan: Montreal and Toronto. Contrasts between a participatory and a centralized approach to urban policy making. *Urban History Review* 17, no. 1: 16–28. https://doi .org/10.7202/1017698ar

Filion, Pierre. 1999. Rupture or continuity: Modern and postmodern planning in Toronto. *International Journal of Urban and Regional Research* 23, no. 3: 421–44. https://doi.org/10.1111/1468-2427.00206

Fingard, Judith. 1977. The relief of the unemployed: The poor in Saint John, Halifax, and St. John's, 1815–1860. In Gilbert A. Stelter and Alan F.J. Artibise, eds., *The Canadian City: Essays in Urban History*, 341–67. Toronto: McClelland and Stewart.

Fingard, Judith. 1989. *The Dark Side of Life in Victorian Halifax*. Porter Lake, NS: Pottersfield Press.

Fischel, William A. 2001. *The Homevoter Hypothesis: How Home Values Influence Local Government Taxation, School Finance and Land Use Policies*. Cambridge, MA: Harvard University Press.

Fischel, William A. 2014. Not by the hand of Horace Mann: How the quest for land values created the American school system. In Gregory K. Ingram and Daphne A. Kenyon, eds., *Education, Land and Location*, 123– 50. Cambridge, MA: Lincoln Institute for Land Policy.

Fischer, Claude S. 1991. Ambivalent communities: How Americans understand their localities. In Alan Wolfe, ed., *America at Century's End*,

79–90. Berkeley, CA: University of California Press. https://doi.org
/10.1525/9780520400252-007

Fischer, Claude S. 2002. Ever-more rooted Americans. *City and Community* 1, no. 2: 177–98. https://doi.org/10.1111/1540-6040.00016

Fischer, Claude S., and Michael Hout. 2006. *Century of Difference: How America Changed in the Last One Hundred Years.* New York: Russell Sage.

Fischler, Raphael. 1998. Health, safety and the general welfare: Markets, politics and social science in early land use regulation and community design. *Journal of Urban History* 24: 675–719. https://doi.org/10.1177 /009614429802400601

Fischler, Raphael. 2007. Development controls in Toronto in the nineteenth century. *Urban History Review* 36: 16–31. https://doi.org/10.7202 /1015817ar

Fischler, Raphael. 2014. Émergence du zonage à Montréal, 1840–1914. In Harold Bérubé, Donald Fyson, and Léon Robichaud, eds., *370 Ans de Gouvernance Montréalaise*, 71–84. Montréal: Éditions Multomondes.

Fischler, Raphael. 2016. Zoning: The past and present of planning as real-estate regulation. *Plan Canada* 5, no. 2: 36–41.

Fisher, Robert. 1982. Community organizations in historical perspective: A typology. *Houston Review* 4: 75–87.

Flanagan, Maureen. 2018. *Constructing the Patriarchal City: Gender and the Built Environments of London, Dublin, Toronto and Chicago, 1870s into the 1940s.* Philadelphia: Temple University Press.

Flanagan, Tom. 2019. *Property Rights and Prosperity: A Case Study of Westbank First Nation.* Vancouver: Fraser Institute, 17 September.

Florida, Richard. 2005. *Cities and the Creative Class.* London: Routledge. https://doi.org/10.4324/9780203997673

Foley, Donald L. 1952. *Neighbors or Urbanites? The Study of a Rochester Residential District.* Rochester: University of Rochester.

Fong, Eric, and Brent Berry. 2017. *Immigration and the City.* Cambridge: Polity.

Foran, Max. 1979. Land development patterns in Calgary, 1884–1945. In Alan F.J. Artibise and Gilbert Stelter, eds., *The Usable Urban Past*, 293–315. Toronto: Macmillan. https://doi.org/10.1515/9780773580640-017

Foran, Max. 2010. *Expansive Discourses: Urban Sprawl in Calgary 1945–1978.* Edmonton: Athabasca University Press. https://doi.org/10.15215 /aupress/9781897425138.01

Ford, Larry. 1994. *Cities and Buildings: Skyscrapers, Skid Rows and Suburbs.* Baltimore: Johns Hopkins University Press.

Forward, Charles N. 1973. The immortality of a fashionable residential district: The Uplands. In Charles N. Forward, ed., *Residential and Neighbourhood Studies in Victoria*, 1–39. Western Geographical Series, vol. 5. Department of Geography, University of Victoria.

Fougères, Dany. 2018. The years of dispersion. In Dany Fougères and Roderick MacLeod, eds., *Montreal: A History of a North American City*, 295–341. Montreal and Kingston: McGill-Queen's University Press. https://doi.org/10.2307/j.ctt2111gbs.17

Franke, Thea, Catherine Tong, Maureen C. Ashe, Heather McKay, Joanie Sims-Goulde, and the Walk, the Talk Team. 2013. The secrets of highly active older adults. *Journal of Aging Studies* 27, no. 4: 398–409. https://doi .org/10.1016/j.jaging.2013.09.003

Frankfurt, Harry G. 2015. *On Inequality*. Princeton, NJ: Princeton University Press.

Fraser, Graham. 1972. *Fighting Back: Urban Renewal in Trefann Court*. Toronto: Hakkert.

Fraser, Sylvia. 1972. *Pandora*. Toronto: McClelland and Stewart.

Frenette, Marc, Garnett Picot, and Roger Sceviour. 2004. When do they leave? The dynamics of living in low-income neighborhoods. *Journal of Urban Economics* 56, no. 3: 484–504. https://doi.org/10.1016/j.jue .2004.06.001

Fried, Marc. 1973. *The World of the Urban Working Class*. Cambridge, MA: Harvard University Press. https://doi.org/10.4159/harvard .9780674189492

Friedman, Avi, and David Krawitz. 2002. *Peeking through the Keyhole: The Evolution of North American Homes*. Montreal and Kingston: McGill-Queen's University Press. https://doi.org/10.1515/9780773570603

Gaetz, Stephen, Erin Dej, Tim Richter, and Melanie Redman. 2016. *The State of Homelessness in Canada*. Toronto: Canadian Observatory on Homelessness Press.

Gagan, Rosemary. 1989. Mortality patterns and public health in Hamilton, Canada, 1900–1914. *Urban History Review* 17, no. 3: 161–76. https://doi .org/10.7202/1017629ar

Gagnon, Robert. 1996. *Histoire de la Commission des Écoles Catholiques de Montrèal*. Montreal: Boréal.

Galois, Robert M. 1979. Social structure in space: The making of Vancouver, 1886–1901. PhD diss., University of British Columbia.

Galster, George C. 2001. On the Nature of Neighbourhood. *Urban Studies* 38, no. 12: 2111–24. https://doi.org/10.1080/00420980120087072

Galster, George C. 2012. The mechanism(s) of neighborhood effects: Theory, evidence and policy implications. In Maarten van Ham, David Manley, Ludi Simpson, Nick Bailey, and Duncan Maclennan, eds., *Neighbourhood Effects Research: New Perspectives*, 23–56. Dordrecht: Springer.

Galster, George C. 2019. *Making Our Neighborhoods, Making Our Selves*. Chicago: University of Chicago Press. https://doi.org/10.7208/chicago /9780226599991.001.0001

Galster, George C., and Sean P. Killen. 1995. The geography of metropolitan opportunity: A reconnaissance and conceptual framework. *Housing Policy Debate* 6, no. 1: 7–43. https://doi.org/10.1080/10511482.1995.9521180

Gans, Herbert J. 1962. *The Urban Villagers*. New York: Free Press.

Gardner, Paula J. 2011. Natural neighborhood networks: Important social networks in the lives of older adults aging in place. *Journal of Aging Studies* 25: 263–71. https://doi.org/10.1016/j.jaging.2011.03.007

Garner, Hugh. 1936. Toronto's Cabbagetown. *Canadian Forum*, 14 (July). Reprinted in Jack L. Granatstein and Peter Stevens, eds., *Forum: Canadian Life and Letters, 1920–1970, Selections from the Canadian Forum*, 145–8. Toronto: University of Toronto Press, 1972.

Garner, Hugh. 1971. *Cabbagetown*. Markham, ON: Simon and Schuster.

Garner, Hugh. 1976. *The Intruders*. Toronto: McGraw-Hill.

Garrioch, David, and Mark Peel. 2006. The social history of neighborhoods. *Journal of Urban History* 32, no. 5: 663–76. https://doi.org/10.1177/0096144206287093

Gaub, Janne E., Danielle Wallace, and Mary E. Hoyle. 2022. The neighborhood according to women: Understanding gendered disorder perceptions. *Crime and Delinquancy* 67, nos. 6–7: 891–915. https://doi.org/10.1177/0011128720968491

Gauvreau, Michael. 2006. Factories and foreigners: Church life in working-class neighbourhoods in Hamilton and Montreal, 1890–1930. In Michael Gauvreau and Ollivier Hubert, eds., *The Churches and Social Order in Nineteenth- and Twentieth-Century Canada*, 225–73. Montreal and Kingston: McGill-Queen's University Press. https://doi.org/10.1515/9780773576001-009

Gee, Marcus. 2020. A shocking map of the city's divisions. *Globe and Mail*, 6 June.

Gee, Marcus. 2021. The buddy system: How drug users became amateur medics in Canada's opioid crisis. *Globe and Mail*, 5 April. https://www.theglobeandmail.com/canada/article-the-buddy-system-how-drug-users-became-amateur-medics-in-canadas/

Germain, Annick. 1984. *Les Mouvements de Reforme Urbaine à Montréal au Tournant du Siècle: Modes de Developpement, Modes d'Urbanisation et Transformations de la Scène Politique*. Montreal: Département de Sociologie, Université de Montréal.

Germain, Annick. 1999. Les quartiers multiethniques. Montréalais: Une lecture urbaine. *Recherches sociographiques* 19, no. 1: 9–32. https://doi.org/10.7202/057242ar

Germain, Annick. 2013. The Montréal School: Urban social mix in a reflexive city. *Anthropologica* 55: 1–11.

Germain, Annick. 2016. The fragmented or cosmopolitan metropolis? A neighbourhood story of immigration in Montreal. *British Journal of Canadian Studies* 29, no. 1: 1–23. https://doi.org/10.3828/bjcs.2016.1

Germain, Annick, and Julie E. Gagnon. 1999. Is neighbourhood a black box? A reply to Galster, Metger and White. *Canadian Journal of Urban Research* 8, no. 2: 172–84.

Germain, Annick, and Damaris Rose. 2000. *Montréal: The Quest for a Metropolis*. Toronto: Wiley.

Gettler, Brian. 2020. *Colonialism's Currency: Money, State and First Nations in Canada, 1820–1950*. Montreal and Kingston: McGill-Queen's University Press. https://doi.org/10.1515/9780228002536

Ghaziani, Amin. 2014. *There Goes the Gayborhood?* Princeton, NJ: Princeton University Press. https://doi.org/10.23943/princeton/9780691158792 .003.0002

Ghosh, Sutama. 2014. Everyday lives in vertical neighbourhoods: A case study of Bangladeshis in Toronto. *International Journal of Urban and Regional Research* 38, no. 6: 2008–2024. https://doi.org/10.1111 /1468-2427.12170

Gidney, R.D. 1999. *From Hope to Harris: The Reshaping of Ontario's Schools*. Toronto: University of Toronto Press. https://doi.org/10.3138 /9781442675087

Gidney, R.D., and W.P.J. Millar. 2012. *How Schools Worked: Public Education in English Canada, 1900–1940*. Montreal and Kingston: McGill-Queen's University Press. https://doi.org/10.1515/9780773587304

Gilbert, Dale. 2015. *Vivre en Quartier Populaire: Saint-Sauveur, 1930–1980*. Quebec: Septentrion.

Gilbert, Daniel. 2007. *Stumbling on Happiness*. New York: Knopf.

Gillette, Howard. 1983. The evolution of neighborhood planning from the Progressive Era to the 1949 Housing Act. *Journal of Urban History* 9: 421–44. https://doi.org/10.1177/009614428300900402

Gilliland, Jason A. 1998. Modeling residential mobility in Montreal, 1860–1900. *Historical Methods* 31, no. 1: 27–42. https://doi.org/10.1080 /01615449809600091

Gilliland, Jason A., and Sherry Olson. 2010. Residential segregation in the industrializing city: A closer look. *Urban Geography* 31, no. 1: 29–58. https://doi.org/10.2747/0272-3638.31.1.29

Gilliland, Jason A., Sherry Olson, and Danielle Gauvreau. 2011. Did segregation increase as the city expanded: The case of Montreal, 1881–1901. *Social Science History* 35, no. 4: 465–503. https://doi.org /10.1215/01455532-1381823

Gladwell, Malcolm. 2000. *The Tipping Point: How Little Things Can Make a Big Difference*. Boston: Little, Brown.

Godfrey, Ray. 2013. The old Elm Tree Tea House. In Nancy Williams and Marie Scott-Baron, eds., *Recollections of a Neighbourhood: Huron-Sussex from UTS to Stop Spadina*, 67–8. Toronto: Wards Indeed Publishing.

Goheen, Peter G. 1970. *Victorian Toronto, 1850–1900: Pattern and Process of Growth*. Department of Geography Research Paper No. 127, University of Chicago.

Gold, Kerry. 2021. The struggle to define the future of Vancouver's False Creek neighbourhood. *Globe and Mail*, 9 April. https://www.theglobeandmail.com/real-estate/vancouver/article-false-creek-south-under-fire/

Gold, Kerry. 2023. Vancouver development community sees uncertainty in multiplex zoning plan. *Globe and Mail*, 1 September.

Goldberg, Michael, and John Mercer. 1986. *The Myth of the North American City: Continentalism Challenged*. Vancouver: UBC Press. https://doi.org/10.59962/9780774857031

Goldman, Gustave. 2014. Canada's Aboriginal population: A unique historical perspective. In Gordon Darroch, ed., *Canada's Century: Hidden Histories*, 124–45. Montreal and Kingston: McGill-Queen's University Press. https://doi.org/10.1515/9780773589391-009

Gordon, David. 2018. Humphrey Carver and the federal government's postwar revival of community planning. *Urban History Review* 46, no. 2: 71–86. https://doi.org/10.7202/1064834ar

Gordon, David. 2021. Building "Canada's Model Town: Oromocto NB, 1950–1969. Unpublished manuscript, Queen's University.

Gorman-Murray, Andrew, and Catherine Nash. 2017. Transformations in LGBT consumer landscapes and leisure spaces in the neoliberal city. *Urban Studies* 54, no. 3: 786–805. https://doi.org/10.1177/0042098016674893

Grabar, Henry. 2020. The year of the neighborhood. Slate, 29 December.

Graff, Harvey. 1979. *The Literacy Myth: Literacy and Social Structure in the Nineteenth Century City*. New York: Academic Press.

Graham, Scott, Stephanie Procyk, and Michelynn Laflèche. 2020. Mapping Canada's social policy space. In Jill Grant, Alan Walks, and Howard Ramos, eds., *Changing Neighbourhoods: Social and Spatial Polarization in Canada's Cities*, 239–51. Vancouver: UBC Press. https://doi.org/10.59962/9780774862042-015

Grainger, Lia. 2020. Next Door in the social network for neighbours. *Globe and Mail*, 15 February. https://www.theglobeandmail.com/life

/home-and-design/article-nextdoor-bets-on-the-power-of
-neighbourhoods/

Grampp, Christopher. 2008. *From Yard to Garden: The Domestication of America's Home Grounds*. Chicago: Center for American Places at Columbia College, Chicago.

Grannis, Rick. 2009. *From the Ground Up: Translating Geography into Community through Neighbor Networks*. Princeton, NJ: Princeton University Press. https://doi.org/10.1515/9781400830572

Grant, Jill. 1994. *The Drama of Democracy: Contention and Dispute in Community Planning*. Toronto: University of Toronto Press. https://doi .org/10.3138/9781442674073

Grant, Jill. 2005. The function of the gates: The social construction of security in gated developments. *Town Planning Review* 76, no. 3: 291–313. https://doi.org/10.3828/tpr.76.3.4

Grant, Jill. 2007. Two sides of the coin? New Urbanism and gated communities. *Housing Policy Debate* 18, no. 3: 481–501. https://doi.org /10.1080/10511482.2007.9521608

Grant, Jill. 2011. Time, scale and control: How New Urbanism (Mis)uses Jane Jacobs. In Max Page and Timothy Mennell, eds., *Reconsidering Jane Jacobs*, 91–103. Chicago: American Planning Association. https://doi.org /10.4324/9781351179775-5

Grant, Jill, Katherine Greene, and Kirstin Maxwell. 2004. The planning and policy implications of gated communities. *Canadian Journal of Urban Research* 13, no. 1: 70–88.

Grant, Jill, and Will Gregory. 2016. Who lives downtown? Neighbourhood change in central Halifax, 1951–2011. *International Planning Studies* 21, no. 2: 176–90. https://doi.org/10.1080/13563475.2015.1115340

Grant, Jill, and Howard Ramos. 2020. Halifax. Scaling inequality. In Jill Grant, Alan Walks, and Howard Ramos, eds., *Changing Neighbourhoods: Social and Spatial Polarization in Canadian Cities*, 171–91. Vancouver: UBC Press. https://doi.org/10.59962/9780774862042-012

Grant, Jill, Alan Walks, and Howard Ramos. 2020. Evaluating neighbourhood inequality and change: Lessons from a national comparison. In Jill Grant, Alan Walks, and Howard Ramos, eds., *Changing Neighbourhoods: Social and Spatial Polarization in Canadian Cities*, 252–79. Vancouver: UBC Press. https://doi.org/10.59962 /9780774862042-016

Grant, Kelly. 2021. High-risk areas have lowest vaccination rates among the elderly, new data show. *Globe and Mail*, 1 April. https://www.theglobeandmail.com/canada

/article-covid-19-immunization-rates-low-in-high-risk-communities
-but-hesitancy/

Grant, Kelly, James Keller, and Les Perreaux. 2021. Warming up to the hot spot strategy. *Globe and Mail*, 24 April. https://www.theglobeandmail .com/canada/article-warming-up-to-the-hotspot-strategy-why -targeting-hard-hit-areas-for/

Gray, James H. 1970. *The Boy from Winnipeg*. Toronto: Macmillan.

Green, David A., W. Craig Riddell, and Francis St. Hilaire, eds. 2016. *Income Inequality: The Canadian Story*. Montreal and Kingston: McGill-Queen's University Press.

Greenhill, Ralph, and A.J. Birrell. 1979. *Canadian Photography, 1839–1920*. Toronto: Coach House Press.

Grimsby Museum. 2021. *Sweat Equity: The Grimsby Homebuilding Cooperative, 1953–1956*. Grimsby: The Museum.

Guest, Avery M., and Barrett A. Lee. 1983. Consensus on locality names within the metropolis. *Sociology and Social Research* 67, no. 4: 375–91.

Guest, Avery M., and Susan K. Wierzbicki. 1999. Social ties at the neighborhood level. *Urban Affairs Review* 35, no. 1: 92–111. https://doi .org/10.1177/10780879922184301

Gunton, Thomas I. 1979. Evolution of urban and regional planning in Canada, 1900–1960. PhD diss., University of British Columbia. https:// open.library.ubc.ca/soa/cIRcle/collections/ubctheses/831 /items/1.0095464

Guppy, Neil, Larissa Sakumoto, and Riona Wilkes. 2019. Social change and gender division of household labour in Canada. *Canadian Review of Sociology* 56, no. 2: 178–203. https://doi.org/10.1111/cars.12242

Haan, Michael. 2005. *The Decline of the Immigrant Homeownership Advantage: Life cycle, Declining Ffortunes and Changing Housing Careers in Montreal, Toronto and Vancouver, 1981–2001*. Ottawa: Statistics Canada. http:// www.researchgate.net/publication/23546119_The_Decline_of_the _Immigrant_Homeownership_Advantage_Life-cycle_Declining _Fortunes_and_Changing_Housing_Careers_in_Montreal_Toronto _and_Vancouver_1981-2001

Hackworth, Jason. 2016. Why is there no Detroit in Canada? *Urban Geography* 37, no. 2: 272–95. https://doi.org/10.1080/02723638.2015.1101249

Hagopian, John. 1999. Galt's "Dickson's Hill": The evolution of a late-Victorian neighbourhood in an Ontarian town. *Urban History Review* 27, no. 2: 25–43. https://doi.org/10.7202/1016580ar

Hampton, Keith N. 2016. Why is helping behavior declining in the United States but not in Canada? Ethnic diversity, new technologies, and other

explanations. *City and Community* 15, no. 4: 380–99. https://doi
.org/10.1111/cico.12206

Hampton, Keith N., and Barry Wellman. 2003. Neighboring in Netville:
How the internet supports community and social capital in a wired
suburb. *City and Community* 2, no. 4: 277–311. https://doi.org/10.1046
/j.1535-6841.2003.00057.x

Harland Bartholemew and Associates. 1929. *A Plan for the City of Vancouver,
British Columbia, Including a General Plan of the Region.* St. Louis, MO:
Harland Bartholemew.

Harney, Robert F. 1975. *Ambiente* and social class in North American Little
Italies. *Canadian Review of Studies in Nationalism* 2, no. 2: 208–24.

Harney, Robert F. 1985. Ethnicity and neighbourhoods. In Robert F. Harney,
ed., *Gathering Place: Peoples and Neighbourhoods of Toronto, 1834–1945*, 1–24.
Toronto: Multicultural History Society of Toronto.

Harris, Amy L. 2010. *Imagining Toronto.* Toronto: Mansfield Press.

Harris, Douglas C. 2011. Condominium and the city: The rise of property in
Vancouver. *Law and Social Enquiry* 36, no. 3: 694–726. https://doi.org
/10.1111/j.1747-4469.2011.01247.x

Harris, Marjorie. 1984. *Toronto: The City of Neighbourhoods.* Toronto:
McClelland and Stewart.

Harris, Richard. 1988. *Democracy in Kingston: A Social Movement in Urban
Politics.* Montreal and Kingston: McGill-Queen's University Press.
https://doi.org/10.1515/9780773561267

Harris, Richard. 1990. Working-class home ownership in the American
metropolis. *Journal of Urban History* 17, no. 1: 46–69. https://doi
.org/10.1177/009614429001700104

Harris, Richard. 1996. *Unplanned Suburbs: Toronto's American Tragedy, 1900–
1950.* Baltimore: Johns Hopkins University Press.

Harris, Richard. 2001. Flattered but not imitated: Co-operative self-help and
the Nova Scotia Housing Commission, 1936–1973. *Acadiensis* 31, no. 1:
103–28.

Harris, Richard. 2004. *Creeping Conformity: How Canada Became Suburban,
1900–1960.* Toronto: University of Toronto Press. https://doi.org
/10.3138/9781442627642

Harris, Richard. 2012a. *Building a Market: The Rise of the Home Improvement
Industry, 1914–1960.* Chicago: University of Chicago Press. https://doi
.org/10.7208/chicago/9780226317687.001.0001

Harris, Richard. 2012b. "Ragged urchins play on marquetry floors": The
discourse of filtering is reconstructed, 1920s–1950s. *Housing Policy Debate*
22, no. 3: 463–82. https://doi.org/10.1080/10511482.2012.680481

Harris, Richard. 2013. The rise of filtering down: The American housing market transformed, 1915–1929. *Social Science History* 37, no. 4: 515–49. https://doi.org/10.1215/01455532-2346879

Harris, Richard. 2020a. Neighborhood upgrading: A fragmented global history. *Progress in Planning* 142. https://doi.org/10.1016/j.progress.2019.04.002

Harris, Richard. 2020b. Hamilton: Poster child for concentrated poverty. In Jill Grant, Alan Walks, and Howard Ramos, eds., *Changing Neighbourhoods: Social and Spatial Polarization in Canadian Cities*, 149–70. Vancouver: UBC Press. https://doi.org/10.59962/9780774862042-011

Harris, Richard. 2021. *How Cities Matter.* Cambridge: Cambridge University Press.

Harris, Richard. 2024. How neighborhoods came to matter more over time: A broad historical sketch. *Journal of Urban Affairs* (forthcoming). https://doi.org/10.1080/07352166.2024.2305128

Harris, Richard, and Chris Hamnett. 1987. The myth of the promised land: The social diffusion of home ownership in Britain and the United States. *Annals of the Association of American Geographers* 77, no. 2: 173–90. https://doi.org/10.1111/j.1467-8306.1987.tb00152.x

Harris, Richard, Greg Levine, and Brian Osborne. 1981. Housing tenure and social classes in Kingston, Ontario, 1881–1901. *Journal of Historical Geography* 7, no. 3: 271–89. https://doi.org/10.1016/0305-7488(81)90003-7

Harris, Richard, and Michael Mercier. 2008. A test for geographers: The geography of educational achievement in Toronto and Hamilton, 1997. *Canadian Geographer* 44, no. 3: 210–27. https://doi.org/10.1111/j.1541-0064.2000.tb00705.x

Harris, Richard, and Charlotte Vorms. 2017. Introduction. In Richard Harris and Charlotte Vorms, eds., *What's in a Name? Talking about Urban Peripheries*, 3–35. Toronto: University of Toronto Press. https://doi.org/10.3138/9781442620643-003

Hasson, Shlomo, and David Ley. 1994. *Neighbourhood Organizations and the Welfare State.* Toronto: University of Toronto Press.

Hauch, Christopher. 1985. *Coping Strategies and Street Life: The Ethnography of Winnipeg's Skid Row Region.* Winnipeg: Institute of Urban Studies.

Heblich, Stephan, Alex Trew, and Yanos Zylberberg. 2021. East-side story: Historical pollution and persistent neighborhood sorting. *Journal of Political Economy* 129, no. 5: 1508–52. https://doi.org/10.1086/713101

Hedman, Lina, and George Galster. 2013. Neighborhood income sorting and the effects of neighborhood income mix on income: A holistic empirical exploration. *Urban Studies* 50, no. 1: 107–27. https://doi.org/10.1177/0042098012452320

Hedman, Lina, and Maarten van Ham. 2012. Understanding neighbourhood effects: Selection bias and residential mobility. In Maarten van Ham, David Manley, Ludi Simpson, Nick Bailey, and Duncan Maclennan, eds., *Neighbourhood Effects Research: New Perspectives*, 79–99. Dordrecht: Springer99. https://doi.org/10.1007/978-94-007-2309-2_4

Heisz, Andrew. 2015. Trends in income inequality in Canada and elsewhere. In David Green, W. Craig Riddell, and Francis St. Hilaire, eds., *Income Inequality: The Canadian Story*, 77–102. Montreal and Kingston: McGill-Queen's University Press.

Helman, Claire. 1987. *The Milton-Park Affair: Canada's Largest Citizen-Developer Confrontation*. Montreal: Véhicule.

Hendler, Sue, with Julie Markovich. 2017. I *Was the Only Woman: Women and Planning in Canada*. Vancouver: UBC Press. https://doi.org/10.59962/9780774825894

Henry, Robert. 2019. "I claim in the name of ..." Indigenous street gangs and the politics of recognition in Prairie cities. In Julie Tomiak, Tyler McCreary, David Hugill, Robert Henry, and Heather Dorries, eds., *Settler City Limits: Indigenous Resurgence and Colonial Violence in the Urban Prairie West*, 222–47. Winnipeg: University of Manitoba Press. https://doi.org/10.1515/9780887555893-011

Heron, Craig. 2015. *Lunch Bucket Lives: Remaking the Workers' City*. Toronto: Between the Lines.

Herring, D. Ann, and Ellen Korol. 2012. The north-south divide: Social inequality and mortality from the 1918 influenza pandemic in Hamilton, Ontario. In Magda Fahrni and Esyllt Jones, eds., *Epidemic Encounters: Influenza, Society and Culture in Canada, 1918–1920*, 97–112. Montreal and Kingston: McGill-Queen's University Press. https://doi.org/10.59962/9780774822145-007

Hess, Paul. 2019. Property, planning and bottom-up pedestrian spaces in Toronto's postwar suburbs. In Mahyar Arefi and Conrad Kickert, eds., *The Palgrave Handbook of Bottom-Up Urbanism*, 287–304. Cham, Switzerland: Palgrave Macmillan304. https://doi.org/10.1007/978-3-319-90131-2_18

Hess, Paul, and Robert Lewis. 2019. Property rights, redevelopment areas, and Toronto ratepayers' associations in the 1950s. *Journal of Urban History* 45, no. 2: 279–99. https://doi.org/10.1177/0096144217696987

Hiebert, Daniel. 1991. Class, ethnicity and residential structure: The social geography of Winnipeg, 1901–1921. *Journal of Historical Geography* 17, no. 1: 56–86. https://doi.org/10.1016/0305-7488(91)90005-G

Hiebert, Daniel. 1993. Integrating production and consumption: Industry, class, ethnicity, and the Jews of Toronto. In Larry S. Bourne and David

Ley, eds., *The Changing Social Geography of Canadian Cities*, 199–213. Montreal and Kingston: McGill-Queen's University Press. https://doi.org/10.1515/9780773563551-011

Hiebert, Daniel. 1995. The social geography of Toronto in 1931: A study of residential differentiation and social structure. *Journal of Historical Geography* 21, no. 1: 55–74. https://doi.org/10.1016/0305-7488(95)90007-1

Hiebert, Daniel. 1999. Immigration and the changing social geography of Vancouver. *BC Studies* 121: 35–82.

Hiebert, Daniel. 2012. *A New Residential Order? The Social Geography of Visible Minority and Religious Groups in Montreal, Toronto and Vancouver in 2031.* Ottawa: Citizenship and Immigration.

Higgins, Donald J.H. 1977. *Urban Canada: Its Government and Politics.* Toronto: Macmillan.

Higgins, Donald J.H. 1986. *Local and Urban Politics in Canada.* Toronto: Gage.

High, Steven. 2019. Little Burgundy: The interwoven histories of race, residence, and work in twentieth-century Montreal. *Urban History Review* 46, no. 1: 23–44. https://doi.org/10.7202/1059112ar

High, Steven. 2022. *Deindustrializing Montreal: Entangled Histories of Race, Residence and Class.* Montreal and Kingston: McGill-Queen's University Press. https://doi.org/10.1515/9780228012313

High, Steven, Lysiane Goulet, Michelle Duchesneau, and Dany Guay-Bélanger. 2020. Interlocking lives: Employment mobility and family fixity in three gentrifying neighbourhoods of Montreal. *International Journal of Urban and Regional Research* 44, no. 3: 505–20. https://doi.org/10.1111/1468-2427.12728

Hill, Donna, ed. 1981. *A Black Man's Toronto, 1914–1980: The Reminiscences of Harry Gairey.* Toronto: Multicultural History Society of Ontario.

Hirschman, Albert O. 1970. *Exit, Voice, and Loyalty: Responses to Decline in Firms, Organizations, and States.* Cambridge, MA: Harvard University Press.

Hobkirk, Alan. 1973. Eastside, westside: Social class images of Vancouver. In David Ley, ed., *Community Participation and the Spatial Order of the City*, 11–23. BC Geographical Series no.19. Vancouver: Tantalus.

Hobsbawm, Eric. 1983. Introduction: Inventing traditions. In Eric Hobsbawm and Terence Ranger, eds., *The Invention of Tradition*, 1–14. Cambridge: Cambridge University Press. https://doi.org/10.1017/CBO9781107295636.001

Hodge, Gerald, and David Gordon. 2008. *Planning Canadian Communities.* 5th ed. Toronto: Thomson.

Holdsworth, Deryck. 1977. House and home in Vancouver: Images of west coast urbanism. In Gilbert A. Stelter and Alan F.J. Artibise, eds., *The*

Canadian City. Essays in Urban History, 196–211. Toronto: McClelland and Stewart.

Hood, Hugh. 1975. *The Swing in the Garden*. Ottawa: Oberon.

Horak, Martin, and Aaron A. Moore. 2015. Policy shift without institutional change: The precarious place of neighborhood revitalization in Toronto. In Clarence N. Stone and Robert Stoker, eds., *Urban Neighborhoods in a New Era. Revitalization Politics in the Postindustrial City* 182–208. Chicago: University of Chicago Press.

Hornstein, Jeffrey M. 2005. *A Nation of Realtors: A Cultural History of the Twentieth-Century American Middle Class*. Durham, NC: Duke University Press.

Horsman, A., and P. Raynor. 1978. Citizen participation in local area planning: Two Vancouver cases. In Len Evenden, ed., *Vancouver: Western Metropolis*, 239–53. Western Geographical Series, vol. 16, Department of Geography, University of Victoria.

Hou, Feng, and John Myles. 2005. Neighborhood inequality: Neighborhood affluence and population health. *Social Science and Medicine* 60, no. 7: 1557–69. https://doi.org/10.1016/j.socscimed.2004.08.033

Hou, Feng, and John Myles. 2008. The changing role of education in the marriage market: Assortative marriage in Canada and the United States since the 1970s. *Canadian Journal of Sociology* 33: 337–66. https://doi.org/10.29173/cjs551

Houston, Susan E. 1982. The "waifs and strays" of a late Victorian city: Juvenile delinquency in Toronto. In Joy Parr, ed., *Childhood and Family in Canadian History*, 129–42. Toronto: McClelland and Stewart.

Howell, Junia. 2019a. The unstudied reference neighborhood: Towards a critical theory of empirical neighborhood studies. *Sociology Compass* 13, no. 1: e12649. https://doi.org/10.1111/soc4.12649

Howell, Junia. 2019b. The truly advantaged: Examining the effects of privileged places on educational attainment. *Sociological Quarterly* 60, no. 3: 420–38. https://doi.org/10.1080/00380253.2019.1580546

Howlett, Karen. 2015. Ontario moves to fast-track process to close "underutilized" schools. *Globe and Mail*, 6 April.

Hugill, David. 2017. What is a settler-colonial city? *Geography Compass* 11, no. 5: e12315. https://doi.org/10.1111/gec3.12315

Hugill, David. 2022. Neeginan: The struggle to build an indigenous "enclave." In Kent Blanchett, Kathleen D. Cahill, and Andrew Needha, eds., *Indian Cities: Histories of Indigenous Urbanization*, 246–66. Norman, OK: University of Oklahoma Press.

Hulbert, François. 1981. Pouvoir municipal et développement urbain: Le cas de Sainte-Foy en banlieue de Québec. *Cahiers de géographie de Québec* 25, no. 66: 361–401. https://doi.org/10.7202/021529ar

Hulchanski, J. David. 1981. The origins of urban land use planning in Ontario, 1900–1946. PhD diss., University of Toronto.

Hulchanski, J. David. 1986. The 1935 Dominion Housing Act: Setting the stage for a permanent federal presence in Canada's housing sector. *Urban History Review* 15, no. 1: 19–40. https://doi.org/10.7202/1018891ar

Hulchanski, J. David. 2010. *The Three Cities within Toronto: Income Polarization among Toronto's Neighborhoods, 1970–2005*. Toronto: University of Toronto Cities Centre.

Hunt, Michael E., and Gail Gunter-Hunt. 1986. Naturally occurring retirement communities. *Journal of Housing for the Elderly* 3, nos. 3–4: 3–22. https://doi.org/10.1300/J081V03N03_02

Hunter, Albert. 1974. *Symbolic Communities: A Study of the Persistence and Change of Chicago's Local Communities*. Chicago: University of Chicago Press.

Hunter, Albert. 1975. The loss of community: An empirical test through replication. *American Sociological Review* 40, no. 5: 537–55. https://doi.org/10.2307/2094194

Hunter, Albert. 1983. The urban neighborhood: Its analytical and social contexts. In Phillip L. Clay and Robert Hollister, eds., *Neighborhood Policy and Planning*, 3–30. Lexington, MA: DC Heath.

Hunter, Albert. 1987. The symbolic ecology of suburbia. Human Behavior and Environment 9: 191–221. https://doi.org/10.1007/978-1-4899-1962-5_6

Hwang, Jackelyn. 2015. The social construction of a gentrifying neighborhood: Reifying and redefining identity and boundaries in inequality. *Urban Affairs Review* 52, no. 1: 603–28. https://doi.org/10.1177/1078087415570643

Iacovetta, Franca. 1992. *Such Hardworking People: Italian Immigrants in Postwar Toronto*. Montreal and Kingston: McGill-Queen's University Press. https://doi.org/10.1515/9780773563155

Iacovetta, Franca. 1999. Gossip, contest, and power in the making of suburban bad girls: Toronto, 1945–60. *Canadian Historical Review* 80, no. 4: 585–623. https://doi.org/10.3138/CHR.80.4.585

Iankova, Katia. 2010. Insertion de la réserve Huronne dans l'espace urban de la ville de Québec. *Recherches Amérindiennes au Québec*. 38, no. 1: 67–78. https://doi.org/10.7202/039748ar

Insler, Michael, and Kurtis Swope. 2016. School quality, residential choice, and the U.S. housing bubble. *Housing Policy Debate* 26, no. 1: 53–79. https://doi.org/10.1080/10511482.2014.956777

Irving, Alan, Harriett Parsons, and Donald Bellamy. 1995. *Neighbours: Three Social Settlements in Downtown Toronto*. Toronto: Canadian Scholars Press.

Jacobs, Dorene. 1971. The Annex Ratepayers' Association: Citizen efforts to exercise social choice in their urban environment. In James A. Draper, ed., *Citizen Participation, Canada: A Book of Readings*, 288–306. Toronto: New Press.

Jacobs, Jane. 1961. *The Death and Life of Great American Cities*. New York: Random House.

James, Cathy. 2001. Reforming reform: Toronto's settlement house movement, 1900–1920. *Canadian Historical Review* 82, no. 1: 55–90. https://doi.org/10.3138/CHR.82.1.55

Janowitz, Morris. 1978. *The Last Half Century: Societal Change and Politics in America*. Chicago: University of Chicago Press.

Jean, Sandrine. 2014. Ville ou banlieue? Les choix residentielle des jeunes familles de classes moyenne dans la grande région de Montréal. *Recherches sociographiques* 55, no. 1: 105–34. https://doi.org/10.7202/1025647ar

Jean, Sandrine. 2015. Personal communication. 29 March.

Jean, Sandrine. 2016. Neighbourhood attachment revisited: Middle-class families in the Montreal metropolitan region. *Urban Studies* 53, no. 12: 2567–83. https://doi.org/10.1177/0042098015594089

Jerrett, Michael, Richard Burnett, Pavlos Kanaroglou, John Eyles, Norm Finkelstein, Chris Giovis, and Jeffrey Brook. 2001. A GIS-environmental justice analysis of particulate air pollution in Hamilton, Canada. *Environment and Planning A* 33: 955–73. https://doi.org/10.1068/a33137

Johnson, Laura, and Robert E. Johnson. 2017. *Regent Park Redux*. New York: Routledge. https://doi.org/10.4324/9781315748993

Jones, Esyllt. 2007. *Influenza 1918: Disease, Death, and Struggle in Winnipeg*. Toronto: University of Toronto Press.

Kahlenberg, Richard D. 2019. Why economic school segregation matters. In Ingrid G. Ellen and Justin P. Steil, eds., *The Dream Revisited: Contemporary Debates about Housing, Segregation, and Opportunity*, 78–80. New York: Columbia University Press80. https://doi.org/10.7312/elle18362-021

Kahneman, Daniel. 2011. *Thinking, Fast and Slow*. New York: Farrar, Straus and Giroux.

Kalbach, Warren. 1990. Ethnic residential segregation and its significance for the individual in an urban setting. In Raymond Breton, Wsevolod Isajiw, and Warren Kalbach, *Ethnic Identity and Equality: Varieties of Experience in a Canadian City*, 92–134. Toronto: University of Toronto Press.

Kaplan, Howard. 1982. *Reform, Planning and City Politics: Montreal, Winnipeg, Toronto*. Toronto: University of Toronto Press.

Kataure, Vispal, and Margaret Walton-Roberts. 2013. The housing preferences and location choices of second-generation South Asians living in ethnic enclaves. *South Asian Diaspora* 5, no. 1: 57–76. https://doi .org/10.1080/19438192.2013.722385

Katz, Michael B. 1975. *The People of Hamilton, Canada West: Family and Class in a Mid-Nineteenth Century Industrial City.* Cambridge, MA: Harvard University Press. https://doi.org/10.4159/harvard.9780674494213

Katz, Michael B., Michael J. Doucet, and Michael Stern. 1982. *The Social Organization of Early Industrial Capitalism.* Cambridge, MA: Harvard University Press. https://doi.org/10.4159/harvard.9780674181533

Keating, Donald. 1975. *The Power to Make It Happen.* Toronto: Green Tree Publishing.

Keating, Peter, ed. 1976. *Into Unknown England, 1866–1913: Selections from the Social Explorers.* London: Fontana.

Keenan, Edward. 2010. A tale of two Torontos. *Eye Weekly,* 3 November.

Keenan, Edward. 2013. *Some Great Idea: Good Neighbourhoods, Crazy Politics and the Invention of Toronto.* Toronto: Coach House Press.

Keillor, Rebecca. 2019. Vancouver implements programs to help connect people experiencing loneliness. *Globe and Mail,* 30 December. https:// www.theglobeandmail.com/canada/british-columbia/article-vancouver -implements-programs-to-help-connect-people-experiencing/

Keller, Suzanne. 1968. *The Urban Neighborhood: A Sociological Perspective.* New York: Random House.

Kelly, Cathal. 2018. *Boy Wonders: A Memoir.* Toronto: Penguin Random House.

Kermoal, Nathalie. 2022. La presence métisse à Edmonton dans les années 1930: De l'éclipse à l'émergence. *Urban History Review* 49, no. 2. https:// doi.org/10.3138/uhr-2021-0001

Kern, Leslie. 2010. *Sex and the Revitalized City: Gender, Condominium Development, and Urban Citizenship.* Vancouver: UBC Press.

Kerr, Don. 1989. Ways of seeing Saskatoon. In Elaine Decourset, Don Kerr, Dan Ring, and Mathew Teitelbaum, eds., *Saskatoon Imagined.* Saskatoon, SK: Mendel Art Gallery.

Kerr, Donald. 1943. Vancouver: A study in urban geography. MA thesis, University of Toronto.

Kerr, Jacqueline, Dori Rosenberg, and Lawrence Frank. 2012. The role of the built environment in healthy aging: Community design, physical activity, and health among older adults. *Journal of Planning Literature* 27, no. 1: 43–60. https://doi.org/10.1177/0885412211415283

Keyfitz, Nathan. 1961. The changing Canadian population. In Samuel D. Clark, ed., *Urbanism and the Changing Canadian Society.* Toronto: University of Toronto Press. https://doi.org/10.3138/9781442652859-002

Kheraj, Sean. 2013. *Inventing Stanley Park: An Environmental History.* Vancouver: UBC Press. https://doi.org/10.59962/9780774824262

Kidd, Bruce. 2015. The Elizabeth Street playground revisited. In Jon Lorinc, Michael McClelland, Ellen Scheinberg, and Tatum Taylor, eds., *The Ward: The Life and Loss of Toronto's First Immigrant Neighbourhood*, 184–7. Toronto: Coach House Press.

Kidd, Kenneth. 2009. The *Star* unveils unique map of of neighbourhoods. *Toronto Star,* 8 March. http://www.thestar.com/news/gta/2009/03/08/the_star_unveils_unique_map_of_neighbourhoods.html. Last accessed 21 Sept. 2021.

Kiernan, Matthew J., and David C. Walker. 1983. Winnipeg. In Warren Magnusson and Andrew Sancton, eds., *City Politics in Canada*, 222–54. Toronto: University of Toronto Press. https://doi.org/10.3138/9781487575908-008

Kirk, Carolyn T., and Gordon W. Kirk. 1981. The impact of the city on home ownership: A comparison of immigrants and native whites at the turn of the century. *Journal of Urban History* 7: 471–87. https://doi.org/10.1177/009614428100700403

Kitchen, Peter, Allison Williams, and Dylan Simone. 2012. Measuring social capital in Hamilton, Ontario. *Social Indicators Research* 108: 215–38. https://doi.org/10.1007/s11205-012-0063-3

Klodawski, Fran. 2006. Landscapes on the margins: Gender and homelessness in Canada. *Gender, Place and Culture* 13, no. 4: 365–81. https://doi.org/10.1080/09663690600808478

Kluch, Sofia, and Jessi Gordon. 2018. G7 women need safety to make gender equality history. Gallup blog, 19 June.

Knelman, Judith. 1983. Riverdale: A neighbourhood of contrasts. *Globe and Mail*, 10 December.

Koch, Tim. 2017. *Disease Maps, History and More: Cartographies of Disease. Maps, Mapping and Medicine.* Redlands, CA: ESRI Press.

Kohen, Dafna, Jeanne Brooks-Gunn, Tama Leventhal, and Clyde Hertzman. 2002. Neighborhood income and physical and social disorder in Canada: Associations with young children's competencies. *Child Development* 73, no. 6: 1844–60. https://doi.org/10.1111/1467-8624.t01-1-00510

Korinek, Valerie. 2018. *Prairie Fairies: A History of Queer Communities and People in Western Canada, 1930–1985.* Toronto: University of Toronto Press. https://doi.org/10.3138/9781487518172

Krishnan, Manisha. 2015. Midtowners battle the rise of the midrise. *Toronto Star,* 25 May.

Krivo, Laure J., Heather M. Washington, Ruth P. Peterson, Christopher R. Browning, Catherine A. Calder, and Mei-Po Kwan. 2013. Social isolation of disadvantage and advantage: The reproduction of inequality in urban space. *Social Forces* 92, no. 1: 141–64. https://doi.org/10.1093/sf/sot043

Krohn, Roger C., Berkeley Fleming, and Marilyn Manzer. 1977. *The Other Economy: The Internal Logic of Local Rented Housing.* Toronto: Peter Martin Associates.

Kuper, Leo. 1951. Social science research and the planning of urban neighbourhoods. *Social Forces* 29, no. 3: 237–43. https://doi.org/10.2307/2572411

Kwan, Mei-Po. 2018. The limits of the neighborhood effect: Contextual uncertainties in geographic environmental health, and social science research. *Annals of the Association of American Geographers* 108, no. 6: 1482–90. https://doi.org/10.1080/24694452.2018.1453777

Kyrck, Hazel. 1929. *Economic Problems of the Family.* New York: Harper and Brothers.

Lafrance, Marc, and David-Thierry Rudell. 1982. Physical expansion and socio-cultural segregation in Quebec City, 1765–1840. In Gilbert A. Stelter and Alan F.J. Artibise, eds., *Shaping the Urban Landscape: Aspects of the Canadian City-Building Process*, 148–72. Ottawa: Carleton University Press. https://doi.org/10.1515/9780773584860-007

Lai, Chuen-Yau. 1973. Socio-economic structures and viability of Chinatown. In Charles Forward, ed., *Residential and Neighbourhood studies in Victoria*, 101–29. Western Geographical Series No. 5. Victoria, BC: Department of Geography, University of Victoria.

Lamarre, Christine. 2010. Quartier. In Christian Topalov, Laurent Coudroy de Lille, Jean-Charles Depaule, and Brigitte Morin, eds., *L'Aventure des Mots de la Ville*, 1013–17. Paris: Laffont.

Lands, LeeAnn. 2008. Be a patriot, buy a home: Re-imagining home owners and home ownership in early 20th century Atlanta. *Journal of Social History* 41, no. 4: 943–65. https://doi.org/10.1353/jsh.0.0029

Lane, Barbara. 2015. *Houses for a New World: Builders and Buyers in American Suburbs, 1945–1965.* Princeton, NJ: Princeton University Press. https://doi.org/10.1515/9780691246420

Lansing, John B., and Robert W. Marans. 1969. Evaluation of neighborhood. *Journal of the American Institute of Planners* 35: 195–99. https://doi.org/10.1080/01944366908977953

Lassiter, Matthew D. 2012. Schools and housing in metropolitan history: An introduction. *Journal of Urban History* 38, no. 2: 195–204. https://doi.org/10.1177/0096144211427111

The last holdout. 2021. *The Economist*, 18 December, 21–3.

Latham, Jim, and Tina Moffatt. 2007. Determinants of variation in food cost and availability in two socioeconomically contrasting neighborhoods of

Hamilton, Ontario, Canada. *Health and Place* 13, no. 1: 273–87. https://doi
.org/10.1016/j.healthplace.2006.01.006

Lauzon, Gilles. 2014. *Pointe-Saint-Charles: L'Urbanisation d'un quartier ouvrier
de Montréal, 1840–1930.* Quebec City: Septentrion.

Lawrence, Jon. 2016. Inventing the "traditional working class": A re-analysis
of interview notes from Young and Wilmott's Family and Kinship in
East London. *Historical Journal* 59, no. 2: 567–93. https://doi.org/10.1017
/S0018246X15000515

League for Social Reconstruction. (1975) 1935. A housing programme. In
Social Planning for Canada, 451–63. Toronto: University of Toronto Press.
https://doi.org/10.3138/9781487583590-023

Lederman, Marsha. 2023. Nextdoor: With friends like these, who needs
frenemies? *Globe and Mail*, 15 April.

Lee, Barrett A., and Karen Campbell. 1997. Common ground: Urban
neighborhoods as survey respondent see them. *Social Science Quarterly*
78, no. 4: 922–36.

Lee, Barrett A., R.S. Oreposa, Barbara J. Metch, and Avery M. Guest. 1984.
Testing the decline of community thesis: Neighborhood organizations
in Seattle, 1929 and 1979. *American Journal of Sociology* 89, no. 5: 1161–88.
https://doi.org/10.1086/227987

Lee, Jo-Anne. 2007. Gender, ethnicity and hybrid forms of community-based
activism in Vancouver, 1957–1978. *Gender, Place and Culture* 14, no. 4: 381–
407. https://doi.org/10.1080/09663690701439702

Legault, Guy R. 2002. *La Ville qu'on a Bâtie: Trente Ans au Service de
l'Urbanisme et de l'Habitation de Montréal, 1956–1986.* Montreal: Liber.

Lemelin, Roger. 1948. *The Town Below.* New York: Reynal and Hitchcock.

Lemelin, Rober. (1942) 1967. *Au Pied de la Pente Douce.* Montreal: Cercle du
livre de France.

Lemon, James. 1985. *Toronto since 1918.* Toronto: Lorimer.

Leslie, Deborah, and Mia Hunt. 2013. Securing the neoliberal city: Discourses
of creativity and priority neighborhoods in Toronto, Canada. *Urban
Geography* 34, no. 8: 1171–92. https://doi.org/10.1080/02723638.2013.823729

Léveillée, Jacques. 1988. Pouvoir local et politiques publiques à Montréal:
Renouveau dans les modalités d'exercice du pouvoir urbain. *Cahiers de
recherché sociologique* 11, no. 2: 37–64. https://doi.org/10.7202/1002048ar

Leventhal, Tama, and Jeanne Brooks-Gunn. 2000. The neighborhoods they
live in: The effects of neighborhood residence on child and adolescent
outcomes. *Psychological Bulletin* 126, no. 2: 309–37. https://doi.org
/10.1037//0033-2909.126.2.309

Leventhal, Tama, Véronique Duprée, and Elizabeth Shuey. 2015. Children
in neighborhoods. In Richard M. Lerner, M.H. Bornstein, and Tama

Leventhal, eds., *Handbook of Child Psychology and Development Science*, 7th ed., 493–533. New York: Wiley. https://doi.org/10.1002/9781118963418 .childpsy413

Lewinburg, Frank. 1983. Will neighbourhoods accept intensification? In John R. Hitchcock, ed., *The Metropolis: Proceedings of a Conference in Honour of Hans Blumenfeld*, 175–82. Toronto: Department of Geography and Centre for Urban and Community Studies, University of Toronto.

Lewinburg, Frank. 1984. Neighbourhood planning: The reform years in Toronto. In Neighbourhood Planning Conference, *Toronto Neighbourhoods: The Next Ten Years: Papers Delivered at the Neighbourhood Planning Conference, Toronto, Nov. 16–17*. Toronto: City of Toronto Planning and Development Department.

Lewis, Nathaniel M. 2013. Ottawa's Le/The Village: Creating a gaybourhood amidst the "death of the village." *Geoforum* 49: 233–42. https://doi .org/10.1016/j.geoforum.2013.01.004

Lewis, Norah L., ed. 2002. *Freedom to Play: We Made Our Own Fun*. Waterloo, ON: Wilfrid Laurier University Press.

Lewis, Pierce F. 1979. Axioms for reading the landscape. In Donald W. Meinig, ed., *The Interpretation of Ordinary Landscapes: Geographical Essays*, 11–32. New York: Oxford University Press.

Lewis, Robert D. 1990. Home ownership reassessed for Montreal in the 1840s. *Canadian Geographer* 34, no. 2: 150–2. https://doi.org/10.1111 /j.1541-0064.1990.tb01260.x

Lewis, Robert D. 1991. The segregated city: Class residential patterns and the development of industrial districts in Montreal, 1861 and 1901. *Journal of Urban History* 17, no. 2: 123–52. https://doi.org/10.1177 /009614429101700201

Lewis, Robert D., and Paul Hess. 2016. Refashioning urban space in postwar Toronto: The Wood-Wellesley redevelopment areas, 1952–1957. *Planning Perspectives* 31, no. 4: 563–84. https://doi.org/10.1177/009614429101700201

Levitan, Seymour, and Carol Miller, eds. 1986. *Lucky to Live in Cedar Cottage: Memories of Lord Selkirk Elementary School and Cedar Cottage Neighbourhood, 1911–1963*. Vancouver: Vancouver School Board.

Ley, David. 1993. Past elites and present gentry: Neighbourhoods of privilege in the inner city. In Larry S. Bourne and David Ley, eds., *The Changing Social Geography of Canadian Cities*, 214–33. Montreal and Kingston: McGill-Queen's University Press. https://doi.org/10.1515 /9780773563551-012

Ley, David. 1994. Social polarization and community response: Contesting marginality in Vancouver's Downtown Eastside. In Frances Frisken, ed., *The Changing Canadian Metropolis*, 699–724. Berkely, CA: Institute of Government Studies Press.

Ley, David. 1995. Between Europe and Asia: The case of the missing sequoias. *Ecumene* 2, no. 2: 185–210. https://doi.org/10.1177 /147447409500200205

Ley, David. 1996. *The New Middle Class and the Remaking of the Central City.* Toronto: Oxford University Press. https://doi.org/10.1093/oso /9780198232926.001.0001

Ley, David, Daniel Hiebert, and Geraldine Pratt. 1992. Time to grow up? From urban village to world city, 1966–1991. In Graeme Wynn and Tim Oke, eds., *Vancouver and Its Region*, 234–66. Vancouver: University of British Columbia Press.

Ley, David, and John Mercer. 1980. Locational conflict and the politics of consumption. *Economic Geography* 56, no. 2: 89–109. https://doi .org/10.2307/142929

Ley, David, and Heather Smith. 2000. Relations between deprivation and immigrant groups in large Canadian cities. *Urban Studies* 37, no. 1: 37–62. https://doi.org/10.1080/0042098002285

Leyshon, Glynn A. 1999. The art of play: Street games in the Depression. *The Beaver* 79 (Aug.–Sept.): 32–6.

Lightbody, James. 1983. Edmonton. In Warren Magnusson and Andrew Sancton, eds., *City Politics in Canada*, 255–90. Toronto: University of Toronto Press. https://doi.org/10.3138/9781487575908-009

Lindgren, April. 2009. News, geography and disadvantage: Mapping newspaper coverage of high-needs neighbourhoods in Toronto, Canada. *Canadian Journal of Urban Research* 18, no. 1: 76–97.

Lippert, Randy K. 2019. *Condo Conquest: Urban Governance and Condoization in New York City and Toronto.* Vancouver: UBC Press. https://doi .org/10.59962/9780774860376

Lithwick, Norman H. 1971. *Urban Poverty.* Central Mortgage and Housing Corporation Research Monograph No. 1. Ottawa: Central Mortgage and Housing Corporation .

Liu, Sikee, and Nicholas Blomley. 2013. Making news and making space: Framing Vancouver's Downtown Eastside. *Canadian Geographer* 57, no. 2: 119–32. https://doi.org/10.1111/j.1541-0064.2012.00453.x

Livingstone, D.W. 2023. *Tipping Point for Advanced Capitalism.* Halifax: Fernwood.

Lloyd, Sheila. 1979. The Ottawa typhoid epidemics of 1911 and 1912. *Urban History Review* 8, no. 1: 66–89. https://doi.org/10.7202 /1019391ar

Lo, Lucia. 2011. Immigrants and social services in the suburbs. In Douglas Young, Patricia Wood, and Roger Keil, eds., *In-between Infrastructure: Urban Connectivity in an Age of Vulnerability*, 131–50. Toronto: Praxis.

Loebach, Janet, and Jason Gilliland. 2019. Examining the social and built environment factors influencing children's independent use of their neighborhood and the experience of local settings as child-friendly. *Journal of Planning Education and Research*. https://doi.org/10.1177/0739456X19828444

Lombardo, Nicholas. 2014. White collar workers and neighbourhood change: Jarvis Street in Toronto. *Urban History Review* 43, no. 1: 1–20. https://doi.org/10.3138/uhr.43.01.01

Loo, Tina. 2010. Africville and the dynamics of state power in postwar Canada. *Acadiensis* 39, no. 2: 23–47.

Loo, Tina. 2019. *Moved by the State: Forced Relocation and Making a Good Life in Postwar Canada*. Vancouver: UBC Press. https://doi.org/10.59962/9780774861021

Looker, Benjamin, 2015. *A Nation of Neighborhoods: Imagining Cities, Communities and Democracy in Postwar America*. Chicago: University of Chicago Press. https://doi.org/10.7208/chicago/9780226290454.001.0001

Lorimer, James, and Myfanwy Phillips. 1971. *Working People: Life in a Downtown City Neighbourhood*. Toronto: Lorimer.

Lorinc, John. 2015. What up and coming looks like: The Junction Triangle at 5. *Globe and Mail*, 28 March.

Lorinc, John, Michael McClelland, Ellen Scheinberg, and Tatum Taylor, eds. 2015. *The Ward: The Life and Loss of Toronto's First Immigrant Neighbourhood*. Toronto: Coach House Press.

Lu, Vanessa. 2010. 416 and 905: We're more alike than we think. *Toronto Star*, 18 April.

Lutz, John, Patrick Dunae, Jason Gilliland, and Don Lafreniere. 2014. Turning space inside out: Spatial history and race in Victoria. In Jennifer Bonnell and Marcel Fortin, eds., *Historical GIS Research in Canada*, 1–26. Calgary: University of Calgary Press. https://doi.org/10.2307/j.ctv6gqt40.5

Macdonald, Bruce. 1992. *Vancouver: A Visual History*. Vancouver: Talonbooks.

MacDonald, Norbert. 1973. A critical growth cycle for Vancouver, 1900–1914. *BC Studies* (Spring): 26–42.

MacDougall, Cynthia. 2015. Growing up on Walton street. In John Lorinc, Michael McClelland, Ellen Scheinberg, and Tatum Taylor, eds, *The Ward: The Life and Loss of Toronto's First Immigrant Neighbourhood*, 112–13. Toronto: Coach House Press.

MacGregor, Roy. 2017. Can Peterborough stand as an inspiration to Québec after mosque attack? *Globe and Mail*, 4 February.

https://www.theglobeandmail.com/news/national/peterborough
-inspiration-to-quebec-mosque-attack/article33893274/

Mackintosh, Phillip G. 2017. *Newspaper City: Toronto's Street Surfaces and the Liberal Press*. Toronto: University of Toronto Press. https://doi .org/10.3138/9781442666566

Mackintosh, Phillip G. 2020. Liberalism underfoot: A micro-geography of street paving and social dissolution – Brunswick Avenue, Toronto, Ontario, 1898–99. In Alida Clemente, Dag Lindström, and Jon Stobart, eds., *Micro-Geographies of the Western City, c1750–1900*, 105–23. London: Routledge. https://doi.org/10.4324/9780429329395-9

MacLennan. 1945. *Two Solitudes*. Toronto: Collins.

Magnusson, Warren. 1981. Community organization and local self-government. In Lionel Feldman, ed., *Politics and Government of Urban Canada: Selected* Readings, 61–86. 4th ed. Toronto: Methuen.

Magnusson, Warren. 1983a. Introduction: The development of Canadian urban government. In Warren Magnusson and Andrew Sancton, eds., *City Politics in Canada*, 3–57. Toronto: University of Toronto Press. https:// doi.org/10.3138/9781487575908-002

Magnusson, Warren. 1983b. Toronto. In Warren Magnusson and Andrew Sancton, eds., *City Politics in Canada*, 94–139. Toronto: University of Toronto Press. https://doi.org/10.3138/9781487575908-004

Making Canadians of little foreigners. 1916. *The Globe*, 10 October.

Mallach, Alan, and Todd Swanstrom. 2023. *The Changing American Neighbourhood: The Meaning of Place in the Twenty-First Century*. Ithaca, NY: Cornell University Press. https://doi.org/10.7591/cornell /9781501770890.001.0001

Maloutas, Thomas, and Nikos Karadimitrou, eds. 2022. *Vertical Cities: Micro-segregation, Social Mix and Urban Housing Markets*. Cheltenham, UK: Edward Elgar.

Maltais, Alexandre. 2014. The evolution of Canadian discourses on urban neighbourhoods. Montréal, 1944–2014. Unpublished manuscript, INRS, Montreal.

Maltais, Alexandre. 2023. *Les Rues qui Changent: Commerce de Détail et Transformation des Quartiers Centraux Montréalais*. Quebec City: Presses de l'Université Laval. https://doi.org/10.1515/9782763749518

Mann, Peter H. 1954. The concept of neighborliness. *American Journal of Sociology*. 60, no. 2: 163–68. https://doi.org/10.1086/221507

Mann, William E. 1961. The social system of a slum: The Lower Ward, Toronto. In Samuel D. Clark, ed., *Urbanism and the Changing Canadian Society* 39–69. Toronto: University of Toronto Press. https://doi .org/10.3138/9781442652859-004

Mann, William E., ed. 1970. *The Underside of Toronto*. Toronto: McClelland and Stewart.

Marchessault, J. 1975. *Like and Child of the Earth*. Trans. by Yvonne Klein. Vancouver: Talonbooks.

Marlyn, John. 1957. *Under the Ribs of Death*. Toronto: McClelland and Stewart.

Marsan, Jean-Claude. 1981. *Montréal in Evolution: Historical Analysis of the Development of Montréal's Architecture and Urban Environment*. Montreal and Kingston: McGill-Queen's University Press.

Marsh, Leonard C. 1950. *Rebuilding a Neighbourhood: Report on a Demonstration Slum-Clearance and Urban Rehabilitation Projec in a Key Central Area in Vancouver*. Research Publication No.1, Vancouver: University of British Columbia.

Martindale, Edith. 1977. *Contemporary Entrepreneurs: The Sociology of Residential Real Estate Agents*. Westport, CT: Greenwood.

Mason, Waye. 2011. *HRM Neighbourhood Map Project*. http://wayemason.ca/hrm-map-project/ (Last accessed 21 Sept. 2021)

Massenkoff, Maxim, and Nathan Wilmers. 2023. Rubbing shoulders: Class segregation in daily activities. *Social Science Research Network*, 22 August. https://doi.org/10.2139/ssrn.4516850

Massey, Doreen. 1995a. Places and their pasts. *History Workshop Journal* 39: 182–92. https://doi.org/10.1093/hwj/39.1.182

Massey, Doreen. 1995b. The conceptualisation of place. In Doreen Massey and Pat Jess, eds., *A Place in the World? Places, Cultures and Globalisation*. Oxford: Open University/ Oxford University Press.

Matheson, Flora, Rahim Moineddin, James R. Dunn, Maria Creatore, Piotr Gozdyra, and Richard H. Glazier. 2006. Urban neighbourhoods, chronic stress, gender and depression. *Social Science and Medicine* 63: 2604–16. https://doi.org/10.1016/j.socscimed.2006.07.001

Matheson, Flora, Heather L. White, Rahim Moineddin, James R. Dunn, and Richard H. Glazier. 2010. Neighborhood chronic stress and gender inequalities in hypertension among Canadian adults: A multi-layer analysis. *Journal of Epidemiology and Community Health* 64, no. 8: 705–13. https://doi.org/10.1136/jech.2008.083303

Mathews, Vanessa. 2019. Lofts in transition: Gentrification in the Warehouse District, Regina, Saskatchewan. *Canadian Geographer* 63, no. 2: 284–96. https://doi.org/10.1111/cag.12495

Mawani, Renisa. 2003. Legal geographies of aboriginal segregation in British Columbia: The making and unmaking of the Songhees reserve. In Carolyn Strange and Alison Bashford, eds., *Isolation: Places and Practices of Exclusion*, 163–80. London: Routledge.

Mayhew, Barry W. 1967. *Local Areas of Vancouver*. Vancouver: Research Department, United Community Services of the Greater Vancouver Area.

Mayne, Alan. 2017. *Slums: The History of a Global Injustice*. London: Reaktion.

McAfee, Ann. 1972. Evolving inner-city residential neighbourhoods: The case of Vancouver's West End. In Julian V. Minghi, ed., *People of the Living Land: Geography of Cultural Diversity in British Columbia*. Vancouver: Tantalus.

McAfee, Ann. 2016. People and plans: Vancouver's CityPlan process. In Ren Thomas, ed., *Planning Canada: A Case Study Approach*. Don Mills, ON: Oxford University Press.

McAfee, Ann, Anne Borooah, Michael Gordon, and Fred Breeze. 1989. Neighbourhood intensification: How to succeed. *Canadian Housing* 5, no. 5: 14–17.

McCann, Larry D. 1996. Planning and building the corporate suburb of Mount Royal, 1910–1925. *Planning Perspectives* 11, no. 3: 259–302. https://doi.org/10.1080/026654396364871

McCann, Larry D. 2017. *Imagining Uplands: John Olmsted's Masterpiece of Residential Design*. Victoria, BC: Brighton Press.

McCormack, A. Ross. 1984. Networks among British immigrants and accommodation to Canadian society: Winnipeg, 1900–1914. *Histoire sociale/ Social History* 17, no. 34: 357–74.

McDonald, Robert A.J. 1996. *Making Vancouver: Class, Status and Social Boundaries, 1863–1913*. Vancouver: UBC Press.

McGinn, Dave. 2013a. Buildings in search of a community. *Globe and Mail,* 2 March.

McGinn, Dave. 2013b. How the 21st century put the Block Parent program into decline. *Globe and Mail,* 28 October.

McGirr, Emily, Andrejs Skaburskis, and Tim S. Danegani. 2015. Expectations, preferences and satisfaction levels among new and long-term residents in a gentrifying Toronto neighbourhood. *Urban Studies* 52, no. 1: 3–19. https://doi.org/10.1177/0042098014522721

McGuigan, Peter T. 2007. *Historic South End Halifax*. Halifax: Nimbus.

McLaughlin, Stephen. 1984. The challenge ahead: Future directions for neighbourhood planning. In *Toronto Neighbourhoods: The Next Ten Years*. Toronto: City of Toronto Planning Department.

McMahon, Michael. 1990. *Metro's Housing Company: The First 35 Years*. Toronto: Metropolitan Housing Company Ltd.

McManus, Ruth, and Phil Ethington. 2007. Suburbs in transition: New approaches to suburban history. *Urban History* 34, no. 2: 317–37. https://doi.org/10.1017/S096392680700466X

Meij, Erik, Tialda Haartsen, and Louise Meijering. 2020. "Everywhere they are trying to hide poverty. I hate it": Spatial practices of the urban poor in Calgary, Canada. *Geoforum* 117: 206–15. https://doi.org/10.1016/j.geoforum.2020.10.002

Mendez, Pablo. 2009. Immigrant Residential Geographies and the "spatial assimilation" debate in Canada, 1997–2006. *Journal of International Migration and Integration* 10: 89–108.

Mercer, Greg. 2021. Membertou's moment: How a Mi'kmaq nation found prosperity and a seafood empire. *Globe and Mail*, 10 January. https://www.theglobeandmail.com/canada/article-membertous-moment-how-a-mikmaq-nation-found-prosperity-and-a-seafood/

Mercure-Jolette. 2015. The "Dozois plan": Lessons learned from urban renewal and the history of urban planning in Montreal, trans. by Oliver Waine. *Metropolitics*, 23 October.

Micallef, Shawn. 2014. *The Trouble with Brunch: Work, Class and the Pursuit of Leisure*. Toronto: Coach House Books.

Micallef, Shawn. 2017. *Frontier City: Toronto on the Verge of Greatness*. Toronto: Signal.

Michelson, William. 1977. *Environmental Choice, Human Behavior and Residential Satisfaction*. New York: Oxford University Press.

Michelson, William. 1985. *From Sun to Sun: Daily Obligations and Community Structure in the Lives of Employed Women and Their Families*. Toronto: Rowman and Allanheld.

Mifflin, Erin, and Robert Wilton. 2005. No place like home: Rooming houses in contemporary urban context. *Environment and Planning A*. 37, no. 3: 403–21. https://doi.org/10.1068/a36119

Miller, John. 2006. *A Sharp Intake of Breath*. Toronto: Dundurn.

Miller, Zane L. 1981. The role and concept of neighborhood in American cities. In Robert Fisher and Peter Romofsky, eds., *Community Organization for Urban Social Change: A Historical Perspective*, 3–32. Westport, CT: Greenwood.

Miller, Zane L. 2001. *Visions of Place: The City, Neighborhoods, Suburbs and Cincinnati's Clifton, 1850–2000*. Columbus: Ohio State University Press.

Mills, Caroline. 1993. Myths and meanings of gentrification. In James Duncan and David Ley, eds., *Place/Culture/Representation*, 149–70. London: Routledge.

Mintz, Corey. 2017. Residents of Kensington Market concerned as Airbnb moves in. *Globe and Mail*, 28 April. https://www.theglobeandmail.com/news/toronto/residents-of-torontos-kensington-market-concerned-as-airbnb-moves-in/article34851601/

Molla, Rani. 2019. The rise of fear-based social media like Nextdoor, Citizen and now Amazon's Neighbors. *Vox*, 9 May.

Montreal Board of Trade, Civic Improvement League. 1935. *A Report on Housing and Slum Clearance for Montreal*. Montreal: Montreal Board of Trade.

Mooney-Melvin, Patricia. 1985. Changing contexts: Neighborhood definition and urban organization. *American Quarterly* 37, no. 3: 357–67. https://doi.org/10.2307/2712662

Moore, Peter W. 1979. Zoning and planning: The Toronto experience, 1904–1970. In Alan F.J. Artibise and Gilbert Stelter, eds., *The Usable Urban Past*, 316–41. Toronto: Macmillan. https://doi.org/10.1515/9780773580640-018

Moore, Peter W. 1982. Zoning and neighbourhood change: The Annex in Toronto, 1900–1970. *Canadian Geographer* 26, no. 1: 21–36. https://doi.org/10.1111/j.1541-0064.1982.tb01437.x

Moore, Peter W. 1983. Public services and residential development in a Toronto neighborhood, 1880–1915. *Journal of Urban History* 9, no. 4: 445–71. https://doi.org/10.1177/009614428300900403

Moos, Markus. 2006. From gentrification to youthification? The increasing importance of young age in delineating high density living. *Urban Studies* 53, no. 14: 2903–20. https://doi.org/10.1177/0042098015603292

Moreno, Carlos. 2024. *The Fifteen-Minute City: A Solution for Saving Our Time and Our Planet*. Hoboken, NJ: Wiley.

Morin, Richard. 1998. Gouvernance locale et développement économique des quartiers de Montréal. *Revue de geographie de Lyon* 73, no. 2: 127–34. https://doi.org/10.3406/geoca.1998.4816

Morley, Alan. 1974. *Vancouver: From Milltown to Metropolis*. 3rd ed. Vancouver: Mitchell Press.

Mormino, Gary R. 1986. *Immigrants on the Hill: Italian-Americans in St. Louis, 1882–1982*. Champaign, IL: University of Illinois Press.

Morris, Howard. 1948. *How to Build a Better Home*. Richmond, VA: Better Homes Publishing.

Morton, Suzanne. 1995. *Ideal Surroundings: Domestic Life in a Halifax Working-Class Suburb in the 1920s*. Toronto: University of Toronto Press. https://doi.org/10.3138/9781487576646

Mumford, Lewis. 1954. The neighbourhood and the neighbourhood unit. *Town Planning Review* 24, no. 4: 256–70. https://doi.org/10.3828/tpr.24.4.d4r60h470713003w

Mumford, Lewis. 1961. *The City in History*. London: Penguin.

Munro, Alice. 1968. The shining houses. In *Dance of the Happy Shades*, 19–29. Toronto: McGraw-Hill Ryerson.

Munro, Alice. 1978. The beggar maid. In *Who Do You Think You Are?*, 67–99. Scarborough, ON: Macmillan-NAL Publishing.

Munro, Sheila. 2001. *Lives of Mothers and Daughters: Growing Up with Alice Munro*. New York: Union Square Press.

Murdie, Robert, and Sutama Ghosh. 2010. Does spatial concentration always mean a lack of integration? Exploring ethnic concentration as integration in Toronto. *Journal of Ethnic and Migration Studies* 36, no. 2: 293–311. https://doi.org/10.1080/13691830903387410

Murdie, Robert, and Carlos Teixeira. 2011. The impact of gentrification on ethnic neighbourhoods of Toronto: A case study of Little Portugal. *Urban Studies* 48, no. 1: 61–83. https://doi.org/10.1177/0042098009360227

Murray, Karen B. 2011. Making space in Vancouver's East End: From Leonard Marsh to the Vancouver Agreement. *BC Studies* 169: 7–49. https://doi.org/10.14288/bcs.v0i169.446

Muzzatti, Stephen. 2012. Si Siamo Italiani! Ethnocultural identity, class consciousness, and anarchic sensibilities in an Italian-Canadian working-class enclave. In Chris Richardson and Hans A. Skott-Myhre, eds., *Habitus of the Hood*, 47–66. Bristol: Intellect.

Nader, George E. 1976. *Cities of Canada*. Vol. 2, *Profiles of Fifteen Metropolitan Centres*. Toronto: Macmillan. https://doi.org/10.2307/3340503

Nasaw, David. 1985. *Children of the City: At Work and at Play*. New York: Oxford University Press.

Nash, Catherine J. 2006. Toronto's gay village (1969 to 1982): Plotting the politics of Toronto's gay identity. *Canadian Geographer* 50, no. 1: 1–16. https://doi.org/10.1111/j.0008-3658.2006.00123.x

Nash, Catherine J., and Andrew Gorman-Murray. 2014. LGBT neighbourhoods and the "new mobilities": Towards understanding transformations in sexual and gendered urban landscapes. *International Journal of Urban and Regional Research* 38, no. 3: 756–72. https://doi.org/10.1111/1468-2427.12104

Navin, Robert B., and Associates. 1934. *An Analysis of a Slum Area in Cleveland*. Cleveland: Cleveland Metropolitan Housing Authority.

Neal, Zachary P., and Jennifer W. Neal. 2012. The public school as a public good: Direct and indirect pathways to community satisfaction. *Journal of Urban Affairs* 34, no. 5: 469–86. https://doi.org/10.1111/j.1467-9906.2011.00595.x

New Methodist church for Don neighborhood. 1922. *The Globe*, 11 December.

Newhouse, David R. 2011. Urban life: Reflections of a middle class Indian. In Heather A. Howard and Craig Proulx, eds., *Aboriginal Peoples in Canadian Cities: Transformations and Continuities*, 22–38. Waterloo, ON: Wilfrid Laurier University Press. https://doi.org/10.51644/9781554583140-002

Nguyen-Huang, Phuong, and John Yinger. 2011. The capitalization of school quality into house values: A review. *Journal of Housing Economics* 20, no. 1: 30–48. https://doi.org/10.1016/j.jhe.2011.02.001

Nolen, Stephanie. 2017. This space is taken. In Stephanie Chambers et al., eds., *Any Other Way: How Toronto Got Queer*, 78–9. Toronto: Coach House Books.

North Toronto a residential area. 1911. *The Globe*, 21 February.

North Toronto going strong for secession. 1920. *The Globe*, 29 March.

Northrup, David. 1979. Saint John, New Brunswick, 1871–1891: The changing residential structure of a slow-growth city. MA thesis, York University.

Olds, Jacqueline, and Richard S. Schwartz. 2009. *The Lonely American: Drifting Apart in the Twenty-First Century*. Boston: Beacon Press.

Olson, Sherry, and Audrey L. Kobayashi. 1993. The emerging ethno-cultural mosaic. In Larry S. Bourne and David Ley, eds., *The Changing Social Geography of Canadian Cities*. Montreal and Kingston: McGill-Queen's University Press. https://doi.org/10.1515/9780773563551-008

Olson, Sherry, and Jean-Claude Robert. 2001. Morphologie de la paroisse urbaine. In Serge Courville and Normand Séguin, eds., *Atlas Historique de Québec.*, vol. 7, *La Paroisse*. Sainte-Foy, QC: Presses de l'Université Laval.

Olson, Sherry, and Patricia Thornton. 2011. *Peopling the North American City: Montreal 1840–1900*. Montreal and Kingston: McGill-Queen's University Press. https://doi.org/10.1515/9780773586000

Onomé, Louise. 2021. *Like Home*. Toronto: Harper Collins.

Ontario. 1934. *Report of the Lieutenant-Governor's Committee on Housing Conditions*. Toronto: The Committee.

Onusko, James. 2021. *Boom Kids: Growing Up in the Calgary Suburbs, 1950–1970*. Waterloo, ON: Wilfrid Laurier University Press. https://doi.org/10.51644/9781771125017

Oreopoulos, Philip. 2008. Neighbourhood effects in Canada: A critique. *Canadian Public Policy* 34, no. 2: 237–58. https://doi.org/10.3138/cpp.34.2.237

Osberg, Lars. 2018. *The Age of Increasing Inequality: The Astonihing Rise of Canada's 1%*. Toronto: Lorimer.

Osofsky, Gilbert. 1963. *Harlem: The Making of a Ghetto, 1890–1930*. New York: Harper and Row.

Ouellet, Valérie, and Dexter MacMillan. 2021. Ontario's pharmacy vaccine rollout leaves some hard-hit communities with limited access to shots. *CBC News*, 16 April.

Pacione, Michael. 1984. Local areas in the city. In David T. Herbert and R.J. Johnston, eds., *Geography and the Urban Environment*, vol. 4, 349–92. London: Wiley.

Page, Max, and Timothy Mennell, eds. 2011. *Reconsidering Jane Jacobs*. Chicago: American Planning Association. https://doi.org/10.4324/9781351179775

Palango, Paul. 1983. Some neighbourhood. In Brightside Reunion Committee, *Brightside Reunion 1983*. Hamilton, ON: The Committee.

Palmer, Bryan, and Gaétan Héroux. 2016. *Toronto's Poor: A Rebellious History*. Toronto: Between the Lines.

Pampalon, Robert, Denis Hamel, Maria DeKoninck, and Marie-Jeanne Disant. 2007. Perception of place and health: Differences between neighborhoods in the Québec city region. *Social Science and Medicine* 65, no. 1: 95–111. https://doi.org/10.1016/j.socscimed.2007.02.044

Paris, Erna. 1976. Ghetto of the mind: Forest Hill in the fifties. In William Kilbourn, ed., *The Toronto Book*, 98–105. Toronto: Macmillan.

Parr, Joy. 1990. *The Gender of Breadwinners: Women, Men and Change in Two Industrial Towns, 1880–1950*. Toronto: University of Toronto Press.

Pask, Jim S. 1981. Myth, money, men and real Estate: The Early Years of Tuxedo, Manitoba, 1903–1929. MA thesis, University of Manitoba.

Pask, Jim S. 1989. *On the East of the River: A History of the East Kildonan – Transcona Community*. Winnipeg: City of Winnipeg.

Paskievich, John. 2017. *The North End Revisited: Photography of John Paskievich*. Winnipeg: University of Manitoba Press. https://doi.org/10.1515/9780887555411

Paterson, Ross. 1985. The development of an interwar suburb: Kingsway Park, Etobicoke. *Urban History Review* 13, no. 3: 225–35. https://doi.org/10.7202/1018104ar

Paterson, Ross. 1989. Creating the packaged suburb: The evolution of planning and business practices in the early Canadian land development industry, 1900–1914. In Barbara Kelly, ed., *Suburbia Re-examined*, 119–32. Westport, CT: Greenwood Press.

Patricios, Nicholas N. 2002. The neighborhood concept: A retrospective of physical design and social interaction. *Journal of Architectual and Planning Research* 19, no. 1: 70–90.

Patterson, Jeffrey. 1993. Housing and community development policies. In John Miron, ed., *House, Home and Community. Progress in Housing Canadians 1945–1986*, 320–38. Montreal and Kingston: McGill-Queen's University Press. https://doi.org/10.1515/9780773563926-020

Pavlic, Dejan, and Zhu Qian. 2014. Declining inner suburbs? A longitudinal-spatial analysis of large metropolitan regions in Canada. *Urban Geography* 35, no. 3: 378–401. https://doi.org/10.1080/02723638.2013.863499

Payne, Keith. 2017. *The Broken Ladder: How Inequality Affects the Way We Think, Live, and Die*. New York: Viking.

Pebley, Anne R., and Narayan Shastry. 2004. Neighborhood, poverty, and children's well-being: A review. In Kathryn Neckerman, ed., *Social Inequality*, 119–46. New York: Russell Sage.

Perry, Clarence A. 1926. The local community as a unit in the planning of urban residential areas. In Clarence A. Perry, ed., *The Urban Community*, 238–41. Chicago: University of Chicago Press.

Perry, Clarence A. 1929. *A Plan for New York and Its Environs*. Vol. 7. New York: New York Regional Plan Association.

Perry, Clarence A. 1930. The tangible aspects of community organization. *Social Forces* 8, no. 4: 558–64. https://doi.org/10.2307/2570374

Peters, Evelyn. 1996. "Urban" and "Aboriginal": An impossible contradiction? In Jon Caulfield and Linda Peake, eds., *City Lives and City Forms: Critical Research and Canadian Urbanism*, 46–62. Toronto: University of Toronto Press. https://doi.org/10.3138/9781442672987-006

Peters, Evelyn. 2005. Indigeneity and marginalisation: Planning for and with urban Aboriginal communities in Canada. *Progress in Planning* 63, no. 4: 327–404. https://doi.org/10.1016/j.progress.2005.03.008

Peters, Evelyn. 2014. Aboriginal peoples in urban areas. In Harry Hillier, ed., *Urban Canada*, 182–205. Don Mills, ON: Oxford University Press205.

Peters, Evelyn, Matthew Stock, and Adrian Werner. 2018. *Rooster Town: The History of an Urban Métis Community, 1901–1961*. Winnipeg: University of Manitoba Press. https://doi.org/10.1515/9780887555688

Pettigrew, Eileen. 1983. *The Silent Enemy: Canada and the Deadly Flu of 1918*. Saskatoon: Western Producer Prairie Books.

Phillipson, Chris, Miriam Bernard, Judith Phillips, and Jim Ogg. 1999. Older people's experiences of community life: Patterns of neighbouring in three urban areas. *Sociological Review* 47: 715–49. https://doi.org/10.1111/1467-954X.00193

Phyne, John. 2014. On a hillside north of the harbour: Changes to the centre of St. John's, 1942–1987. *Newfoundland and Labrador Studies* 29, no. 1: 1719–26. https://doi.org/10.7202/1062244ar

Phyne, John, and Christine Knott. 2018. Schools, streets and stores: Childhood geographies of the inner city of St. John's, 1935–1966. Paper presented at the Atlantic Canada Studies Conference, Acadia University, Wolfville, NS, 4–6 May.

Piketty, Thomas. 2014. *Capital in the Twenty-First Century*. Cambridge, MA: Belknap Press. https://doi.org/10.4159/9780674369542

Pinault, Lauren, Daniel Crouse, Michael Jerrett, Michael Brauer, and Michael Tjepkema. 2016. Spatial associations between socioeconomic groups and NO2 air pollution exposure within three large Canadian cities.

Environmental Research 147: 373–82. https://doi.org/10.1016/j.envres
.2016.02.033

Pitsula, James M. 1979. The emergence of social work in Toronto. *Journal of Canadian Studies* 14, no. 1: 35–42. https://doi.org/10.3138/jcs.14.1.35

Plane, Jocelyn, and Fran Klodawski. 2013. Neighbourhood amenities and health: Examining the significance of a local park. *Social Science and Medicine* 99: 1–8. https://doi.org/10.1016/j.socscimed.2013.10.008

Plant, Monica. 2016. Undermount's hero with a snowblower. *Hamilton Spectator*, 20 April.

Plant, Monica. 2017. An urban barn-raising. *Hamilton Spectator*, 13 May.

Pocius, Gerald L. 1991. *A Place to Belong: Community, Order and Everyday Space in Calvert, Newfoundland*. Montreal and Kingston: McGill-Queen's University Press. https://doi.org/10.1515/9780773562707

Podmore, Julie. 1998. (Re)reading the "loft living" habitus in Montreal's inner city. *International Journal of Urban and Regional Research* 22, no. 2: 283–302. https://doi.org/10.1111/1468-2427.00140

Podmore, Julie. 2006. Gone "underground"? Lesbian visibility and the consolidation of queer space in Montreal. *Social and Cultural Geography* 7, no. 4: 595–625. https://doi.org/10.1080/14649360600825737

Podmore, Julie, and Alison Bain. 2021. Whither queer suburbanisms? Beyond heterosuburbia and queer metronormatives. *Progress in Human Geography* 23, no. 2: 175–87.

Polanyi, Michael, Lesley Johnston, Anita Khama, Said Dirie, and Michael Kerr. 2014. *The Hidden Epidemic: A Report on Child and Family Poverty in Toronto*. Toronto: Social Planning Council. http://www.torontocas.ca /app/Uploads/documents/cast-report2014-final-web71.pdf

Porteous, J. Douglas. 1973. The Burnside teenage gang: Territoriality, social space, and community planning. In Charles N. Forward, ed., *Residential and Neighbourhood Studies in Victoria*, 130–48. Western Geographical Studies, Vol. 5. Victoria, BC: Department of Geography, University of Victoria.

Porteous, Douglas. 1977. *Environment and Behavior: Planning and Everyday Life*. Reading, MA: Addison-Wesley.

Porter, John. 1965. *The Vertical Mosaic: An Analysis of Social Class and Power in Canada*. Toronto: University of Toronto Press. https://doi.org/10.3138 /9781442683044

Portolese, Marisa. 2023. *Goose Village*. Montreal: Marisa Portolese.

Pothier, Melanie, Nishan Zewge-Abubaker, Madelaine Cahuas, Carla B. Klassen, and Sarah Wakefield. 2019. Is "including them" enough? How narratives of race and class shape participation in a resident-led

neighbourhood revitalization initiative. *Geoforum* 98: 161–9. https://doi
.org/10.1016/j.geoforum.2018.11.009

Poulton, Michael. 1995. Affordable homes at an affordable (social) price.
In George Falliset al., eds., *Home Remedies: Rethinking Canadian Housing
Policy*, 50–122. Toronto: C.D. Howe Institute.

Poutanen, Mary A., and Jason Gilliland. 2017. Mapping work in early
twentieth-century Montreal: A Rabbi, a neighbourhood and a
community. *Urban History Review* 45, no. 2: 7–24. https://doi.org
/10.3138/uhr.45.02.01

Poutanen, Mary A., and Sherry Olson. 2020. Public houses and hidden
networks: Roles of women in mid-19th-century Montreal. In Alida
Clemente, Dag Lindström, and Jon Stobart, eds., *Micro-Geographies of
the Western City, c1750–1900*, 105–24. London: Routledge. https://doi
.org/10.4324/9780429329395-13

President's Conference on Home Building and Home Ownership. 1932.
Home Ownership, Incomes and Types of Dwellings. Washington, DC: The
Conference.

Preston, Valerie, and Brian Ray. 2020. Placing the second generation: A case
study of Toronto. *Canadian Geographer* 64, no. 2: 215–31. https://doi.org
/10.1111/cag.12597

Prouse, Victoria, Howard Ramos, Jill Grant, and Martha Radice. 2014.
How and when scale matters: The modifiable areal unit problem and
income inequality in Halifax. *Canadian Journal of Urban Research* 23, no. 1
(Supplement): 61–82.

Pucher, John, and Ralph Buehler. 2006. Why Canadians cycle more than
Americans: A comparative analysis of bicycling trends and policies.
Transport Policy 13, no. 3: 265–79. https://doi.org/10.1016/j.tranpol
.2005.11.001

Putnam, Robert D. 2000. *Bowling Alone: The Collapse and Revival of American
Community*. New York: Simon and Schuster. https://doi.org/10.1145
/358916.361990

Putnam, Robert D. 2015. *Our Kids: The American Dream in Crisis*. New York:
Simon and Schuster.

Putnam, Robert D., and Shaylyn R. Garrett. 2020. *The Upswin:. How America
Came Together a Century Ago and How We Can Do It Again*. New York:
Simon and Schuster.

Qadeer, Mohammed, Sandeep Agrawal, and Alexander Lovell. 2010.
Evolution of ethnic enclaves in the Toronto metropolitan area, 2001–2006.
International Journal of Migration and Integration 11, no. 3: 315–19. https://
doi.org/10.1007/s12134-010-0142-8

Qadeer, Mohammed, and Sandeep Kumar. 2006. Ethnic enclaves and social cohesion. *Canadian Journal of Urban Research* 15, no. 2: 1–17.

Quercia, Robert G., and George C. Galster. 2000. Threshold effects and neighborhood change. *Journal of Planning Education and Research* 20, no. 2: 146–62. https://doi.org/10.1177/0739456X0002000202

Ramadier, Thierry, and Carole Deprés. 2004. Les territoires de mobilité et les representations d'une banlieue vieillissante de Québec. *Recherches Sociographiques* 45, no. 3: 521–48. https://doi.org/10.7202/011468ar

Rankin, Katherine, and Heather McLean. 2015. Governing the commercial streets of the city. New terrains of disinvestment and gentrification in Toronto's inner suburbs. *Antipode* 47, no. 1: 216–39. https://doi.org/10.1111/anti.12096

Ray, Brian, and Valerie Preston. 2009. Are immigrants socially isolated? An assessment of neighbours and neighbouring in Canadian cities. *Journal of International Migration and Integration* 10, no. 3: 219–44. https://doi.org/10.1007/s12134-009-0104-1

Reeder, David. 2010. Neighbourhood. In Christian Topalov, Laurent Coudroy de Lille, Jean-Charles Depaule, and Brigitte Morin, eds., *L'Aventure des Mots de la Ville*, 811–25. Paris: Laffont.

Religion in Canada. 2022. *Wikipedia.* https://en.wikipedia.org/wiki/Religion_in_Canada

Remiggi, Frank W. 1998. Le village gai de Montréal: Entre le ghetto et l'espace identitaire. In Irène Demczuk and Frank W. Remiggi, eds., *Sortis de l'Ombre: Histoires des Communautés Lesbiennes et Gaie de Montréal*, 267–89. Montreal: VLB Éditeur.

A residential district. 1905. *The Globe*, 8 June.

Reynolds, Graham, and Wanda Robson. 2016. *Viola Desmond's Canada: A History of Blacks and Racial Segregation in the Promised Land.* Halifax: Fernwood.

Richardson, Chris. 2014. Orientalism at home: The case of "Canada's toughest neighbourhood." *British Journal of Canadian Studies* 27, no. 1: 75–95. https://doi.org/10.3828/bjcs.2014.5

Richler, Mordecai. 1969. *The Street.* Toronto: McClelland and Stewart.

Richler, Mordecai. (1959) 1970. *The Apprenticeship of Duddy Kravitz.* Harmondsworth, UK: Penguin.

Riis, Jacob. 1890. *How the Other Half Lives.* New York: Scribner.

Roberts, Elizabeth. 1993. Neighbours: North-west England, 1940–1970. *Oral History* 21, no. 2: 37–45.

Robertson, Angus. 1977. The pursuit of power, profit and privacy: A study of Vancouver's West End elite, 1886–1914. MA thesis, University of British Columbia.

Robick, Brian. 2011. Blight: The development of a contested concept. MA thesis, Carnegie Mellon University.

Robinson, Keith, and Angel Harris. 2014. *The Broken Compass: Parental Involvement with Children's Education.* Cambridge, MA: Harvard University Press. https://doi.org/10.4159/harvard.9780674726291

Rohe, William. 2009. From local to global: One hundred years of neighborhood planning. *Journal of the American Planning Association* 75, no. 2: 209–30. https://doi.org/10.1080/01944360902751077

Rojc, Phillip. 2017. How Seattle is curbing the power of neighborhood groups. *Planetizen,* 8 April. https://www.planetizen.com/node/92163/how-seattle-curbing-power-neighborhood-groups

Rollwagen, Heather. 2015. Constructing renters as a threat to neighbourhood safety. *Housing Studies* 30, no. 1: 1–21. https://doi.org/10.1080/02673037.2014.925099

Rose, Albert. 1958. *Regent Park: A Study in Slum Clearance.* Toronto: University of Toronto Press. https://doi.org/10.3138/9781487584160

Rose, Damaris. 2015. Gender, sexuality and the city. In Pierre Filion, Markus Moos, Tara Vinodrai, and Ryan Walker, eds., *Canadian Cities in Transition: Perspectives for an Urban Age,* 155–74. Toronto: Oxford University Press.

Rose, Damaris, Annick Germain, Marie-Hélène Bacqué, Gary Bridge, Yankel Fijalkow, and Tom Slater. 2013. "Social mix" and neighbourhood revitalization in a transatlantic perspective: Comparing local policy discourses and expectations in Paris (France), Bristol (UK) and Montréal (Canada). *International Journal of Urban and Regional Research* 37, no. 2: 430–50. https://doi.org/10.1111/j.1468-2427.2012.01127.x

Rose, Félix. 2020. *Les Rose* [The Rose Family]. Montreal: National Film Boardof Canada.

Rosenfeld, Jean. 2000. A noble house in the city. Domestic architecture as elite signification in late 19th-century Hamilton. PhD diss., University of Guelph.

Rosenfeld, Michael J., and Reuben J. Thomas. 2012. Searching for a mate: The rise of the internet as a social intermediary. *American Sociological Review* 77, no. 4: 523–47. https://doi.org/10.1177/0003122412448050

Rosow, Irving. 1948. Home ownership motives. *American Sociological Review* 13: 751–6.

Ross, Nancy A., Lisa N. Oliver, and Paul J. Villeneuve. 2013. The contribution of neighbourhood material and social deprivation to survival: A 22-year follow-up of more than 500,000 Canadians. *International Journal of Environmental Research and Public Health* 10: 1378–91. https://doi.org/10.3390/ijerph10041378

Rothwell, Jonathan T., and Douglas S. Massey. 2014. Geographic effects on intergenerational income mobility. *Economic Geography* 91, no. 1: 83–106. https://doi.org/10.1111/ecge.12072

Rowe, Emma E. 2017. *Middle-Class School Choice in Urban Spaces: The Economics of Public Schooling and Globalized Education Reform.* London: Routledge. https://doi.org/10.4324/9781315651736

Rowley, Gwyn. 1978. "Plus ça change ..." A Canadian skid row. *Canadian Geographer* 22, no. 3: 211–24. https://doi.org/10.1111/j.1541-0064.1978.tb01013.x

Roy, Gabrielle. (1947) 1989. *The Tin Flute.* Toronto: McClelland and Stewart.

Roy, Gabrielle. 1945. *Bonheur d'Occasion.* Montreal: Société d'Éditions Pascal.

Rudin, Phyllis. 2022. Why did I walk along every street in Montreal? C'est le fun. *Globe and Mail,* 22 February. https://www.theglobeandmail.com/life/first-person/article-why-did-i-walk-along-every-street-in-montreal-cest-le-fun/

Rueck, Daniel. 2011. When bridges become barriers: Montreal and Kahnawake Mohawk Territory. In Stéphane Castonguay and Michèle Dagenais, eds., *Metropolitan Natures: Environmental Histories of Montreal,* 228–44. Pittsburgh: University of Pittsburgh Press. https://doi.org/10.2307/j.ctv111jfg2.18

Rutherford, Paul. 1974. Introduction. In Paul Rutherford, ed., *Saving the Canadian City: The First Phase, 1880–1920,* ix–xxiii. Toronto: University of Toronto Press. https://doi.org/10.3138/9781487583446-001

Rutherford, Paul. 1977. Tomorrow's metropolis: The urban reform movement in Canada, 1880–1920. In Gilbert A. Stelter and Alan F.J. Artibise, eds., *The Canadian City: Essays in Urban History,* 368–92. Toronto: McClelland and Stewart.

Rutland, Ted. 2018. *Displacing Blackness: Planning, Power and Race in Twentieth-Century Halifax.* Toronto: University of Toronto Press. https://doi.org/10.3138/9781487518233

Sabourin, Joanne. 1994. The process of gentrification: Lessons from an inner-city neighbourhood. In Frances Frisken, ed., *The Changing Canadian Metropolis: A Public Policy Perspective,* vol. 1, 259–92. Toronto: Canadian Urban Institute.

Sager, Eric. 2014. Canada's immigrants in 1911: A class analysis. In Gordon Darroch, ed., *Canada's Century: Hidden Histories,* 428–51. Montreal and Kingston: McGill-Queen's University Press. https://doi.org/10.1515/9780773589391-018

Sampson, Robert J. 2012. *Great American City: Chicago and the Enduring Neighbourhood Effect.* Chicago: University of Chicago Press. https://doi.org/10.7208/chicago/9780226733883.001.0001

Sancton, Andrew. 1983. Montreal. In Warren Magnusson and Andrew Sancton, eds., *City Politics in Canada*, 58–93. Toronto: University of Toronto Press. https://doi.org/10.3138/9781487575908-003

Sandel, Michael. 2012. *What Money Can't Buy: The Moral Limits of Markets.* New York: Farrar, Straus and Giroux.

Sandercock, Leonie, and Giovanni Attili. 2009. *Where Strangers Become Neighbours: Integrating Immigrants in Vancouver, Canada.* Dordrecht: Springer.

Saunders, Doug. 2010. *Arrival City: The Final Migration and the Next World.* Toronto: Knopf.

Saunders, Doug. 2013. We're a nation of suburban apartment dwellers but afraid to admit it. *Globe and Mail*, 30 March.

Savage, Mike, Gaynor Bagnall, and Brian Longhurst. 2005. *Globalization and Belonging.* London: Sage. https://doi.org/10.4135/9781446216880

Scherzer, Kenneth A. 1992. *The Unbounded Community: Neighborhood Life and Social Structure in New York City, 1830–1875.* Durham, NC: Duke University Press. https://doi.org/10.2307/2075753

Schultz, Patricia V. 1975. *The East York Workers' Association: A Response to the Great Depression.* Toronto: New Hogtown Press.

Scott, F.R. 1981. Calamity. In *The Collected Poems of F.R. Scott.* Toronto: McClelland and Stewart.

Seeley, John R., R. Alexander Sim, and Elizabeth W. Loosley. 1956. *Crestwood Heights: A Study of the Culture of Suburban Life.* Toronto: University of Toronto Press.

Séguin, Ann-Marie, Philippe Apparacio, and Mylene Riva. 2012. The impact of geographical scale in identifying areas as possible sites for area-based interventions to tackle poverty: The case of Montreal. *Applied Spatial Analysis and Policy* 5, no. 3: 231–51. https://doi.org/10.1007/s12061 -011-9068-6

Seligman, Amanda. 2016. *Chicago's Block Clubs: How Neighbors Shape the City.* Chicago: University of Chicago Press. https://doi.org/10.7208/chicago /9780226385990.001.0001

Sewell, J. 1993. *The Shape of the City: Toronto Struggles with Modern Planning.* Toronto: University of Toronto Press. https://doi.org/10.3138 /9781442628106

Shareck, Martine, Yan Kestens, and Katherince L. Frohlich. 2014. Moving beyond the residential neighborhood to explore social inequalities in exposure to area-level disadvantage. *Social Science and Medicine* 108 (special issue): 106–14. https://doi.org/10.1016/j.socscimed.2014.02.044

Sharkey, Patrick. 2017. A fragile urban consensus. *Chronicle of Higher Education*, 30 July.

Sharpe, Christopher A., and A.J. Shawyer. 2016. *Sweat Equity: Cooperative House-building in Newfoundland, 1920–1974*. St. John's: Institute of Social and Economic Research.

Sharpe, Christopher A., and A.J. Shawyer. 2021. *Corner Windows and Cul-de-Sacs: The Remarkable Story of Newfoundland's First Garden Suburb*. St. John's, NL: Memorial University Press. https://doi.org/10.1515/9781990445033

Sharpe, Christopher A. 2024. Personal communication, 29 February.

Shields, Rob, Kieran Moran, and Dianne Gillespie. 2020. Edmonton, Amiskevaciy Wâskahikan, and a Papaschase suburb for settlers. *Canadian Geographer* 64, no. 1: 105–19. https://doi.org/10.1111/cag.12562

Shragge, E. 2013. *Activism and Social Change: Lessons for Community Organizing*. Toronto: University of Toronto Press.

Silver, Jim, Parvin Ghorayshi, Joan Hay, and Darlene Klyne. 2006. *In a Voice of Their Own. Urban Aboriginal Community Development*. Ottawa: Canadian Centre for Policy Alternatives.

Silver, Jim, Joan Hay and Peter Gorzen. 2006. In but not of: Aboriginal people in an inner-city neighbourhood. In Jim Silver, ed., *In Their Own Voices. Building Urban Aboriginal Communities*, 40–69. Halifax: Fernwood.

Simpson, Michael. 1985. *Thomas Adams and the Modern Planning Movement: Britain, Canada and the United States, 1900–1940*. London: Mansell.

Slater, Tom. 2013. Your life chances affect where you live: A critique of the "cottage industry" of neighbourhood effects research. *International Journal of Urban and Regional Research* 37, no. 2: 367–87. https://doi.org/10.1111/j.1468-2427.2013.01215.x

Small, Mario L., and Laura Adler. 2019. The role of space in the formation of social ties. *Annual Review of Sociology* 45: 111–32. https://doi.org/10.1146/annurev-soc-073018-022707

Smith, Garry. 1969. What's in a (neighborhood) name? *Hamilton Spectator*, 26 July.

Smith, Peter J., and Larry D. McCann. 1981. Residential land use change in Edmonton. *Annals of the Association of American Geographers* 71, no. 4: 534–51. https://doi.org/10.1111/j.1467-8306.1981.tb01373.x

Social Planning Council of Metropolitan Toronto. 1979. *Metro's Suburbs in Transition, Part I: Evolution and Overview*. Toronto: The Council.

Soderstrom, Mary. 2008. *The Walkable City*. Montreal: Véhicule.

Speisman, Stephen. 1985. St. John's Shtetl: The Ward in 1911. In Robert F. Harney, ed., *Gathering Place: People and Neighbourhoods of Toronto, 1834–1945*, 107–20. Toronto: Multicultural History Society of Ontario.

Spence-Sales, Harold. 1950. *How to Subdivide for Housing Developments*. Ottawa: Community Planning Association of Canada.

Stamp, Robert M. 1977. The response to urban growth: The bureaucratization of public education in Calgary, 1884–1914. In Gilbert A. Stelter and Alan F.J. Artibise, eds., *The Canadian City: Essays in Urban History*, 282–99. Toronto: McClelland and Stewart.

Stamp, Robert M. 1978. Canadian high schools in the 1920s and 1930s: The social challenge to the academic tradition. *Canadian Historical Association Historical Papers* 13, no. 1: 76–93. https://doi.org/10.7202/030478ar

Stanger-Ross, Jordan. 2008. Municipal colonialism in Vancouver: City planning and the conflict over Indian reserves, 1928–1950. *Canadian Historical Review* 89, no. 4: 544–80. https://doi.org/10.1353/can.0.0113

Stanger-Ross, Jordan. 2009. *Staying Italian: Urban Change and Ethnic Life in Postwar Toronto and Philadelphia*. Chicago: University of Chicago Press. https://doi.org/10.7208/chicago/9780226770765.001.0001

Stanger-Ross, Jordan, and Hildy Stanger-Ross. 2012. Placing the poor: The ecology of poverty in postwar Canada. *Journal of Canadian Studies* 46, no. 1: 213–44. https://doi.org/10.3138/jcs.46.1.213

Stapleton, John, Brian Murphy, and Yue Xing. 2012. *The 'Working Poor' in the Toronto Region*. Toronto: Metcalf Foundation.

Stewart, Bryce. 1913. The housing of our immigrant workers. *Papers and Proceedings of the Canadian Political Science Association* 1: 98–111. Reprinted in Paul Rutherford, ed., *Saving the Canadian City: The First Phase, 1880–1920*, 137–54 (Toronto: University of Toronto Press, 1974).

Stewart, Matthew. 2021. *The 9.9 Percent: The New Aristocracy That Is Entrenching Inequality and Warping Our Culture*. New York: Simon and Schuster.

Strange, Carolyn. 1995. *Toronto's Girl Problem: The Perils and Pleasures of the City, 1880–1930*. Toronto: University of Toronto Press. https://doi.org/10.3138/9781442682696

Strong-Boag, Veronica. 1982. Intruders in the nursery: Childcare professionals reshape the years from one to five, 1920–1940. In Joy Parr, ed., *Childhood and Family in Canadian History*, 160–78. Toronto: McClelland and Stewart.

Strong-Boag, Veronica. 1988. *The New Day Recalled: The Lives of Girls and Women in English Canada 1919–1939*. Toronto: Penguin.

Strong-Boag, Veronica. 1991. Home dreams: Women and the suburban experiment in Canada, 1945–60. *Canadian Historical Review* 72, no. 4: 471–504. https://doi.org/10.3138/CHR-072-04-03

Stross, Randall. 2012. Meet your neighbors, if only online. *New York Times*, 12 May.

Stubbs, Todd R. 2018. "Efficiency and evangelism": Peter Bryce and the making of Liberal Protestantism at Toronto's Earlscourt Methodist church. *Histoire sociale/Social History* 51, no. 104: 231–51. https://doi.org /10.1353/his.2018.0025

Sturino, Franc. 1978. A case study of a South Italian family in Toronto. *Urban History Review* 2: 38–57. https://doi.org/10.7202/1019424ar

Sundaram, Maria et al. 2021. The individual and social determinants of COVID-19 diagnosis in Ontario, Canada: A population-wide study. Preprint. https://www.cmaj.ca/content/193/20/E723

Survivors of the Assiniboia Indian Residential School. 2021. *Did You See Us? Reunion, Remembrance, and Reclamation at an Urban Residential School.* Winnipeg: University of Manitoba Press.

Sutherland, Neil. 1986. "The triumph of formaism": Elementary schooling in Vancouver from the 1920s to the 1960s. *BC Studies* 69–70: 175–210. https:// doi.org/10.59962/9780774857079-008

Sutherland, Neil. 1997. *Growing Up: Childhood in English Canada from the Great War to the Age of Television.* Toronto: University of Toronto Press. https:// doi.org/10.3138/9781442675520

Suttles, Gerald. 1972. *The Social Construction of Communities.* Chicago: University of Chicago Press.

Sweeney, Robert. 1995. Aperçu d'un effort collectif Québécois: La création, au début du xxe. siècle, d'un marché privé et institutionnalisé de capitaux. *Revue d'histoire de l'Amérique française* 49, no. 1: 35–72. https:// doi.org/10.7202/305399ar

Sweeney, Robert. 2014. Property, gender and popular class housing in turn of the century Montréal. Paper presented at the Annual Meetings of the Social Science History Association, Toronto, November. https://www .academia.edu/9248710/Property_gender_and_popular_class_housing _in_turn_of_the_century_Montr%C3%A9al

Sweeney, Robert. 2020. Divvying up space: Housing segregation and national identity in turn of the century Montreal. In Robert Sweeney, ed., *Sharing Spaces: Essays in Honour of Sherry Olson,* 111–27. Ottawa: University of Ottawa Press. https://doi.org/10.2307/j.ctvz0h9hc.13

Sweetser, Frank L. 1942. A new emphasis for neighborhood research. *American Sociological Review* 7, no. 4: 525–33.

Synge, Jane. 1976. Immigrant communities – British and continental European – in early twentieth century Hamilton, Ontario. *Oral History* 4, no. 2: 38–51.

Synge, Jane. 1978. Self help and neighbourliness: Patterns of life in Hamilton. In Irving Abella and David Millar, eds., *The Canadian Worker in the Twentieth Century,* 97–104. Toronto: Oxford University Press.

Synge, Jane. 1979. The transition from school to work: Growing up working class in early 20th. century Hamilton, Ontario. In Kent Ishwaran, ed., *Childhood and Adolescence in Canada*, 249–69. Toronto: McGraw-Hill Ryerson.

Talbot, Carol. 1984. *Growing Up Black in Canada*. Toronto: Williams-Wallace.

Talbot, Marion, and Sophonisba Breckenridge. 1912. *The Modern Household*. Boston: Whitcomb and Barrows.

Talen, Emily. 2018. *Neighborhood*. New York: Oxford University Press. https://doi.org/10.1093/oso/9780190907495.001.0001

Talen, Emily, and Julia Koschinsky. 2013. The walkable neighborhood: A literature review. *International Journal of Sustainable Land Use and Urban Planning* 1, no. 1: 42–63. https://doi.org/10.24102/ijslup.v1i1.211

Taylor, John H. 1986. *Ottawa: An Illustrated History*. Toronto: Lorimer.

Teixeira, Carlos. 2006. Residential experiences and the culture of suburbanisation: A case study of Portuguese homebuyers in Mississauga. *Housing Studies* 22, no. 4: 58–86. https://doi.org/10.1080/02673030701387622

Teixeira, Carlos. 2014. Living on the "edge of the suburbs" of Vancouver: A case study of the housing experiences and coping strategies of recent immigrants in Surrey and Richmond. *Canadian Geographer* 58, no. 2: 168–87. https://doi.org/10.1111/j.1541-0064.2013.12055.x

Thompson, Sara K., and Rosemary Gartner. 2014. The spatial distribution and social context of homicide in Toronto's neighbourhoods. *Journal of Research in Crime and Delinquancy* 51, no. 1: 88–118. https://doi.org/10.1177/0022427813487352

Thornton, Patricia, and Sherry Olson. 2001. A deadly discrimination among Montreal infants, 1860–1900. *Continuity and Change* 16, no. 1: 95–135. https://doi.org/10.1017/S0268416001003733

Tolfo, Giuseppe, and Brian Doucet. 2021. Gentrification in the media: The eviction of critical class perspective. *Urban Geography* 42, no. 10: 1418–39. https://www.tandfonline.com/doi/abs/10.1080/02723638.2020.1785247

Tomalty, Ray, and Alan Mallach. 2015. *America's Urban Future: Lessons from North of the Border*. Washington, DC: Island Press. https://doi.org/10.5822/978-1-61091-597-7

Tomiak, Julie. 2017. Contesting the settler city: Indigenous self-determination, new urban reserves, and the neoliberalization of colonialism. *Antipode* 49, no. 4: 928–45. https://doi.org/10.1111/anti.12308

Tong, Catherine E., Heather A. McKay, Anne Martin-Matthews, Atiya Mahmud, and Joanie Sims-Gould. 2020. "These few blocks, these are my village": The physical activity and mobility of foreign-born older adults. *The Gerontologist* 60, no. 4: 638–50. https://doi.org/10.1093/geront/gnz005

Too liberal mortgages condemned by Reid. 1948. *Globe and Mail*, 6 May.

Topalov, Christian. 2003. "Traditional working-class neighborhoods": An enquiry into the emergence of a sociological model in the 1950s and 1960s. *Osiris* 18: 212–33. https://doi.org/10.1086/649385

Topalov, Christian. 2017. The naming process. In Richard Harris and Charlotte Vorms, eds., *What's in a Name? Talking about Urban Peripheries*, 36–67. Toronto: University of Toronto Press. https://doi.org/10.3138/9781442620643-004

Topalov, Christian, Laurent Coudroy de Lille, Jean-Charles Depaule, and Brigitte Marin, eds. 2010. *L'Aventure des Mots de la Ville*. Paris: Laffont.

Topsy Turvy. 2023. *The Economist*, 2 December, p.66.

Toronto City Planning Board. 1943. *The Master Plan for the City of Toronto and Environs* (31 Dec.). Author.

Toronto City Planning Board. 1949. *Third Report and Official Plan* (21 June). Author.

Toronto Foundation. 2018. *Toronto Social Capital Study*. Author. https://torontofoundation.ca/wp-content/uploads/2018/11/TF-Social CapitalStudy-Final-Clean-min.pdf

Toronto Public Health. 2012. *The Walkable City: Neighbourhood Design and Preferences, Travel Choices and Health*. Author.

Town, Harold. 1984. Introduction. In Marjorie Harris, *Toronto: The City of Neighbourhoods*, 6–9. Toronto: McClelland and Stewart.

Townshend, Ivan. 2006. From public neighbourhood to multi-tier private neighbourhoods: The evolving ecology of neighbourhood privatization in Calgary. *GeoJournal* 66: 103–20. https://doi.org/10.1007/s10708-006-9010-7

Townshend, Ivan, Byron Miller, and Derek Cook. 2020. Neighbourhood change in Calgary: An evolving geography of income inequality and social difference. In Jill Grant, Alan Walks, and Howard Ramos, eds., *Changing Neighbourhoods: Social and Spatial Polarization in Canadian Cities*, 193–213. Vancouver: UBC Press. https://doi.org/10.59962/9780774862042-013

Townshend, Ivan, and Robert Murdie. 2020. Using social dimensions and neighbourhood typologies to characterize neighbourhood change. In Jill Grant, Alan Walks, and Howard Ramos, eds., *Changing Neighbourhoods: Social and Spatial Polarization in Canadian Cities*, 53–74. Vancouver: UBC Press. https://doi.org/10.59962/9780774862042-007

Tremblay, Michel. 1981. *The Fat Woman Next Door Is Pregnant*. Vancouver: Talonbooks.

Tremblay, Michel. 1989. *The Heart Laid Bare*. Toronto: McClelland and Stewart.

Troper, Harold. 1987. Jews and Canadian Immigration Policy, 1900–1950. In Moses Rischin, ed., *The Jews of North America*, 235–46. Detroit: Wayne State University Press.

Trudelle, Catherine, Juan-Luis Klein, Jean-Marc Fontan, Diane-Gabrielle Tremblay, and Christophe Bocquin. 2016. Conflits urbains, compromise et cohésion socioterritoriale: Le cas de Tohu à Montréal. *Revue d'Économie Régionale et Urbaine* 2: 417–43. https://doi.org/10.3917/reru.162.0417

Turcotte, Geneviève, and Marie Lavigne. 1980. Différentiation Sociale et Accessibilité à l'Espace Neuf. Études et Documents #18, INRS-Urbanisation, Montreal.

Turcotte, Martin. 2009. Are suburban residents really less physically active? *Canadian Social Trends* No. 87. Ottawa: Statistics Canada.

Turok, Ivan. 2004. The rationale for area-based policies: Lessons from international experience. In Peter Robinson, Jeff McCarthy, and Clive Foster, eds., *Urban Reconstruction in the Developing World: Learning through an International Best Practice*. Sandown, South Africa: Heinemann.

Underhill, Frank H. (1910/11) 1974. Commission government in cities. In Paul Rutherford, ed., *Saving the Canadian City: The First Phase 1880–1920*, 325–34. Toronto: University of Toronto Press. https://doi.org/10.3138/9781487583446-032

Usiskin, Roz. 1980. Continuity and change: The Jewish experience in Winnipeg's North End, 1900–1914. *Canadian Jewish Historical Society Journal* 4, no. 1: 71–94.

Valeriote, Richard. 2010. *Alice Street: A Memoir.* Montreal and Kingston: McGill-Queen's University Press.

Vallières, Pierre. 1971. *White Niggers of America.* Trans. by Joan Pinkham. Toronto: McClelland and Stewart. https://doi.org/10.14452/MR-022-07-1970-11_4

Valverde, Mariana. 2008. *The Age of Light, Soap and Water: Moral Reform in English Canada, 1885–1925.* 2nd ed. Toronto: McClelland and Stewart. https://doi.org/10.3138/9781442689268

Valverde, Mariana. 2016. A tale of two- or three- cities: Gentrification and community consultations. In Jay Pitter and John Lorinc, eds., *Subdivided City: Building in an Age of Hyper-Diversity*, 199–207. Toronto: Coach House Press.

Vancouver City Planning Department. 1975. *Vancouver Local Areas.* Author.

Van Ham, Maarten, Sanne Boschman, and Matt Vogel. 2018. Incorporating neighborhood choice in a model of neighborhood effects on income. *Demography* 55, no. 3: 1069–90. https://doi.org/10.1007/s13524-018-0672-9

Van Ham, Maarten, David Manley, Nick Bailey, Ludi Simpson, and Duncan Maclennan, eds. 2013. *Understanding Neighbourhood Dynamics.* Dordrecht: Springer. https://doi.org/10.1007/978-94-007-4854-5

Van Horssen, Jessica. 2016. *A Town Called Asbestos*. Vancouver: UBC Press. https://doi.org/10.59962/9780774828437

Van Nus, Walter. 1979. Toward the city efficient: The theory and practice of zoning, 1909–1939. In Alan F.J. Artibise and Gilbert Stelter, eds., *The Usable Urban Past: Planning and Politics in the Modern Canadian City*, 226–45. Toronto: Macmillan. https://doi.org/10.1515/9780773580640-014

Van Nus, Walter. 1998. A community of communities: Suburbs in the development of "Greater Montreal." In Isabelle Gournay and France Vanlaethem, eds., *Montreal Metropolis 1880–1930*, 59–67. Montreal: Canadian Centre for Architecture.

Vassanji, M.G. 1991. *No New Land*. Toronto: McClelland and Stewart.

Vaughan, Laura. 2018. *Mapping Society: The Spatial Dimensions of Social Cartography*. London: University College London Press. https://doi.org/10.2307/j.ctv550dcj

Vaz, Edmund. 1965. Middle-class adolescents: Self-reported delinquency and youth culture activities. *Canadian Review of Sociology and Anthropology* 2, no. 1: 52–70. https://doi.org/10.1111/j.1755-618X.1965.tb01329.x

Verma, Ann R. 2017. Mississauga meetup: Queer organizing in the 905. In Stephanie Chambers et al., eds., *Any Other Way: How Toronto Got Queer*, 226–8. Toronto: Coach House Press.

Vermette, Katherena. 2016. *The Break*. Toronto: House of Anansi.

Vertovec, Steven, and Daniel Hiebert. 2021. *Superdiversity* (website). https://superdiv-canada.mmg.mpg.de

Vickers, Simon. 2020. Trefann Court revisited: The activist afterlives of John Sewell and Edna Dixon. *Histoire sociale/Social History* 53, no. 108: 351–71. https://doi.org/10.1353/his.2020.0017

Ville de Montréal. 2004. *Plan de l'Urbanisme de Montréal*. Author.

Ville de Montréal, Bureau du Plan d'Urbanisme. 1992. *Plan d'Urbanisme de la Ville de Montréal*. Montreal: Service de l'Habitation et du Développement Urbain.

Ville de Montréal, Direction des Transports. 2013. *Quartiers Verts: Guide d'Aménagement Durables des Rues de Montréal*. Author.

Ville de Montréal, Service de l'Urbanisme. 1944. *Planning for Montreal, Master Plan: Preliminary Report*. Author.

Ville de Montréal, Service de l'Urbanisme. 1979. *Opération 10000 Logements*. Author.

Ville de Montréal, Service des Infrastructures. 2008. *Transportation Plan*. Montreal: Division du Développement des Transports.

Ville de Montréal, Service du Développement Culturel. 2005. *Montreal, Cultural Metropolis. Cultural Development Policy, 2005–2015*. Montreal: Direction du Développement Culturel et des Bibliothèques.

Vipond, Robert C. 2017. *Making a Global City: How One Toronto School Embraced Diversity.* Toronto: University of Toronto Press. https://doi.org/10.3138/9781442624429

Vitiello, Domenic, and Zoe Blickenderfer. 2020. The planned destruction of Chinatowns in the United States and Canada since c. 1900. *Planning Perspectives* 35, no. 1: 143–68. https://doi.org/10.1080/02665433.2018.1515653

Volunteer workers take leading part in assisting needy. 1925. *The Globe*, 15 April.

Von Hoffman, Alexander. 1994. *Local Attachments: The Making of an American Urban Neighborhood.* Baltimore: Johns Hopkins University Press.

Wade, Jill. 1994. *Houses for All: The Struggle for Social Housing in Vancouver, 1919–50.* Vancouver: UBC Press. https://doi.org/10.59962/9780774856478

Wakefield, Sarah, and Colin McMullan. 2005. Healing in places of decline: (Re)imagining everyday landscapes in Hamilton, Ontario. *Health and Place* 11, no. 4: 299–312. https://doi.org/10.1016/j.healthplace.2004.05.001

Walker, James W.St.G. 1997. *"Race," Rights, and the Law in the Supreme Court of Canada: Historical Case Studies.* Toronto: Osgoode Society and the University of Waterloo Press.

Walks, R. Alan. 2006. The causes of city-suburban political polarization: A Canadian case study. *Annals of the Association of American Geographers* 96, no. 2: 390–414. https://doi.org/10.1111/j.1467-8306.2006.00483.x

Walks, R. Alan. 2010. New divisions: Social polarization and neighbourhood inequality in the Canadian city. In Trudi Bunting, Pierre Filion, and Ryan Walker, eds. *Canadian Cities in Transition. New Directions in the Twenty-First Century*, 125–59. Toronto: Oxford University Press.

Walks, R. Alan. 2014. Gated communities, neighbourhood selection and segregation: The residential preferences and demographics of gated community residents in Canada. *Town Planning Review* 85, no. 1: 39–66. https://doi.org/10.3828/tpr.2014.5

Walks, R. Alan. 2020. Urban divisions: Inequality, neighbourhood poverty and homelessness in the Canadian city. In Markus Moos, Tara Vinodrai, and Ryan Walker, eds., *Canadian Cities in Transition: Understanding Contemporary Urbanism*, 175–94. Toronto: Oxford University Press.

Walks, R. Alan, and Larry S. Bourne. 2006. Ghettos in Canada's cities? Racial segregation, ethnic enclaves and poverty concentration in Canadian urban areas. *Canadian Geographer* 50, no. 3: 273–97. https://doi.org/10.1111/j.1541-0064.2006.00142.x

Walls, Martha. 2016. The disposition of the ladies: Mi'kmaw women and the removal of the King's Road reserve, Sydney, Nova Scotia. *Journal of Canadian Studies* 50, no. 3: 538–65. https://doi.org/10.3138/jcs.50.3.538

Wang, Lu. 2014. Immigrant health, socioeconomic factors and residential neighbourhood characteristics: A comparison of multiple ethnic groups in Canada. *Applied Geography* 51: 90–8. https://doi.org/10.1016/j.apgeog .2014.03.010

Wang, Shuguang, and Jason Zhong. 2013. Delineating ethnoburbs in Metropolitan Toronto. CERIS Working Paper No 100. Toronto: CERIS – Ontario Metropolis Centre.

Ward, David. 1976. The Victorian slum: An enduring myth. *Annals of the Association of American Geographers* 66, no. 2: 323–36. https://doi.org /10.1111/j.1467-8306.1976.tb01093.x

Ward, David. 1989. *Poverty, Ethnicity and the American City, 1840–1925: Changing Conceptions of the Slum and the Ghetto.* Cambridge: Cambridge University Press.

Ward, Stephen V. 2002. *Planning the Twentieth-Century City: The Advanced Capitalist World.* Chichester, UK: Wiley.

Ward, W. Peter. 1999. *A History of Domestic Space: Privacy and the Canadian Home.* Vancouver: UBC Press.

Weaver, John. 1977. Tomorrow's metropolis revisited: A critical assessment of urban reform in Canada, 1890–1920. In Gilbert A. Stelter and Alan F.J. Artibise, eds., *The Canadian City: Essays in Urban History*, 393–418. Toronto: McClelland and Stewart.

Weaver, John. 1979. The property industry and land use controls: The Vancouver experience, 1910–1945. *Plan Canada* 19, nos. 3–4: 211–25.

Weaver, John. 1988. From land assembly to social maturity: The suburban life of Westdale (Hamilton), Ontario, 1911–1951. *Histoire sociale/Social History* 21: 411–40.

Weaver, Robert D., Suzanne M. McMurphy, and Nazim N. Habibov. 2013. Analyzing the impact of bonding and bridging social capital on economic well-being: Results from Canada's general social survey. *Sociological Spectrum* 33, no. 6: 566–83. https://doi.org/10.1080/02732173.2013.836149

Webster, Chris. 2003. The nature of the neighbourhood. *Urban Studies* 40, no. 13: 2591–612. https://doi.org/10.1080/0042098032000146803aaa

Webster, Douglas. 1970. Neighbourhood planning. *Globe and Mail*, 16 February.

Wehrwein, George A., and Coleman Woodbury. 1930. Tenancy versus ownership in urban land utilization. *Annals of the American Academy of Political and Social Science* 148: 184–98. Supplement. https://doi.org /10.1177/000271623014800126

Weightman, Barbara. 1978. *The Musqueam Reserve: A Cast Study of the Indian Milieu in an Urban Environment.* Seattle, WA: University of Washington Press.

Weisburd, David, et al. 2016. *Place Matters: Criminology for the Twenty-First Century.* New York: Cambridge University Press.

Weisburd, David, Elizabeth R. Groff, and Sue-Ming Yang. 2018. *The Criminology of Place: Street Segments and Our Understanding of the Crime Problem.* New York: Oxford University Press.

Weiss, M. 1987. *The Rise of the Community Builders: The American Real Estate Industry and Urban Land Planning.* New York: Columbia University Press.

Wekerle, Gerda. 1976. Vertical village: Social contacts in a singles high-rise complex. *Sociological Focus* 9, no. 3: 299–315. https://doi.org/10.1080/00380237.1976.10570939

Wellesley Instute. 2010. *Precarious Housing in Canada.* Toronto: Wellesley Institute. https://www.wellesleyinstitute.com/wp-content/uploads/2010/08/Precarious_Housing_In_Canada.pdf

Wellman, Barry. 1979. The community question: The intimate networks of East Yorkers. *American Journal of Sociology* 84: 1201–31. https://doi.org/10.1086/226906

Wellman, Barry. 1992. Men in networks: Private communities, domestic friendships. In Peter Nardi, ed., *Men's Friendships*, 74–114. Newbury Park, CA: Sage. https://doi.org/10.4135/9781483325736.n5

Wellman, Barry, and Barry Leighton. 1979. Networks, neighborhoods and communities. Approaches to the study of the community question. *Urban Affairs Quarterly* 14, no. 13: 363–90. https://doi.org/10.1177/107808747901400305

Westley, Margaret W. 1990. *Remembrance of Grandeur: The Anglo-Protestant Elite of Montreal 1900–1950.* Montreal: Libre Expression.

Weston, Julia. 1982. Gentrification and displacement: An inner city dilemma. *Habitat* 25, no. 1: 10–19.

Wetherell, Donald G., and Irene Kmet. 1995. *Town Life: Main Street and the Evolution of Small Town Alberta, 1880–1947.* Edmonton: University of Alberta Press.

White, Richard. 2016. *Planning Toronto, 1940–1980: The Planners, the Plans and Their Legacies.* Vancouver: UBC Press. https://doi.org/10.59962/9780774829373

Whitney, Keith. 1970. Skid row. In William E. Mann, ed., *The Underside of Toronto*, 65–74. Toronto: McClelland and Stewart.

Whitzman, Carolyn. 2009. *Suburb, Slum, Urban Village. Transformations in Toronto's Parkdale Neighbourhood, 1875–2002.* Vancouver: UBC Press. https://doi.org/10.59962/9780774815376

Wideman, Trevor, and Jeffrey Masuda. 2018. Assembling "Japantown": A critical toponymy of urban dispossession in Vancouver, Canada. *Urban*

Geography 39, no. 4: 493–518. https://doi.org/10.1080/02723638.2017.1360038

Wilkinson, Richard G. 2005. *The Impact of Inequality: How to Make Sick Societies Healthier.* New York: New Press.

Wilkinson, Richard, and Kate Pickett. 2019. *The Inner Level: How More Equal Societies Reduce Stress, Restore Sanity and Improve Everyone's Well-Being.* New York: Allen Lane.

Williams, Allison, and Peter Kitchen. 2012. Sense of place and health in Hamilton, Ontario: A case study. *Social Indicators Research* 108: 257–76. https://doi.org/10.1007/s11205-012-0065-1

Williams, Dorothy W. 1997. *The Road to Now: A History of Blacks in Montreal.* Montreal: Véhicule.

Wills, Jacqueline G. 1995. *Marriage of Convenience: Business and Social Work in Toronto, 1918–1957.* Toronto: University of Toronto Press.

Wirth, Louis. 1938. Urbanism as a way of life. *American Journal of Sociology* 44: 3–24. https://doi.org/10.1086/217913

Wolch, Jennifer, Jason Byrne, and Joshua P. Newell. 2014. Urban green space, public health, and environmental justice: The challenge of making cities "just green enough." *Landscape and Urban Planning* 125: 234–44. https://doi.org/10.1016/j.landurbplan.2014.01.017

Wolter, Hugo W. 1948. Living in our community. In Citizens Advisory Committee, *Working and Living in Toronto: Prospects and Problems, Research Conference Dec. 1948.* Toronto.

Wood, Andy. 2020. *Faith, Hope and Charity, 1500–1640.* Cambridge: Cambridge University Press. https://doi.org/10.1017/9781108886765

Woodsworth, James S. 1909. *Strangers within Our Gates: Or, Coming Canadians.* Toronto: Missionary Society of the Methodist Church.

Woodsworth, James S. (1911) 1972. *My Neighbor.* Toronto: University of Toronto Press.

Worley, William S. 1993. *J.C. Nichols and the Shaping of Kansas City: Innovation in Planned Residential Communities.* Columbia: University of Missouri Press.

Xu, Xiao. 2024. Inspired by the Chinese trend, U of T graduate rents out her time to the social anxious and isolated. *Globe and Mail,* 4 March.

Young, Michael. 1958. *The Rise of the Meritocracy, 1870–2033.* Harmondsworth, UK: Penguin.

Young, Michael and Peter Wilmott. 1957. *Family and Kinship in East London.* Harmondsworth, UK: Penguin.

Young, Phyllis B. (1960) 2007. *The Torontonians.* Montreal and Kingston: McGill-Queen's University Press.

Zaami, Mariam. 2015. "I fit the description": Experiences of social and spatial exclusion among immigrant youth in the Jane and Finch neighbourhood of Toronto. *Canadian Ethnic Studies* 47, no. 3: 69–89. https://doi.org/10.1353/ces.2015.0032

Zask, Joëlle. 2021. Montreal's alleyways: A laboratory for democratic life. *Metropolitics*, 12 February. https://doi.org/10.3917/nect.012.0021

Zembrzycki, Stacey. 2007. "There were always men in our house": Gender and the childhood memories of working-class Ukrainians in depression-era Canada. *Labour/Le Travail* 60: 77–105.

Zucchi, John E. 1985. Italian hometown settlements and the development of an Italian community in Toronto, 1875–1935. In Robert Harney, ed., *Gathering Place: People and Neighbourhoods of Toronto, 1834–1945*, 121–46. Toronto: Multicultural History Society of Ontario.

Zucchi, John E. 1988. *Italians in Toronto: Development of a National Identity, 1875–1935*. Montreal and Kingston: McGill-Queen's University Press. https://doi.org/10.1515/9780773561687

Zucchi, John E. 2007. A history of ethnic enclaves in Canada. Canada's Ethnic Group Series Booklet No. 31. Ottawa: Canadian Historical Association.

Zunz, Olivier. 1982. *The Changing Face of Inequality: Urbanization, Industrial Development and Immigrants in Detroit 1880–1920*. Chicago: University of Chicago Press.

Index of People and Subjects

The only authors included in this index are writers of fiction or memoirs. Book titles are not listed. References to tables and figures are in bold. There is a separate index of places.

Aboriginal peoples. *See* First Nations

Adams, Thomas, 170, 171

Airbnb, 143

Ames, Herbert: on poverty, 40, 89, 94, 146

apartment buildings, 35, 87; construction booms of, 111, 138; immigrant enclaves in, 120; middle class, 107, 208, 210; neighbouring in, 18, 21–3, 229; opposition to, 138, 160, 162, 173, 182; public health in, 122

"at homes," 45, 46, **253**

backyards: enclosure of, 42; play in, 42, 74, **253**; use of, 24, 42, 59, 230

Baillie, David, 41; on poverty in Hamilton, 55–6, 129, 136

Beauchemin, Yves: on neighbouring, 39–40

Black people, 60, 149; Canada and U.S. compared, 105, 232; discriminated against, 103, 198; ghettoes of, 27, 105, 108–9; neighbourhoods of, 40–1, 53, 108–10, 115, **261, 262.** *See also* visible minorities

blocks, city, 10, 27, 49, 75, 109; children in, 42, 67, 71–8, **257**; evidence on, 88–93; First Nations in, 54, 61–7; the homeless in, 53–5; immigrants in, 58, 59, 60–1, 102, **259**; maps of, 88–91; neighbouring, importance for, 17–21, 25–7, 117, 194; organizing, 82, 134, 140, 159, 164, 203; segregation by, 90–1, 93, 100; in suburbs, 27. *See also* neighbourhoods; neighbouring; *voisinage*

Boschman, Robert: on working-class Prince Albert, 41, 74, 80

boys. *See under* children

Bradley, Jean, 41; on Newtonbrook, 34, **252**; on working-class neighbouring, 34

British immigrants:
    neighbourhoods of, 57–8, 105–6;
    segregation of, 57–8, 100; in
    suburbs, 33, 34, 215–16; working
    class, 33, 34, 106, 215–16
Bruce, H.A. (Lieutenant-Governor,
    Ontario), 156–7
brunch: as marker of class, 31

Callaghan, Morley, 208
Canada Mortgage and Housing
    Corporation (CMHC), 151, 179
Carver, Humphrey, 176, 178–9,
    180
Catholics, 92, 163, 201, 215. See also
    religion
Cauchon, Noulon, 171
census: on neighbourhoods, 87,
    88; tracts, character of, 100;
    dissemination areas, 111
Chesterton, G.K.: on neighbours, 17
children, 8, 11, **235**; block,
    importance for, 29, 71–2; boys, 71,
    73–4, 76–8, 81–2, 210, 217; of elite,
    46, 48; friendships, 71, 72, 75, 77,
    80; girls, 71, 73–6, 124, 132; health
    of, 87, 130–1; neighbourhood,
    importance of, for, 34–5, 42, 67,
    70, 71–8, 120–9, 131–3; peers, 75,
    85, 121, 128, 132, 136, 219–22;
    playing outside, 22, 42, 67,
    71–5, **253, 257**. See also education;
    playgrounds; schools
Chinatowns: Vancouver, 58, 101,
    104, 140, 148, **255**; Victoria, 58
Chinese immigrants, 60, 64, 82;
    affluent, 61–2; discriminated
    against, 49, 58, 103, 172, 196;
    health of, 130; neighbourhoods
    of, 58, 82, 101–2, 104

churches, 201; Catholic, 91–2, 163,
    201–2, 215; cultural identity and,
    110, 114, 130, 163; decline of,
    201–2; as neighbourhood
    hubs, 43, 47, 58, 174, 179, 181;
    Protestant, 40, 168, 175, 201; social
    supports and, 163, 167, 168. See
    also religion, Christian
Clarke, Austin: on Black
    immigrants, 47
Clarke, Mary, 169
class, social: categories, 29–31;
    defined, 29–31; and education,
    214, 223; and homeownership,
    206–14; intersections, 83–4,
    107–12; and neighbourhood
    politics, 142, 169; neighbourhoods
    shaped by, 5, 29, 50, 88–99, 105–6,
    129, 199–200; and neighbouring,
    29–49, 52–6, 79. See also creative
    class; elite, social; inequality,
    social; low-income households;
    middle class; working class
Code Red series: Hamilton's public
    health, 127, 150
Commission of Conservation, 170
common interest developments
    (CIDs), 68–9, **263**
community, 44, 47, 54, 63, 66, 69,
    109, 143; absence of, 116–17;
    beyond neighbourhood, 65,
    107–8, 202, 230; building of, 25,
    51, 133–4, 140, 174; churches
    create, 163; in cooperatives, 35;
    decline of, 196, 200–6; design
    for, 87; in gangs, 77; of gays,
    107; of gentrifiers, 44; imagined,
    36, 44, 60, 204n2, 204–6; of
    immigrants, 5, 11, 58, 60; and
    neighbourhoods, 3–4, 8, 19, 26,

33, 179, 223; in public housing,
35; working-class, 33–4, 36–7; and
women, 82. *See also* neighbouring;
neighbourhoods
Community Planning Association,
178
condominiums, 23, 211; investment
in, 213; neighbouring in, 23, 44;
neighbours in, 20, 229
confirmation bias, 36
Coopérative d'habitation de
Montréal, 28
cooperatives: building by, 28, 34–5;
neighbouring in, 28, 34–5
covenants. *See* restrictions, deed
COVID-19: neighbourhood
geography, 120–1, 122, 124. *See
also* public health
creative class, 31. *See also* middle class
crime, 76, 128, 150; avoidance of, 133,
233; Canada and U.S. compared,
233; geographic concentration
of, 21, 35, 66, 129, 149–50, 190;
as neighbourhood effect, 123; in
public housing, 35, 66, 110, 190.
See also gangs; public safety
Curtis, Clifford, 178

Dalzell, A.G., 171
deed restrictions. *See* restrictions,
deed
Depression, the Great, 47, 71, 196,
199; and homeownership, 136,
207; and housing policy, 7, 176,
177; and neighbourhood decline,
151, 172, 175–6; neighbouring
during, 20, 33, 34, 38, 80; poverty
during, 52, 53, 175; and school
attendance, 216; social services
during, 161

developers: Canada and U.S.
compared, 197; neighbourhoods
named by, 10, 113, 203–4, 206;
neighbourhoods planned by,
43, 49, 155, 172, 173–5, 179, 180;
political influence of, 175–6, 193;
proposals resisted, 117, 138–9,
186; segregation created by, 88,
123. *See also* National Association
of Real Estate Boards
Dobson, Kathy: on neighbours, 140;
on poverty, 41, 55; on schools,
136; on women, 80
dogs: neighbouring and, 22, 67;
noise of, 17
do-it-yourself (DIY), 73, 210, 211
Dominion Housing Act, (DHA),
177. *See also* housing policy,
federal
"double ghetto," 78. *See also*
women
Dozois, Paul, 181
Drake (Aubrey D. Graham): Forest
Hill, in, 47
Drapeau, Jean (Mayor of Montreal),
185, 187
drugs: gangs and, 77; and
neighbourhood stereotyping, 149;
and poverty, 56

education, 75; Canada and U.S.
compared, 231; Catholic Church
and, 215; class attitudes to, 31,
131–2, 214–18, 235; importance
of, 9, 214, 218–23, 229, 235;
neighbourhood influence on, 11,
128, 217–23
elderly, the: neighbourhood,
importance for, 5, 29, 51, 67;
neighbouring by, 67–70

Electors' Action Movement, The (TEAM), 187
elite, social, 45; defined, 29–30; and homeownership, 209–11, 213; neighbourhoods of, 45–9, 50–1, 61–2, 99, 110, 145; neighbouring by, 45–8, 229–30; schooling of, 133, 214–15. *See also* class, social
enclaves, ethnic, 6, 59–60, 105–6, 108–9, 196–8, **264**. *See also* ghettoes; neighbourhoods
environmental determinism: social reformers, of, 149–51. *See also* slums
ethnoburbs, 112, 197–8
"eyes on the street," 21

family: extended, as neighbours, 20, 25, 32, 66–7, 79, 101, **252**; importance of, 38, 39, 53. *See also* children; men; women
fiction, as a source, 12, 41. *See also* *specific authors*
"15-minute city": best cities for, 205–6; examples, 114, **264**; ideal of, 178, 205, 229. *See also* intensification residential; walkability
filtering, of housing: belief in, 151, 176–7, 180; contradicted, 186–7; described, 50–1
financial institutions: Canada and U.S. compared, 177
First Nations, 13, 63, 82; and crime, 66; discriminated against, 62; displacement of, 62, 63; neighbourhoods with, 51, 53, 63, 64–7; neighbouring by, 65–7; poverty among, 51, 53, 67, 190; segregation of, 5, 62, 63, 103, 104; urbanization of, 62, 63

Ford, Doug (Ontario Premier): on the homeless, 30
Fraser, Sylvia: on neighbouring, 38, 79–80

gangs, 77, 110; of children, 71, 75, 76, 137; of First Nations people, 54, 66, 77; girls in, 76, **258**; racist, 76–7; territory of, 77–8, **258**
Garner, Hugh, 150, 152; on Cabbagetown, 33, 41, 84, 113–14; on schools, 218; on working-class neighbouring, 33
gated neighbourhoods, 44, 205, 233
gays: discriminated against, 106; neighbourhoods of, 6, 106–8, 148; segregation of, 4, 6
gentrification, 148, 186; built environment and, 10, 220; Canada and U.S. compared, 231–2; and house prices, 26–7, **251**; and planning, 186–7; resistance to, 141–2. *See also* urban redevelopment
gentrifiers, 7; built environment preferences, 122–3; middle class, 26, 43–4; neighbouring by, 26–7, 43–4
ghettoes: Black, 27, 105, 108, 232–3; Canada and U.S. compared, 105, 232; defined, 88, 105; gay, 74; Jewish, 88. *See also* enclaves; neighbourhoods
girls. *See under* children
Globe (and Mail): as a source, 7, 12, 158, 237–40; on usage, 176, 180, 181, 185, 189, 193, 212–13, 215, 219
Gray, James, 41: on children, 76; on inequality, 137; on residential mobility, 37

greenery: importance in neighbourhoods, 5, 70, 87, 129, 131. *See also* parks; public health

Hastings, Charles (Medical Officer of Health, Toronto), 168
Hernandez, Catherine: on Scarborough, 41
high-rises. *See* apartment buildings; condominiums
historical perspective: value of, vii, 9–12, 231
homeless, the, 4; defined, 30, 52; concentrations of, 53–6; numbers of, 52
homeowners, 248, 249; attitude to change, 141; income of, 200; middle class, 143, 187, 213, 219; mobility of, 9, 142–3; as neighbourhood activists, 142, 143, 169, 184, 207, 228; neighbouring by, 24; speculation by, 229. *See also* gentrifiers; homeownership
homeownership: Canada and U.S. compared, 13, 206–14; financial motives for, 9, 210–14; growth of, 8–9, 14, 206–8; immigrant attitudes towards, 207–9; impact on neighbourhoods, 142–3, 206, 212–14, 223; and income inequality, 213–14; middle-class attitudes towards, 9, 209–10; price of, 83, 213; research on, xii; working-class attitudes towards, 207–9. *See also* homeowners; "not in my back yard"
Hood, Hugh: on Rosedale, 47
"housing market": concept of, 7, 177

housing policy, federal: Canada and U.S. compared, 175, 232; capital gains tax, 211, 213; homeownership promoted by, 210–11; mortgage programs, 177; public housing, 25, 182

immigrant origins: Afghanistan, 60–1; Armenia, 59; Bangladesh, 59, 100; Barbados, 47; Caribbean, 60, 107, 197; Croatia, 39; Eastern Europe, 59, 105–6, 196; Finland, 130; Hong Kong, 49, 57, 60, 103; Hungary, 32; Jamaica, 80, 109, 149; India, 57, 60; Pakistan, 60, 104; Poland, 104, 146; Portugal, 59, 101–2, 103, 104, 130; Scotland, 101, 111, 57; Somalia, 80; South Asia, 102, 103, 130; Ukraine, 73, 146; United States, 60, 196, 197. *See also* British immigrants; Chinese immigrants; Irish immigrants; Italian immigrants; Jewish immigrants; visible minorities
immigrants: discriminated against, 49, 51, 57, 58–9, 97, 102–3, 172; divisions within groups, 60; homeownership aspirations of, 208–9; neighbourhoods of, 6, 28–9, 51, 57–61, 105–6, 108–9, 196–8, **259;** neighbouring by, 57–63; reasons to congregate, 57–61; segregation of, 5, 6, 28–9, 59, 88–9, 91, 97, 98–102, 105–6, 108–12; in suburbs, 23, 35, 101–2, 190. *See also* British immigrants; Chinese immigrants; immigrant origins; Irish immigrants; Italian immigrants; Jewish immigrants; visible minorities

immigration: Canada and U.S.
compared, 232; effect on
neighbourhoods, 8, 196, 197;
services, 61, 167; sources of, 103,
196–7; waves of, 8, 167, 196–8,
200, 232. *See also* immigrant
origins
inequality, social: Canada and
U.S. compared, 13, 199–200,
233; education, effect on, and
house prices, 213–14, 230–1; and
neighbourhoods, 91, 122, 132,
199–200, 214, 223; segregation,
effect on, 8, 199–200; waves
of, 8, 13, 198–9. *See also under*
neighbourhoods
Innu, 61
intensification, residential:
promoted by planners, 7–8, 188;
resisted by residents, 138. *See also*
urban redevelopment
Irish immigrants, 129; discriminated
against, 97, 102; neighbourhoods
of, 36, 53, 97, 100; segregation of,
97–8, 99–100, 101
Italian immigrants, 60, 130, 197, 208;
architecture of, 86; discriminated
against, 59, 103; health of, 130;
neighbourhoods of, 60, 101–2,
104, 113, 146; neighbouring by, 57,
59; segregation of, 101; working-
class, 110

Jacobs, Jane: and neighbourhood
organizations, 183–4; on
neighbouring, 21; on planning,
183
Jews, 110; education as priority,
208; discriminated against, 49, 58,
103, 148, 172; health of, 125, 129;

neighbourhoods of, 27, 41, 58–9,
88, 114, 264 (*see also* ghettoes);
segregation of, 90, 101, 104, 105

Kelly, Cathal: on misdemeanours, 77
Kennedy, John F. (U.S. president),
147

landlords: legislation affecting,
136; and neighbourhood
activism, 169; as neighbours, 32,
59, 68, 114
Lemelin, Roger 41; on
slummers, 147–8; use of
"*quartier*," 164
lesbians: discriminated against,
106; in neighbourhoods, 106–8,
148; playing football, 74. *See also*
gays
loneliness: building design
mitigates, 22; in multi-unit
dwellings, 68. *See also* mental
health
Lorimer, James: on Cabbagetown,
18, 26, 30, 37; on class, 30,
31, 37; on gentrifiers, 26; on
importance of block, 20, 115; on
neighbouring, 18, 26, 37, 72; on
women's role, 79, 80
low-income households:
definition of, 52–3; health of
125–7; of immigrants, 51, 57, 190;
neighbourhoods of, 8, 28, 55–6,
92, 117, 145–8, 167–8, 192, 199;
neighbours of, 53, 56; residential
choice, lack of, 5, 50–1, 175;
segregation of, 6, 49, 53, 92, 192;
supports for, 167–8. *See also* class,
social; homeless, the; poverty;
slums; tenants

MacDougall, Cynthia: on children, 74; on neighbouring, 59

MacLennan, Hugh: on neighbourhood decline, 50; on public schools, 72, 215

Marchessault, Jovette: on poverty, 55

Marlyn, John: on neighbouring, 41

marriage: assortative mating, 133; neighbourhood influence on, 45–6, 75

Marsh, Leonard, 151

media, 64, 77, 116, 234; community, affecting, 8, 25, 73, 202; neighbourhood image, affecting, 7, 55, 119, 149–50. *See also Globe and Mail*; newspapers; Nextdoor

melting pot: model of assimilation, 102

men, 54, 74, 149; as builders, 28, 33, 34, 73; and do-it-yourself, 73, 210; and domestic work, 74, 78, 79, 83, 210, 229; gangs of, 76; as migrants, 54, 59; as neighbours, 19, 28, 29, 33, 34, 80, 82–3, 229. *See also* gangs; gays

mental health: geography of, 127–9; and neighbouring, 68; and poverty, 56. *See also* loneliness; public health

Métis: discriminated against, 51; displacement of, 62; in neighbourhoods, 40, 51, 54, 61–3, 65, 67, **256**; on skid row, 54

Micallef, Shawn: on class, 31; on inner suburbs, 147; as a neighbour, 22, 23; on neighbouring, 39, 44

middle class, 7, 40, 81; bias of, 12, 40, 150; defined, 30–1; education as priority, 61, 133,

217–23; family roles in, 80; homeownership aspirations of, 9, 209–14; motives of, 6–7, 11, 43; as neighbourhood activists, 140, 142; neighbourhoods, importance for, 43–5; neighbourhoods of, 47, 49, 50, 75, 123, 145, 152, 186, 199; neighbouring by, 27, 40, 42–5. *See also* creative class; homeownership; schools; segregation: class, social; suburbs

Miller, John: on Rosedale, 47–8

mobility: Canada and U.S. compared, 233; everyday, 8, 39–40, 50, 69–70, 202; residential; 21, 111, 231; social, 5, 29, 102, 108, 131, 133, 233

Montreal Citizens' Movement (MCM), 186

mosaic: Canada as a, 102; across neighbourhoods, 110

Mumford, Lewis, 178, 179, 203

municipal government: amalgamation, 184, 191; annexation, 41, 134, 160, 169; Canada and U.S. compared, 166, 232, 233; neighbourhood strategies, 8, 78, 143, 188, 190–3, 214, 220–1; role of, 166, 192–4. *See also* planning, municipal; services, municipal; urban reform; zoning

Munro, Alice: on class, 30; as a neighbour, 24

Munro, Sheila: on Alice Munro, 24; on children playing, 71

Muslims, discriminated against, 198; supported, 234

National Association of Real Estate
Boards (NAREB), 173–4. *See also*
developers
National Housing Act, Canada
(1938), 179. *See also* housing
policy, federal
naturally occurring retirement
communities (NORC), 70
"neighbourhood": associations of
word, 155–62, 212–13, 215; and
"community," 4; convergence
with *"quartier,"* 7, 164–5; meaning
of, 4, 5, 155–62, 172, 181, 193;
spelling, 156, 238; usage of, 7,
155–62, 238–40, **240, 245.** *See also*
"quartier"; "residential area";
"residential district"
neighbourhood associations:
emergence of term, 169, 238,
247; examples, 117, 134. *See
also* ratepayers' associations;
residents' associations
neighbourhood effects: indirect,
119, 122–3; long-term, 6, 123,
131–3, 222–3; physical, 125–7;
policy responses, 189–92; short-
term, 6, 120–31, 133, 144; social,
127–33, 144, 222–3. *See also*
neighbourhoods: significance of;
selection bias
Neighbourhood Houses, 137, 161,
167, 196
Neighbourhood Improvement
Program (NIP): impact of,
187; introduction of, 164, 187;
purpose, 187
neighbourhood organizations:
building community, 134;
demanding attention, 134–8;
resisting change, 138–44; shaping

urban government, 184–8,
192. *See also* neighbourhood
associations; "not in my back
yard"; ratepayers' associations;
residents' associations
neighbourhoods: boundaries of, 6,
85–6, 100, 112–18, 221; Canada
and U.S. compared, 232, 234;
"city of," 8, 188, 192, 193, 214;
common interest developments
(CIDs), 113, 145, 147, **263;** decline
of, 50–1, 145, 147, 175–7, 186, 190,
232; evolution of, 3, 27; gated,
44, 205, 233; naming of, 151–2,
159; outsiders' perceptions of,
140, 144–52; physical influences
of, 125–7; physical variety, 86–7;
scales of, 5–6, 88–112, 117, 171,
229; significance of, 8–9, 10–11,
119–33, 195–223, 227–9; social
influences of, 127–133; social
variety, 88–112; tipping points
in, 227–8; types, 117–8. *See also*
filtering; "neighbourhood";
neighbourhood effects;
segregation; slums
neighbourhood unit: Canada and
U.S. compared, 175, 232; concept
of, 174, 203; influence of idea,
174–5, 178–80, 181, 203–4; origin
of concept, 174, 218
Neighbourhood Watch, 21, 69, 164
neighbourhood work, 192
Neighbourhood Workers'
Association, Toronto, 161, 168,
196
neighbouring, 17–49; bad, 17, 20;
conflictual, 20, 234; decline of,
3, 39, 202–3; latent, 24, 43, 44;
manifest, 19, 24, 27, 35, 50, 67, 82;

mixed types, 24–5, 38–9, 43–4; nature of, 18–21

neighbours. *See* neighbouring

Newhouse, David: on being Indigenous, 65

newspapers: big city bias of, 13; delivery of, 76; influence of, 204; as a source, 12, 150, 158, 237–40. *See also Globe and Mail*; media; Nextdoor

New Urbanist neighbourhoods: neighbouring in, 44–5

Nextdoor: Canada and U.S. compared, 233; neighbouring influence, 20, 203, 231; as a source, 20, 233

Ngram Viewer, Google's: as a source, 156, 237

nostalgia: deceptive memory and, 36, 109; for past neighbourhoods, 3–4, 36, 109

"not in my back yard" (NIMBY): history, 9, 138, 189; impact on neighbourhoods, 138, 139, 214, 229

Onomé, Louise: on Mississauga, 41, 76; on neighbouring, 76; on neighbourhood stereotyping, 140, 146, 150

owner-building: by cooperatives, 28, 34; and neighbouring, 28, 33, 34, 208

Paris, Erna: on Forest Hill, 41

parks: children in, 72, 73, 132; as a neighbourhood feature, 83, 84, 173, 174, 181; value of, 83, 87, 131, 165, 173. *See also* greenery; public health

paroisse: as neighbourhood hub, 91, 163, 164; usage of word, 164, 165

Perry, Clarence, 174, 178, 179, 202–4, 231

Peterson, Oscar, 147

planning, municipal: Canada and U.S. compared, 170, 178, 232; of city neighbourhoods, 7, 176, 180–1, 182, 186–94; history of, 170–2, 176–94; homeowners, influence of, 184, 186–7, 189, 206, 227; humility required for, 12; participation in, 162, 184–8, 191; in Quebec, 163–5, 179–80, 185–6, 191–2; planning districts, 116–17, 162, 181; for social mix, 220–1; of suburbs, 7, 176, 178–80

planning acts, 179

Plant family, 19, 78

poor, the. *See* homeless; low-income households; poverty

poverty, 65, 67, 110, 130; advantages of concentration, 27, 79–80; associations of, 56, 108, 125–6, 149; "culture of," 148–9; cyclical, 52; definitions of, 52; descriptions of, 146–50; disadvantages of concentration, 65, 127–9, 130–3; geography of, 53, 55, 89, 91, 147, 157, 190, 228, 232; incidence of, 89, 91; invisibility of, 147; "neighbourhood," associated with, 162; policies to reduce, 120, 167–8, 191–2, 228; in public housing, 110, 190; shame of, 38, 151. *See also* homeless; low-income households; skid rows; "slum"; slums

privilege: Canada and U.S. compared, 233; class, 5, 12, 48,

112, 214–15; male, 214–15; white, 5, 51, 58, 62–3, 115, 198. *See also* discrimination; inequality, social

Protestants, 40, 41, 103. *See also* religion

public health: and disease, 120–1, 124–5, 147, 167; geography of, 150; initiatives, 168, 188–9, 228; and pollution, 129. *See also* COVID-19; mental health

public housing, 77, 100; concentrations of, 190, 232; as expression of class, 29, 198; federal policy towards, 25, 182; neighbouring in, 25–6, 35, 80; as self-contained, 112. *See also specific projects*

"*quartier*": meaning, convergence with "neighbourhood," 4, 14, 155, 164–5; power of, 186

race: meaning of, 102–3

"ratepayer": meaning of, 143, 169

ratepayers' associations, 139, 172; purposes of, 134–51, 169, 172, 182–3, 219; renaming, 182–3, **247**; women in, 189. *See also* neighbourhood associations; residents' associations

redevelopment. *See* urban redevelopment

reformers, social, 41; and neighbourhoods, 162; research by, 146–7; use of "neighbourhood," 162

refugees: in neighbourhoods, 54, 108

religion, Christian: Canada and U.S. compared, 202; Catholics, 92, 163, 201, 215; discrimination among churches, 103, 130, 175; and

ethnicity, 101, 110; in Montreal, 97; Protestants, 40, 41, 103; segregation by, 99–100, 103–4, 129, 130. *See also* churches; Jews; Muslims; WASP

reserves of First Nations, 54; Coast Salish, 62; displacement to, 62, 63, 64; Kahnawake, 64; Membertou, 64; Mi'kmaq, 63; Musqueam, 64; neighbouring in, 65–6; Papaschase, 64; Six Nations, 65; Songhees, 63; Squamish, 63; Welamukotuk, 64; Wendake, 64; Westbank, 64

"residential area": meaning, 7, 160, 162, 237–8; regulation of, 160, 170; usage of term, 155, 156–7, 158–9, 160, 181, **244, 245**. *See also* "residential district"; "neighbourhood"

"residential district": meaning, 7, 159–60, 162, 192, 237–8; regulation of, 159–60; usage of term, 155, 157, 158–9, **244, 245**. *See also* "residential area"; "neighbourhood"

residential mobility. *See under* mobility

residents' associations: emergence of, 169, 183, **247**; purposes of, 82, 138; tenants in, 4–5, 135, 142; women in, 189. *See also* neighbourhood associations; ratepayers' associations

restrictions, deed: architectural, 61; on ethnicity or race, 49, 123, 172–3; on land use, 49, 172, 173; on value, 47, 49, 51, 61, 175

Richler, Mordecai, 41; on gangs, 76; on Outremont, 145

rooming houses, 51, 56;
concentrations of, 51, 53, 56,
75, 124; neighbouring in, 56;
regulation of, 134; tenants in, 56,
124
Rose, Félix, 137
Roy, Gabrielle: on children
playing, 72; on neighbouring,
32; on working class, 32; use of
"*quartier*," 164

safety, public: Canada and U.S.
compared, 233; "eyes on the
street," 21; in neighbourhood, 4,
35, 56, 63, 66–7, 68, 87. *See also*
crime; public health
schools, private, 45, 46, 215; Canada
and U.S. compared, 231; in
England, 214–15
schools, public, 9, 43, 248;
attendance, 216–19; Canada and
U.S. compared, 231; catchment
areas, 9, 61, 117, 221–3, 235, **267**;
class in, 215–18; closures, 9,
133, 221–2; 228; districts of, 9,
117; immigrants in, 58, 114, 215;
importance for neighbourhoods,
8, 121, 131–3, 214–17, 219–23, 229,
231; house price effects, 214, 222;
as neighbourhood hubs, 43, 47, 84,
85, 174, 175, 178–9; peers in, 121,
128, 132, 137, 219–22; poverty and,
37, 72, 131–2, 136, 235; proximity
to, 205–6; research on, xii
Scott, F.R.: on neighbours, 48
segregation, 23; and city size, 6,
94–101, 230; of classes, 89, 90–1,
92–9, 105–6, 108–12; of ethnic
groups, 88–9, 91, 97, 98–102,
105–6, 108–12, **260, 264**; of gays

and lesbians, 106–8; information
on, 88, 199; levels of, 88–94;
meaning of, 88; measures of,
89, 93, 95; of racialized groups,
102–5; scales of, 6, 88–118, **259**;
significance of, 92, 146
selection bias, 11, 21–2, 25, 121, 126,
131. *See also* neighbourhood effect
servants: of middle class, 209; in
elite neighbourhoods, 46, 47–8,
145
services, municipal: importance
of, 130–1, 166; lobbying for, 9,
82, 134, 137, 170, 172; provision
of, 30, 92, 93–4, 130–1, 166, 168.
See also municipal government;
schools
settlement houses, 161, 167, 196
single-family homes: in
Montreal, 179–80, 211–12; and
neighbouring, 24, 32, 42, 59;
status of, 192, 210; zoning for, 160,
173, 189
skid rows: Winnipeg's, **254;**
neighbouring in, 53–5
"slum": meaning of, 37, 150–1, 176,
181; usage of term, 37, 151, 152,
156–7, 161, 168, 181, 191. *See also*
slums
slum clearance. *See* urban
redevelopment
slums: association with
"neighbourhood," 7; clearance
of, 152, 176, 181, 232; "culture
of," 148; descriptions of, 33,
53–7, 145; neighbouring in, 53–7;
stereotyping of, 37, 53, 84, 148–52;
suburban, 137. *See also* poverty;
skid rows; "slum"
slum tourism, 147–8

snow shovelling: and neighbouring, 67, 68, 73
social explorers, the, 146–7
social mobility. *See under* mobility
sources of information: gaps in, 121, 229; geographical biases of, 13; historical, 88–94, 229–30; physical, 86–7; qualitative, 12–13, 237–40; social biases of, 12; statistical, 87–94
suburbs, **248, 249, 253**; becoming city neighbourhoods, 10, 27; branding of, 206; Canada and U.S. compared, 232, 233; cooperative development in, 28, 34; elite, 113; ethnoburbs, 197; garden, 151; gays and lesbians in, 107–8; and homeownership, 211–12; immigrant, 23, 35, 101–2, 190; inner, 22, 23, 116–17, 186, 190, 197–8; middle class, 42–3, 49, 71, 81–3, 145, 197; neighbouring in, 20, 24–7, 43–5, 71–2, 81–3; planning of, 7, 43, 171, 175, 177–80; regulation of, 7; schools in, 219, 221–2; unplanned, 33–4, 43, 137; way of life in, 25–6, 43–4, 81, 122–3, 205; women in, 80, 81–3; working class, 33–5, 39–40, 72, 106. *See also specific suburbs*; common interest developments; gated neighbourhoods; neighbourhood unit; New Urbanist neighbourhoods
synagogues: as neighbourhood hubs, 47, 85, 114, 201, 234

tenants: activism of, 135–6, 142, 143, 182–3, 184, 228; evictions of, 33; franchise of, 169; immigrant, 59; income of, 200; mobility of, 4, 21–2, 32, 37, 38, 54, 142–3, 228; as neighbours, 21–2, 32, 33, 37, 56, 230; as a proportion of households, 8–9, 227; status of, 4–5, 20, 222. *See also* apartment buildings; landlords
tipping points: in neighbourhoods, 93, 227–8
transit: Canada and U.S. compared, 231
Tremblay, Michel, 41; on neighbouring, 32, 39–40, 41, 48, 144

urban context: Canada and U.S. compared, 166; city-size effects, 94–6, 99–101; effect on neighbourhoods, 12; trends in city size, 96–9
urbanization. *See* urban context
urban redevelopment: Canada and U.S. compared, 231–2; gentrification and, 7, 141–2, 186, 204; inner city, 27, 51, 162, 177, 185–6, 188; opposition to, 7, 20, 58, 134, 140–4, 162, 181–3, 212; precondition for, 51; slum clearance, 151–2, 176–7; urban renewal, 12, 108, 140, 181–3, 186–7. *See also* "not in my back yard"
urban reform: Canada and U.S. compared, 166, 170; early twentieth century, 7, 166–71, 184–5, 199; 1960s–1970s, 184–8, 192. *See also* municipal government
urban village: ethnic, 146; gay village, 106–8, 148; neighbourhood unit as, 203; stereotype of, 27, 36–7, 113–14, 145, 148

Valeriote, Richard: on
neighbouring, 32
Vassanji, M.G.: on immigrant
enclaves, 23, 41, 60
Vermette, Katherena: on Winnipeg, 41
visible minorities: discrimination
against, 4, 57, 58–9, 102–3; income
of, 200; meaning of term, 4, 69,
198; segregation of, 6, 91, 103–6.
See also Black people; immigrant
origins
voisinage: meaning, 4, 5, 21, 164, 165;
usage of term, 164

WASP: culture, 41, 46–7, 51;
prejudice, 58, 103, 196; elite
neighbourhoods, 84
whitepainting. *See* gentrification
women, 74, 91; class differences, 43,
45, 54, 79–81, 209; conflicts among,
80–1; constraints faced by, 69, 78–9,
81–2; domestic responsibilities of,
34, 78–83, 209; "double ghetto," 78;
immigrant, 69, 80, 101; in labour
force, 5, 8, 42, 78, 83, 195; mental
health of, 128; as neighbourhood
organizers, 19, 45, 78, 81–2, 109,
202, 229; as neighbours, 5–6, 21, 24,
32, 34, 38–9, 68, 72, 78–83, 103; as

planners, 189–90; safety of, 54,
66, 233; single, 22, 23; in suburbs,
81–2. *See also* children: girls;
lesbians
Woodsworth, J.S.: on immigrants,
146; on neighbours, 17; on
segregation, 146; and settlement
house movement, 167
working class: definition of,
30; family roles in, 79–81;
homeownership aspirations,
207–9; importance of
neighbourhood for, 27, 31–40;
middle-class studies of, 40;
neighbourhoods discovered,
36–7; neighbourhoods of, 31–40,
95, 97, 99, 126; neighbouring by,
31–40; suburbs, 39. *See also* class,
social; segregation: class

Young, Phyllis, 41, 43, 47; on girls,
75; on the middle class, 42

zoning: Canada and U.S. compared,
172; comprehensive, 159, 177, 185;
developers promote, 155, 174;
origins of, 7, 46, 159, 166, 168–71,
173–5, **266**; residential, 46, 135,
160, **266**

Index of Places

Abbotsford (BC), 103
Africville (Halifax), 40, 108, 144; neighbouring in, 109; redeveloped, 140
Alwington Place (Kingston), 49, 91
Annex, the (Toronto), 13; neighbourhood organizations in, 133–4, 162, 169, 170–1; neighbouring in, 17
Asbestos (QC), 96
Atlanta (GA), 209

Banff Trail (Calgary), 73, 76
Beach/Beaches, the (Toronto), 44, 152, 205
Beasley (Hamilton), 117
Bethnal Green (London, England), 36
Brampton (ON), 102, 112
Brightside (Hamilton), 126

Cabbagetown (Toronto), 10, 39, 41, 77, 113, **251**; gentrification of, 152, 187; naming of, 152; neighbouring in, 18, 33, 37; redevelopment of, 25, 180–1; as a slum, 33, 84, 151

Calgary, 80, 99, **243**; neighbourhoods in, 105, 110–11, 113; planned neighbourhoods in, 179, **263**. *See also specific neighbourhoods*
Calvert, (NF), 94, **259**
Cedar Cottage (Vancouver), 81–2
Charlottetown (PE), **243**
Church and Wellesley (Toronto), 107
Cité-jardin-du-Tricentaire (Montreal), 180
Columbus (OH), 142

Don Mills (Toronto), 13; as a planned suburb, 179, 180
Downtown Eastside (Vancouver), 13, 127, neighbourhood organization in, 135; poverty in, 55, 150–1, 190; stereotyped, 55, 150
Dresden (ON), 109
Duberger (Québec City), 27
Durand (Hamilton), 138, 145, 170

Earlscourt (Toronto), 106, 168, 215; immigrants in, 106; neighbourhood organizations in, 169; neighbouring in, 33; services in 168–9

East End (London, UK), 151–2; neighbouring in the, 36, 37

East York (Toronto): neighbouring in, 20, 33–4, 38

Edmonton, 49, 64, **243**; neighbourhood decline in, 27, 51; urban reform in, 183, 184. *See also* Mount Royal

Esquimalt (BC), 63

Forest Hill (Toronto), 41, 147, 121, 133; homeownership valued, 210, 211; as a Jewish enclave, 41, 59, 144; schools in, 121, 133

Fort Victoria (BC). *See* Victoria

Galt (ON), 94–5

Gander (NF): building cooperative, 28

Glasgow (UK), 116

Golden Square Mile (Montreal), 48; as an anglophone area, 103–4; as an elite area, 46, 48, 93

Goose Village (Montreal), 36; neighbouring in, 35, 75

Greater Forest Lawn (Calgary), 55

Griffintown (Montreal), 89; as immigrant enclave, 36

Grimsby (ON), children playing in, **253**; neighbouring in, 34; owner-building in, 28, 34

Guelph (ON), 32

Halifax (NS), 53, 167, 187, **243**; neighbourhood movement in, 185; planners' neighbourhoods

of, 116; neighbourhoods, scale of, 115; neighbourhoods in, 50, 91, 95, 112. *See also specific neighbourhoods*

Hamilton (ON), 117, 175, **243**, education and schools in, 216–17, 219–20; gentrification opposed in, 141–2; homeownership in, 207, 208, 209, 216; immigrants in, 57–8, 59, 106; neighbourhood organizing in, 78, 135, 138, 170, 182, 184, 185; neighbourhood policy in, 143, 190–1; neighbourhoods in, 46, 55–9, 105, 111, 145; neighbouring in, 37–8, 41, 79–80; segregation in, 89, 93, 94, 95, 98; urban redevelopment in, 140, 180, 181. *See also specific neighbourhoods*

Hanover (ON), 95, 99, 243

Harlem (New York), evolution of, 27; slumming in, 147

High Park (Toronto neighbourhood), 68, 70

Humber Valley Village (Toronto), 179

Jamaica Plain (Boston), 25

Jane-Finch (Toronto), 161n1; residents stereotyped, 55, 149–50

Kelowna (BC), 64

Kingston (ON), 92, 103, **243**; homeownership in, 207, 209; neighbourhood organizations in, 135, 140, 149; neighbourhoods in, 49, 55, 90–1, 105, 115; schools in, 133, 221–2. *See also specific neighbourhoods*

Kingsway Park (Toronto), 49, 173

Kirkendall (Hamilton), 117

Kitchener (ON), 169–70
Kitsilano (Vancouver), 127; neighourhood organizations in, 134–5, 138, 139, 170

Lawrence Heights (Toronto), 110, 113; neighbouring in, 35, 80
Lawrence Manor (Toronto), 179
Leaside (Toronto), 41, 47
Little Burgundy (Montreal), 77; Black history in, 60, 109, 147, 148; gentrification of, 141, 152, 205; redevelopment of, 140, 186; school catchments in, 133. *See also* Saint-Antoine
Little Italy, 60; label, accuracy of, 105, 146; neighbouring in, 59
Liverpool (UK): neighbouring in, 24
London (UK): East End, 36, 37, 151–2; gentrification in, 220; neighbouring in, 3–4; slumming in, 147
London (ON), 73, 138, 203, **243**
Longeuil (Montreal), 39. *See also* Ville Jacques-Cartier
Lord Selkirk Park (Winnipeg), 65
Lower Town (Ottawa), 125
Lower Town (Quebec City), 163; working class in, 163
Lower Ward (Toronto), 37

Markham (ON), 112
Milton-Park (Montreal), 186
Minden (ON) 96
Mississauga (ON), 76, 108, 140; neighbourhoods in, 101–2, 140; public health in, 131
Montreal, 48, 76, 152, 187, 188, 204, 233, **243**; amalgamation in, 191; anglophones in, 46–7, 48–9, 57;

as a "city of neighbourhoods," 155–6, 192; francophones in, 32, 129; gay villages in, 106–7; gentrification in, 44, 186, 192, 205; homeownership in, 14, 179–80, 207–8; immigrants in, 57, 97, 99–100, 114; Jews in, 129, 143; names for neighbourhoods in, 152, 157–8, 164n3, 164–5, neighbourhood decline in, 50, 181; neighbourhood organizing in, 140, 143, 185–6; neighbourhoods in, 89, 90, 92, 97–8, 99–100, 103–4, 110, 112, 126, 145–6, 159; neighbouring in, 32, 144; parishes in, 91–2, 163; poverty in, 55, 89, 125; public health geography, 125, 129, **265**; religious groups in, 97, 99–100, 129; schools in, 72, 133, 220; segregation in, 90, 92, 95, 97, 99–100, 103, 105, **260**; social services in, 91, 124, 163; tenants in, 114, 197, 207–8, 228, 230, **252**; urban redevelopment in, 181, 186; urban reform in, 166–7, 170, 185–6, 228. *See also specific neighbourhoods*
Mount Dennis (Toronto), 143
Mount Royal (Edmonton), 49

Nepean (ON), 134–5
Newtonbrook (Toronto), 34; map of, **252**
New Westminster (BC), 108
New York City, 37. *See also* Harlem
Niagara neighbourhood (Toronto), 104, 187
North End (Hamilton), 55, 136
North End (Kingston), 76, 135;

neighbourhoods in, 90, 115; poverty in, 91; schools in, 222

North End (Winnipeg), 41, 108, 137; First Nations in, 65–7; immigrants in, 32, 37, 99, 105, 215; neighbouring in, 32, 37, 66; public health in, 124, 147; working class in, 93, 105

North Vancouver: Alice Munro in, 24

Oromocto (NB), 64

Oshawa (ON): homeless in, 54–5

Ottawa, 56, 87, 218, **243**; gentrification in, 187; neighbourhood organization in, 183; neighbourhoods in, 112, 148; neighbouring in, 25; planning in, 171; public health geography, 125; urban reform in, 185. *See also* public health; *specific neighbourhoods*

Outremont (Montreal), 48, 126; as an enclave, 144–5; neighbouring in, 48

Paris (ON), 97, 99, 243; neighbourhoods in, 95

Parkdale (Toronto), 145; evolution of, 27

Peterborough (ON), 65; good neighbours in, 234

Plateau-Mont-Royal, 32, 55, 144

Pleasant Hill (Saskatoon), 66

Point-Saint-Charles. *See* Pointe-Saint-Charles

Pointe-Saint-Charles, 114, 141, 201; neighbourhood organizing in, 140; neighbouring in, 33, 35, 114; poverty in, 55; schools in, 133, 136

Port Arthur (ON): immigrants, 89, 130. *See also* Thunder Bay

Port Moody (BC), 22

Prince Albert (SK), 41, 80; children playing in, 72, 74

Quebec: British immigrants in, 57; Catholic Church in, 163, 201, 215; cooperatives in, 28; distinctiveness of, 14; Quiet Revolution in, 137; schools in, 215; separatism in, 136–7, 191. *See also specific cities, neighbourhoods, and quartiers*

Quebec City, 41, 64, 100, 164, **243**; immigrants in, 100–1, 130; neighbourhoods in, 105; school catchments in, **267**; segregation in, 97, 100. *See also specific neighbourhoods*

Quebeckers: familiarity with neighbours, 21

Regent Park (Toronto): neighbouring in, 25–6, 35; as public housing, 25, 180–1

Regina (SK), 101, 102, **243**; gentrification in, 204

Renfrew Heights (Vancouver), 40

Richmond (BC), 112

Richmond Heights (Halifax), 70; neighbouring in, 39, 80

Riverdale (Toronto), 77; neighbourhood organizations in, 135, 141, 183; South Riverdale, 141

Rochester (New York), 204, 205; neighbouring in, 39, 202

Rooster Town (Winnipeg): a Métis enclave, 40, 63, 256

Rosedale (Toronto), 152; mansions in, 48, 150; servants in, 47–8

Saint-Antoine (Montreal), 60;
    renamed, 152; as working class,
    97. *See also* Little Burgundy
Sainte-Foy (Quebec City), **267**
Saint John (NB), 98
Saint-Léonard-de-Port-Maurice
    (Montreal), 28, 180
Saint-Michel (Montreal), 141
Saint Roch (Quebec City), 97
Saint-Sauveur (Quebec), 39
Sandy Hill (Ottawa), 46
Saskatoon (SK), 103; children in, 76,
    128; public health in, 128. *See also*
    Pleasant Hill
Seattle, 143, 202, 206
Shaughnessy (Vancouver), 152,
    173; conflict in, 61–2; influence of,
    170, 189
Singapore, 22
South Korea: noise
    complaints, 18
Spadina expressway (Toronto):
    opposition to, 134, 140–1,
    183–4
St. Catharines (ON), **243**
St. Henri (Montreal), 72, 144
St. John's (NF), 73, 151; co-ops in,
    28; neighbourhoods in, 105, 151,
    206
Strathcona (Vancouver): image
    of, 144, 181; neighbourhood
    organization in, 82, 144; poverty
    in, 151, 181
St. Urbain (Montreal), 13
Sudbury (ON), 73, **243**; immigrants
    in, 105, 146
Sydenham Ward (Kingston): school
    in, 221–2
Sydney (NS), 109; First Nations in,
    63, 64; immigrants in, 89, **259**

Thorncliffe Park (Toronto):
    immigrant enclave in, 60–1
Thunder Bay (ON), 89, 243. *See also*
    Port Arthur
Toronto, 41, 150, 232, **243**; census
    tracts created in, 100; "city
    of homes," 138; "city of
    neighbourhoods," 192–3; crime
    geography, 21, 76–7, 78, 128,
    150, 190; education and schools
    in, 215–16, 218–19; gays and
    lesbians, 107; gentrification
    in, 141, 206; homeowners in,
    142–3, 200, 208; immigrants in,
    33, 34, 59, 86, 101–6, 109–11,
    148, 190, 197–8; inner suburbs
    of, 22, 23, 190, 197–8, 232;
    "neighbourhood," use of, 156–7,
    158–62, 181; neighbourhood
    boundaries in, 115, 116–17;
    neighbourhood organizations
    in, 138–9, 140–1, 142, 168–9, 183;
    neighbourhood strategy of City,
    168, 181–2, 187, 190, 191, 193;
    neighbourhoods in, 47–8, 50, 53,
    70, 94, 105–6, 111–12, 147, 157,
    168, 180, 199; neighbouring in, 19,
    27, 37, 44, 203, 205, 229; planners'
    neighbourhoods in, 116–17, 180,
    181; planning of, 184–5; poverty
    in, 37, 53, 190, 197–8; public
    health in, 126, 168; segregation
    in, 92, 94, 96, 98, 100, 103, 105;
    social services in, 124, 167–8;
    suburbs of, 25, 33, 34, 39, 43;
    urban redevelopment in, 13, 180,
    186–7, 189; urban reform in, 167,
    188; walkability in, 189, 205–6;
    zoning, 135, 172. *See also specific*
    neighbourhoods

Trefann Court (Toronto), 141, 143, 149, 183, 184
Trois-Rivières (QC): British immigrants in, 57

United States, compared with Canada: Black experience, 105, 232; concentrated poverty, 128, 232; crime, 233; development industry, 197; education and schools, 217–20, 222, 231; financial institutions, 177; First Nations, geography of, 62; gentrification, 231–2; homeowner priorities, 142, 210–14; homeownership rates, 206–10; house prices, 213, 231; housing policy, federal, 175, 232; immigration mix, 232; inequality, 199–200, 213, 233; municipal government, 166, 232, 233; "neighbourhood," use of, 156–7; neighbourhood decline, 232; neighbourhood unit, use of, 178, 232; neighbourhoods, prominence of, 234; "niceness," 234; planning, 170, 178, 232; religious attendance, 202; suburbs, 232, 233; transit, 231; urbanization, 166; urban reform, 166; zoning, 172
United States research: usefulness for current study, 13–14, 199, 213, 230
University Hill (Vancouver), 40
Uplands (Victoria): deed restrictions, 49, 51, 173

Vancouver, 40, 70, 75, 79, 139, 200, **243**; Chinese immigrants in, 58, 82, 101, 148; education, 218; First Nations in, 63, 64; gays and lesbians in, 107; immigrants in, 69, 102, 104, 200; neighbourhoods in, 104; neighbouring in, 69, 70, 71; neighbouring promoted in, 68, 187–8; physical geography of, 96; public health geography, 127, 128. *See also specific neighbourhoods.*
"Vancouverism," 188, 192
Victoria: Chinatown, 58; First Nations in, 62, 63; immigrants in, 100–1; segregation in, 100–1
Vienna: building design in, 22
Ville Jacques-Cartier (Montreal): influence on separatists, 137. *See also* Longeuil

Ward, the (Toronto), 13, 60, 74, 104; as a Jewish enclave, 59, 114; neighbouring in, 32; poverty in, 175–6; as a slum, 84, 151; stereotyped, 84, 151, 176
Westdale (Hamilton), 136; deed restrictions in, 49, 175; elite section of, 49
West End (Boston): neighbouring in, 37; as an "urban village," 113
West End (Kingston), 90–1
West End (Vancouver): elite in 45, 61, 215, **253**; gays in, 107; redeveloped, 51
Westmount (Montreal), 172; anglophone, 103, 177; exclusive 84, 113; image of, 152; neighbouring in, 46–7, 48, 49; public health in, 125; schools in, 133, 136; zoning in, 46, 49, 170, 173

West Vancouver: Alice Munro in, 24; children in, 71

Windsor (ON): neighbouring in, 39; neighbourhoods in 31, 109, **262**

Winnipeg, **243**; First Nations in, 54, 65, 82, 103; immigrants in, 57, 105, 148; Métis in, 65; neighbourhood decline in, 51; neighbouring in, 65–6, 136, 143; neighbourhood organizations in, 188; neighbourhood policy for, 168; neighbourhoods in, 53–4, 55, 65–6, 98–9, 111, 172–3, 254; planning movement in, 170; public health geography, 124, 147; segregation in, 92–3, 94, 98–9, 103, 105, 147; urban reform in, 167. *See also specific neighbourhoods*

Yaletown (Vancouver), 99

York Township (ON), 34. *See also* Earlscourt

www.ingramcontent.com/pod-product-compliance
Lightning Source LLC
Chambersburg PA
CBHW031116020426
42333CB00012B/101